T0291632

MODERN RADIO TECHNIQUE

General Editor: J. A. RATCLIFFE

A SURVEY OF
THE PRINCIPLES & PRACTICE
OF WAVE GUIDES

A SURVEY OF
THE PRINCIPLES & PRACTICE
OF WAVE GUIDES

BY

L. G. H. HUXLEY, M.A., D.PHIL.

Reader in Electromagnetism in the University of Birmingham
Formerly Principal Scientific Officer, Ministry of
Aircraft Production

CAMBRIDGE
AT THE UNIVERSITY PRESS
1947

To

E. M. H.

CAMBRIDGE
UNIVERSITY PRESS

University Printing House, Cambridge CB2 8BS, United Kingdom

Cambridge University Press is part of the University of Cambridge.

It furthers the University's mission by disseminating knowledge in the pursuit of education, learning and research at the highest international levels of excellence.

www.cambridge.org
Information on this title: www.cambridge.org/9781316509814

First published 1947
First paperback edition 2015

A catalogue record for this publication is available from the British Library

ISBN 978-1-316-50981-4 Paperback

PREFACE

Although the propagation of electromagnetic waves in metal tubes —or wave guides as they are now called—has been studied for some fifty years, until recently the subject was in the main of theoretical, rather than of practical, interest. However, with the development of the first microwave radar equipment during 1940–1 the subject was suddenly transformed to one of prime practical importance, and in the following years was developed at a phenomenal rate, both in Britain and the United States. This book is written to provide an introductory survey of these recent developments.

The treatment in the first six chapters is based on courses on microwave techniques which were given during the war at the Radar School of the Telecommunications Research Establishment (T.R.E.), and it is believed that the book has lost nothing of importance in content and rigour through this elementary and physical approach to the subject. Chapter 7 has been included for those readers who may prefer a more formal treatment.

The physical interpretation, given in § 5·4, of the normalized admittance or impedance of an obstacle or other discontinuity in a wave guide, in terms of scattering coefficient, has the advantage of relating these quantities immediately to the experimental data obtained with a standing wave indicator. It was the subject of a paper that I read before a technical colloquium at T.R.E. during the war. The treatment of Babinet's Principle in Chapter 7 is also believed to be original.

When this book was written, the greater proportion of the technical developments described were hidden in secret reports and memoranda which are not available to the general reader. No reference is made to this literature, and very few of the many contributors to the development of the subject have been mentioned by name. Accounts of confidential work done during the war are now appearing in technical journals, and the reader can obtain more detailed information on particular matters from these. Attention is drawn in particular to *The Proceedings of the Radio-location Convention*, March–May 1946, *J. Instn Elect. Engrs*, vol. 93, part III A, nos. 1, 3 and 4, 1946.

I welcome this opportunity of expressing my appreciation of the considerable assistance that I received from numerous colleagues at the Telecommunications Research Establishment without which this book could not have been written, but I wish to record in particular my indebtedness to Dr G. G. Macfarlane and Dr W. Cochrane with whom I engaged in many stimulating discussions on the subject of wave guides. I am also indebted to Mr J. A. Ratcliffe, who first aroused my interest in the subject, for helpful criticism of the text.

I wish also to express my thanks to the Director-General of Scientific Research (Air), Ministry of Supply, for permission to publish this book, which follows closely a monograph written by me as a contribution to the Scientific War Records of the Ministry of Supply (Air).

It is recorded, in conclusion, that, although the book has received official scrutiny before publication, I accept full responsibility for all opinions and statements in it. Further, I acknowledge Crown Copyright in respect of all illustrations in the book.

L. G. H. HUXLEY

DEPARTMENT OF ELECTRICAL ENGINEERING
THE UNIVERSITY, EDGBASTON
BIRMINGHAM 15

28 *June* 1946

CONTENTS

Chapter 5. WAVE-GUIDE IMPEDANCE AND
FURTHER TECHNIQUES

Chapter 6. CAVITY RESONATORS

Chapter 7. MATHEMATICAL TREATMENT OF SELECTED TOPICS

CORRIGENDA

p. 4, equation (3).   *For* 10^{-8} *read* 10^8.

p. 5, 1·3, line 8.   *For* E/q *read* F/q.

p. 10, equation (7).   *Omit 2 under radical.*

p. 18, line 16.   *For* smallest *read* largest.

p. 22, line 22.   *For* D *read* B.

p. 61, 6 lines from bottom.   *For* fig. 4·29 (c) *read* fig. 4·29 (a).

p. 62, line 9.   *For* oscillation *read* oscillator.

p. 70, line 3 from bottom.   Cadmium plating is used externally, not internally.

p. 83, line 6 from bottom.   *For* not used *read* not extensively used.

p. 87, lines 8 and 13.   *For* G *read* C.

p. 88.   In the account of *choke couplings and plungers*, it should have been mentioned that loose contacts between the wall and the plunger or between the two portions of the coupler, are rendered innocuous since they are placed at a node of current in the half wave recess.

p. 93, line 22.   *For* (a) *read* (c).

p. 115, line 5 from bottom.   *For* 5·2·2 (1) and (2) *read* of 5·2·2.

p. 119, line 6.   *For* § 7·16 *read* § 7·17.

p. 133, line 3 from bottom.   *For* 10 (a) *read* 5·10 (a).

p. 137.   Equation (2) *should read* $y_1 = jb_1 = -j\dfrac{\lambda_g}{W}\cot^2\left(\dfrac{\pi c}{2W}\right)$.

It is given correctly on the following page.

p. 141, fig. 5·16 (9).   *Include term* -2 *within the square bracket.*

p. 141, fig. 5·16 (10).   *For* R *read* r *and for* $2n-1\lambda$ *read* $(2n-1)\lambda$.

p. 144, line 8.   *Correction as in* p. 83, line 6.

p. 145, fig. 5·18 (a).   *Reflected components should read* (E_r, H_r).

p. 146.   In formula following (1) *read* $(1-\rho)$.

p. 146, line 14 from bottom.   *For* Z_H *read* $(Z_H)_2$.

p. 148, line 21.   *For* fig. 5·16 *read* fig. 5·14.

p. 155, fig. 5·26 (a).   *For* $\frac{4}{3}$ *read* $\frac{3}{4}$.

pp. 164, 165, near bottom of page. *For* $\dfrac{j(1+f_1f_2)}{f_3}$ *read* $\dfrac{j(1+f_1f_2)}{2f_2}$;

also make corresponding correction on fig. 5·30 (*a*).

p. 167, line 5.   *For*   right-hand   *read*   left-hand.

p. 179, line 23.   *For*   DE   *read*   BE.

p. 181, line 2.   *For*   DCHR   *read*   DCHK.

p. 190, line 9.   *For*   cell   *read*   cells.

p. 191, line 2 from bottom.   *For*   mean   *read*   peak.

p. 192, bottom line.   *For*   power   *read*   peak power.

p. 196.   *For*   intrinsic impedance   *read everywhere*   total impedance. (Ref. p. 119.)

p. 198, line 3 from bottom.   *For* $v=\sqrt{\dfrac{L}{C}}$ *read* $v=\dfrac{1}{\sqrt{LC}}$.

p. 202, line 5 from bottom.   *For*   Cuttler   *read*   Cutler.

p. 203, first formula.   *For*   D   *read*   d   *within the round bracket.*

p. 214, line 4.   *For*   (e)   *read*   (a).

p. 224.   The E-mode with the lowest frequency is the E_{110} not the E_{111} as stated. This correction is required also on p. 258. The component U of the Hertz vector for the E_{110} mode is

$$U=\sin\left(\frac{\pi x}{a}\right).\sin\left(\frac{\pi y}{b}\right)e^{j\omega t}.$$

p. 229, line 2.   *Omit words*   C opposite to B.

p. 235, centre of page, paragraph beginning 'the input impedance...'. Cause and effect are interchanged.

p. 248, § 7·3.   In equation above (3) *for*   eE   *read*   e$\dot{\text{E}}$.

p. 253, equation (3).   *For* $\dfrac{\partial}{\partial\mu_3}$ *read* $\dfrac{\partial}{\partial\mu_2}$ for first symbol within square bracket.

p. 275, end of § 7·10·7.   The correct values of the roots are: $\rho_{11}=4\cdot49$, $\rho_{21}=5\cdot8$, $\rho_{12}=7\cdot64$, $\sigma_{11}=2\cdot75$.

p. 276, second line from bottom.   *For*   electric   *read*   magnetic.

p. 291, bottom line.   *For* v *read* σ.

Chapter 1

THE ELECTROMAGNETIC FIELD
OF A TEM-WAVE

1·1. Introduction

At frequencies less than 3000 Mc./sec., the transmission line has long been the standard device for conveying power at high frequencies from a transmitter to an aerial, or from an aerial to a receiver. When, however, during 1940 and 1941, radar devices were first developed to operate on frequencies of 3000 Mc./sec. and greater, it was apparent that at these frequencies transmission lines possess two serious and inherent limitations.

First, the only feasible form of transmission line in a compact equipment is a screened feeder line such as a coaxial cable in which the central conductor is retained in position by dielectric filling. It is found that even with polythene (power factor 0·0005) as the dielectric, the attenuation produced by such a cable is sufficiently great to impair the overall efficiency of a microwave radar equipment. For instance, the attenuation produced by Uniradio 21—a good-quality polythene cable designed for use at ultra-high frequencies—is about 0·6 db./m. at a frequency of 3000 Mc./sec., and at frequencies of 9000 Mc./sec. the loss is several times as great.

Cables are therefore not employed in microwave radar devices except in short lengths.

The second limitation of a polythene cable is its low power-handling capacity. When the mean power carried in the cable exceeds about 200 W. the ohmic heating of the inner conductor is sufficient to soften the polythene dielectric. On the other hand, cables possess the advantage of flexibility and therefore find extensive use in the laboratory and in test equipment where wastage of a proportion of the power is of no consequence, and where the level of the power is low.

Evidently, the success of microwave radar depended, among other factors, on the development of a more efficient device than the feeder line to carry power to and from the aerial.

The principal loss in a cable is attributable to the presence of the inner conductor. Not only is the inherent ohmic loss of power (copper loss) in this conductor an appreciable fraction of the total, but this conductor is responsible for the presence of the dielectric whose main function is to support it in position. If, therefore, it were possible to carry the power along an empty metal tube devoid of inner conductor and dielectric it would be anticipated that the attenuation would merely be that due to the outer conductor of a cable of the same size.

Such a tube, called a wave guide, was adopted as the practical substitute for the high-frequency cable in radar equipments, and in what follows we are concerned with the practical applications of wave guides, and the underlying principles of their operation.

In comparison with cables, practical wave guides possess the double advantage of small attenuation coefficient (about 0·05 db./m. at 3000 Mc./sec.) and of high-power handling capacity (2 MW. or more peak power, according to design).

On the other hand, wave guides have the disadvantage of inflexibility and greater weight.

Although a rigorous study of the propagation of electromagnetic waves along transmission lines should be based on the differential equations of the electromagnetic field, yet the radio engineer prefers to describe the relevant properties of a transmission-line system in the language of circuit theory which employs the concepts of voltage, current and impedance. Because many wave-guide techniques correspond closely to similar techniques of transmission-line practice, it has been found profitable to study wave guides, first, from the standpoint of the electromagnetic field, and then to translate the results into the language of circuits. In this way it is often possible to represent a wave-guide system symbolically by an equivalent transmission-line system.

In the remaining sections of this chapter, it is shown how the field description and the circuit approach give completely equivalent accounts of the propagation of the principal wave on a transmission line. A more critical examination of the meaning of the term impedance as applied to a transmission line or wave guide is given in Chapter 5. The present chapter also serves to summarize some basic facts about electromagnetic waves in free space or guided along transmission lines, of which use will be made in the sequel.

1·2. Field of a plane-polarized electromagnetic wave in free space

In this section we summarize the chief features of an electromagnetic plane wave in an unbounded medium. Here and throughout the book, the rationalized metre-kilogram-second (m.k.s.)* system of units is used because of its practical convenience.

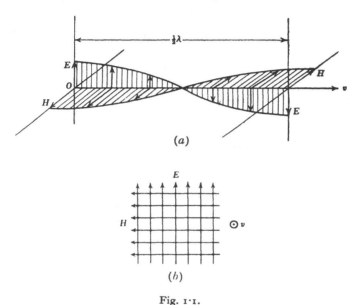

Fig. 1·1.

Fig. 1·1 (a) shows the relative orientations of the electric and magnetic vectors and the direction of propagation in a plane-polarized sinusoidal progressive wave in an unbounded isotropic medium. The electromagnetic field of this wave train has the following properties:

(a) The electric field **E** and the magnetic field **H** are at right angles to each other and to the direction of propagation. An electromagnetic wave of this type, in which both **E** and **H** are perpendicular to the direction of propagation, is called a TEM-wave (transverse electric-magnetic). Fig. 1·1 (b) shows the **E** and **H** fields in the wave front through O (fig. 1·1 (a)).

---

* Stratton, *Electromagnetic Theory*, p. 16.

(*b*) The velocity of propagation (phase velocity) is

$$v = \frac{1}{\sqrt{(\epsilon\mu)}} \text{ m./sec.,} \tag{1}$$

where ϵ is the electric inductive capacity of the medium and μ its magnetic inductive capacity. Further,

$$\epsilon = \epsilon_0 K_e \quad \text{and} \quad \mu = \mu_0 K_m,$$

where ϵ_0 and μ_0 are the inductive capacities of free space and K_e and K_m are respectively the specific inductive capacities of the medium. In a homogeneous medium K_e and K_m are called, respectively, the dielectric constant and the magnetic permeability. In the m.k.s. system of units

$$\mu_0 = 4\pi \times 10^{-7} \text{ henry m.}^{-1}. \tag{2}$$

Also, from (1) the velocity of electromagnetic TEM-waves *in vacuo* ($K_e = K_m = 1$),

$$v = c = \frac{1}{\sqrt{(\mu_0 \epsilon_0)}}$$
$$= 2 \cdot 998 \times 10^{-8} \text{ m.sec.}^{-1}$$
$$\doteqdot 3 \times 10^{-8} \text{ m.sec.}^{-1}. \tag{3}$$

From (2) and (3) we find

$$\epsilon_0 = 8 \cdot 854 \times 10^{-12} \text{ farad m.}^{-1}$$
$$\doteqdot \frac{1}{36\pi \times 10^9}. \tag{4}$$

The ratio

$$\sqrt{\frac{\mu_0}{\epsilon_0}} = 376 \cdot 6 \text{ ohms}$$
$$\doteqdot 120\pi \text{ ohms}. \tag{5}$$

(*c*) At a fixed point such as O (fig. 1·1), **E** and **H** vibrate in phase in a non-conducting medium. Thus, if at O,

$$E = E_0 \cos \omega t,$$

then

$$H = H_0 \cos \omega t.$$

(*d*) The amplitudes E_0 and H_0, when the medium is non-conducting, are related as follows:

$$\sqrt{\epsilon}\, E_0 = \sqrt{\mu}\, H_0,$$

or
$$\frac{E_0}{H_0} = \frac{E}{H} = \sqrt{\frac{\mu}{\epsilon}}$$

$$= \sqrt{\frac{K_m}{K_e}} \sqrt{\frac{\mu_0}{\epsilon_0}} = 120\pi \sqrt{\frac{K_m}{K_e}}$$

$$= 376 \cdot 6 \sqrt{\frac{K_m}{K_e}} \text{ ohms.} \qquad (6)$$

Because the ratio E/H, as indicated in (6), has the dimensions of a resistance, and frequently appears in theoretical discussions, it is called the wave impedance of the TEM-wave. In the m.k.s. system the electric-field strength E is expressed in volts per metre, and the magnetic-field strength H in amperes per metre.

These properties of the wave follow from the fact that the fields **E** and **H** satisfy Maxwell's equations of the electromagnetic field. It remains to state some additional properties of an electromagnetic field that relate to a postulated flux of power in the field, and to the behaviour of **E** and **H** at the surface of an ideal perfect conductor.

1·3. Poynting flux

In practice we do not directly observe the fields **E** and **H** but forces between stationary charges, forces between moving charges and the heat generated when currents flow in conductors. The field representation provides a convenient pictorial synthesis of the electromechanical actions, although the fields themselves are not directly observed. For instance, when a force of **F** newtons ($10^5 F$ dynes) acts on a charge of q coulombs it is concluded that the field **E** at the charge is $\mathbf{E}/q$ V./m. Similarly, the relation between the magnetic-field strength and the force on a charge q moving at velocity **V** is

$$F = qVB \sin\theta \text{ newtons,}$$

where the magnetic induction $\mathbf{B} = \mu\mathbf{H}$. The force **F** is at right angles to the plane containing the directions of the vectors **V** and **B**. The angle θ is the angle between **B** and **V**. In the language of vector algebra

$$\mathbf{F} = q(\mathbf{V} \times \mathbf{B}) \quad (\text{V in m.sec.}^{-1}).$$

When power is delivered to one electrical system by the action of another distant electrical system on it, we may suppose that the power flows through the intervening electromagnetic field.

Poynting's theorem is a postulate concerning the local flux of power at any point in the field. It states that if **E** and **H** are the values of the electric- and magnetic-field strengths at the field point and θ the angle between them, then there is a flux density of power W normal to the plane containing directions of **E** and **H**, and equal to

$$P = EH \sin\theta \text{ W./sq.m.,}$$

or in notation of vector algebra

$$\mathbf{P} = (\mathbf{E} \times \mathbf{H}) \text{ W./sq.m.} \tag{1}$$

The relative directions of **E**, **H** and **P** are shown in fig. 1·2. It may be observed that according to (1), if either one of **E** or **H** is reversed in direction then **P** is reversed, but that if both **E** and **H** are reversed together then **P** preserves its original sense.

The actual flux of power across the area of **A** in fig. 1·2 is

$$W = (\mathbf{E} \times \mathbf{H}).d\mathbf{A} \text{ W.}$$

The energy-flow vector **P** defined in (1) is only one of many which can be used to describe the total transfer of power through the field, but for the

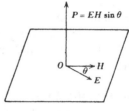

Fig. 1·2.

study of power flux associated with electromagnetic waves the vector **P** is the most useful and convenient.* We shall find that it is valuable in leading to an estimate of the attenuation of waves in wave guides due to loss of power in the walls.

According to Poynting's theorem there is a flux density of power in the direction of propagation of the electromagnetic wave whose instantaneous value is EH W./sq.m. ($\sin\theta = 1$). If $E = E_0 \cos\omega t$ and $H = H_0 \cos\omega t$, this becomes

$$E_0 H_0 \cos^2\omega t,$$

and the mean flux density averaged over a cycle is $\frac{1}{2}E_0 H_0$.

According to 1·2 (6) this mean flux density of power is

$$\frac{E_0^2}{240\pi}\sqrt{\frac{K_e}{K_m}} \text{ W./sq.m.} \tag{2}$$

---

* See Stratton, *Electromagnetic Theory*, p. 133; Slepian, *J. Appl. Phys.* 1942, vol. 13, p. 512; J. J. Thomson, *Recent Researches*, 1893, p. 313.

1·4. Behaviour of electric and magnetic fields at the surface of a conductor

Consider first the relation between the components of the electric-field strength on opposite sides of a sheet of electric charge whose density is s coulombs/sq.m. Let the electric inductive capacities of the media above and below the sheet be respectively ϵ_1 and ϵ_2, and

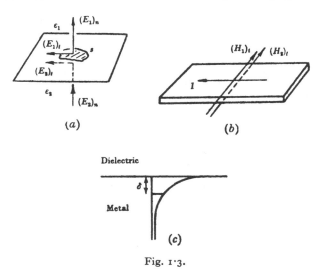

Fig. 1·3.

suppose the tangential and normal components of **E** to be $(E_1)_t$ and $(E_1)_n$ above the sheet (fig. 1·3 (a)) and $(E_2)_t$ and $(E_2)_n$ below it. Then

$$(\epsilon_1 E_1)_n - (\epsilon_2 E_2)_n = s, \quad (E_1)_t = (E_2)_t. \tag{1}$$

These relations are valid both for static and for time-dependent fields.

We next consider the connexion between the components of **H** at opposite faces of a sheet of current I amp./m. We have (fig. 1·3 (b))

$$(\mu_1 H_1)_n = (\mu_2 H_2)_n, \quad (H_1)_t - (H_2)_t = I, \tag{2}$$

where μ_1 and μ_2 are the magnetic inductive capacities of the media on either side of the sheet.

We may summarize the contents of (1) and (2) as follows:

In crossing a sheet of charge the tangential component of the electric vector **E** of an electromagnetic field is continuous, but the normal components of the electric induction $\mathbf{D} = \epsilon\mathbf{E}$ changes discontinuously by an amount equal to the surface-charge density.

In crossing a sheet of current the normal component of the magnetic induction $\mathbf{B} = \mu\mathbf{H}$ is continuous, but the tangential component of the magnetic field $\mathbf{H}$ changes discontinuously by an amount I, the current per unit length in the sheet.

Ideal perfect conductors—boundary conditions. A perfectly conducting material is one whose coefficient of conductivity is infinite. Within such an ideal conductor the electromagnetic field would be everywhere zero. Suppose that of the two media on opposite sides of the sheet of charge in fig. 1·3 (a), the upper is a dielectric with inductive capacity ϵ, and the lower a perfect conductor. It follows from 1·3 (1) with $E_2 = 0$, that

$$(\epsilon E_1)_n = s, \quad (\epsilon E_1)_t = 0. \tag{3}$$

Thus the tangential component of the electric field in a dielectric vanishes at a perfectly conducting boundary.

Similarly, if in fig. 1·3 (b) we suppose the lower medium to be a perfect conductor then $H_2 = 0$, and, according to (2),

$$(\mu H_1)_n = 0, \quad (H_1)_t = I. \tag{4}$$

Thus the normal component of $\mathbf{H}$ vanishes at a conducting boundary. Further, the current I flows entirely on the surface in an infinitely thin layer. The relative directions of I and $(H_1)_t$ are shown correctly in fig. 1·3 (b).

We shall later be concerned with investigating the field patterns of electromagnetic fields within wave guides and cavities with metal boundaries which we shall suppose to behave as perfect conductors. Whatever forms these patterns may take, they must be such that the electric field meets the boundary at right angles and the magnetic field touches the boundary tangentially.

The perfect conductor is an idealization of actual metallic conductors which possess very large but finite conductivities.

In practice the electromagnetic field penetrates from the dielectric into the conductor, but the field strengths and current density diminish exponentially with depth below the surface when the surface is plane (§7·14). For instance, the tangential component of $\mathbf{H}$ reaches its maximum value at the boundary and decays exponentially with depth as shown in fig. 1·3 (c). The depth δ at which H_t

decays to $\dfrac{1}{e} = \dfrac{1}{2\cdot718}$ of its surface value is called the *skin depth*. At the depth $2\pi\delta$ the amplitude falls to about $1/500$ of its surface value.

The skin depth is related to the frequency f of the oscillating electromagnetic field and the conductivity of the metal by the following formula (§7·14):

$$\delta = \sqrt{\dfrac{1}{\pi\mu f\sigma}}\ \text{m.,} \tag{5}$$

where μ is the magnetic inductive capacity of the metal which, except for the special instance of iron, is equal to $\mu_0 = 4\pi \times 10^{-7}$. The frequency f is in cycles per second and σ is the specific conductivity of the metal in mhos per metre cube.

The distance δ provides a convenient index of the degree of penetration of the electromagnetic field into the conductor.

To fix ideas, consider a particular application of formula (5) which is pertinent to our theme: we consider the penetration of waves whose free-space wave-length is $\lambda = 10$ cm., into the walls of a copper wave guide. In formula (5) put $f = 3 \times 10^9$, $\sigma = 5\cdot8 \times 10^7$ mhos/m., $\mu = \mu_0 = 4\pi \times 10^{-7}$ to find

$$\delta = 1\cdot17 \times 10^{-6}\,\text{m.}$$
$$= 1\cdot17 \times 10^{-4}\,\text{cm.}$$

Evidently all currents and fields within the metal are effectively concentrated into a very narrow layer near the surface.

The surface currents are accompanied by the conversion of electrical power into heat. It can be shown that the power dissipated in heat per square metre of the surface is

$$\dfrac{1}{2}\dfrac{I^2}{\sigma\delta}\,\text{W.m.}^{-2}, \tag{6}$$

where I is the amplitude of surface current in amperes per metre. This wastage of power is that which would occur if the current, instead of being distributed exponentially with depth, flowed uniformly in a surface layer of depth δ. This fact further stresses the usefulness of the quantity δ. The *equivalent surface resistance* is therefore $1/\sigma\delta$ ohms/sq.m. These facts are important for estimating the attenuation produced by the walls of a wave guide on progressive waves in it.

Whereas at the surface of a perfect conductor the tangential component of the electric field would be zero, at the surface of a good conductor, such as a metal, there is a small tangential component of the electric field.

The ratio of the tangential electric field to the tangential magnetic field at the conducting surface is

$$\frac{E_t}{H_t} = \sqrt{\frac{2\pi\mu f}{\sigma}}. \tag{7}$$

Consider again the case of copper at a frequency of 3×10^9 c./sec. Formula (7) gives $(E_t/H_t)_{\text{copper}} = 2 \times 10^{-2}$. This may be compared with the corresponding ratio for a plane wave in free space which according to equation 1·2 (6) is 120π. The tangential electric field necessary to drive a surface current of 1 amp./sq.m. on copper is therefore $(H_t = 1) 2 \times 10^{-2}$ V./m. at $f = 3 \times 10^9$. We conclude that the tangential component of the electric field at the surface of a metal is relatively small and that we may treat the metal as a perfect conductor when investigating the geometrical form of the electromagnetic field pattern in its vicinity.

1·5. Field pattern of a principal wave on parallel conductors —transmission-line formulae

We proceed to discuss (using the facts summarized in the previous sections) the properties of electromagnetic fields in regions bounded by metal surfaces, and we begin with the simple case of the principal wave in a parallel strip transmission line.

We consider the field pattern of fig. 1·1 (b) and note that it is one that can exist between a pair of parallel metal strips (except near the edges). This is, in fact, the field configuration of the *principal wave or TEM-wave* on a parallel-strip transmission line (neglecting edge effects). This wave is shown in fig. 1·4 (a). The electric lines stand perpendicular to, and the magnetic lines lie tangentially against, the metal walls and therefore do not violate the boundary conditions formulated in equations 1·4 (3) and (4). It is assumed that in discussing field patterns we may consider metals to behave as perfect conductors.

At the sections A and B, a distance half a wave-length ($\frac{1}{2}\lambda$) apart (fig. 1·4 (a)), the fields are shown with their maximum values and at the section C midway between their instantaneous value is zero.

The whole pattern travels from left to right at speed $v = 1/\sqrt{(\mu\varepsilon)}$. The wave-length on the line is the same as that in free space for the same frequency.

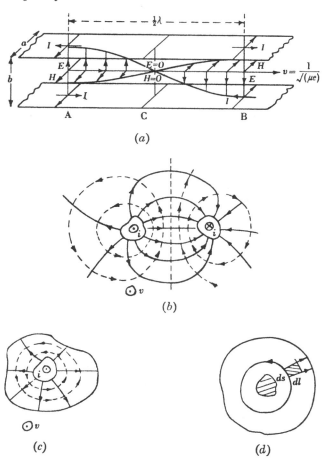

Fig. 1·4.

The properties of this electromagnetic field are identical with those of the unbounded field described in §§ 1·2 and 1·3.

Where the field **H** lies against the surface of the conductor a surface current $I = H$ flows either parallel to or against the direction of propagation. Suppose the width of each metal strip to be a m. and the distance between them b m. At a section where the instantaneous field strengths are E and H, the corresponding total current flowing in opposite senses on each plate is

$$i = aI = aH.$$

The voltage between the plates at this section is

$$V = bE.$$

The ratio of this voltage V to the associated current i in a single progressive principal wave is called the *characteristic impedance* Z_0 of the transmission line. Therefore, according to equation 1·2 (6),

$$Z_0 = \frac{V}{i} = \frac{b}{a}\frac{E}{H} = 120\pi\frac{b}{a}\sqrt{\frac{K_m}{K_e}} \text{ ohms.} \tag{1}$$

This formula neglects edge effects.

The impedance Z_0 does not depend on the amplitudes of **E** and **H** but is determined by the geometry of the line and the nature of the medium between the plates. In this particular case, the pattern of the lines of force in the section is the same as that of the electrostatic field which develops when the plates are maintained at a fixed potential difference V. Similarly, the magnetic field **H** in the section is the same as the magnetostatic field that results when the plates are joined by a conductor at one end and a steady current $i = aI$ is sent along one strip and back again along the other.

It is suggested that the field patterns of a principal (i.e. TEM) wave guided by a pair of parallel conducting cylinders with arbitrary geometrical cross-sections are also the same as the corresponding electrostatic and magnetostatic field configurations. This is, indeed, the case, and in figs. 1·4 (b) and (c) an indication is given of the field patterns of principal waves on transmission lines with cross-sections of arbitrary form. These are the electromagnetic field patterns of TEM or principal waves.

We next derive a general formula for the characteristic impedance of any pair of parallel cylinders such as those of figs. 1·4 (b) and (c). Let dl and ds be elements of length measured respectively along an electric and a magnetic line of force (fig. 1·4 (d)). The voltage across a section of the field is

$$V = \int E\,dl,$$

taken along a line of force from one conductor to the other, and the current on the surface of the conductors at this same section is

$$i = \oint H\, ds,$$

where $\oint$ is taken completely round a magnetic line of force.

The characteristic impedance Z_0, being the ratio V/i in a progressive wave, is therefore

$$Z_0 = \frac{\int E\, dl}{\oint H\, ds}. \tag{2}$$

If q is the charge per unit length of one of the conductors at the section in question, then, by Gauss's flux theorem,

$$q = \epsilon \oint E\, ds.$$

The capacity C per unit length is

$$C = \frac{q}{V} = \frac{\epsilon \oint E\, ds}{\int E\, dl}. \tag{3}$$

The self-inductance per unit length L is defined as follows: The flux of magnetic induction across a strip whose edge is an electric line of force and whose width is unity and parallel to the axes of the cylinders is

$$\mu \int H\, dl.$$

Since H is proportional at each point to the total current i we have

$$\mu \int H\, dl = Li = L \oint H\, ds.$$

The coefficient of proportionality L is called the self-inductance of the transmission-line system per unit length. Consequently,

$$L = \frac{\mu \int H\, dl}{\oint H\, ds}. \tag{4}$$

The ratio L/C is, from (3) and (4),

$$\frac{L}{C} = \frac{\mu \int H\,dl}{\oint H\,ds} \times \frac{\int E\,dl}{\epsilon \oint E\,ds}. \tag{5}$$

But at each point of the wave front of a TEM-wave

$$\sqrt{\epsilon}\,E = \sqrt{\mu}\,H.$$

Consequently equation (5) may be written

$$\frac{L}{C} = \frac{\sqrt{(\mu\epsilon)} \left[\int E\,dl\right]^2}{\sqrt{(\mu\epsilon)} \left[\oint H\,ds\right]^2} = Z_0^2,$$

that is,

$$Z_0 = \sqrt{\frac{L}{C}}. \tag{6}$$

Further, from (3) and (4)

$$\frac{1}{\sqrt{(LC)}} = \frac{1}{\sqrt{(\mu\epsilon)}} = v, \tag{7}$$

whence

$$Z_0 = \frac{\sqrt{(\mu\epsilon)}}{C} \text{ ohms}. \tag{8}$$

These are standard formulae of transmission-line theory.

In the particular instance of the parallel-strip transmission line of fig. 1·4 (a),

$$C = \frac{\epsilon a}{b} \text{ farads/m.,}$$

$$L = \frac{\mu b}{a} \text{ henries/m.,}$$

$$Z_0 = \sqrt{\frac{L}{C}} = \frac{b}{a}\sqrt{\frac{\mu}{\epsilon}} = 120\pi\,\frac{b}{a}\sqrt{\frac{K_m}{K_e}} \text{ ohms,}$$

which is formula (1).

Power carried by a progressive principal wave on a transmission line. As an example of the application of Poynting's theorem we calculate the power carried by a progressive wave on a transmission line.

The instantaneous flux of power across the elementary area $ds\,dl$ shaded in fig. 1·4 (d) is

$$dW = EH\,ds\,dl \text{ watts,}$$

and the total flux of power across the whole section is

$$W = \int\!\!\int EH\,ds\,dl$$
$$= \int E\,dl \oint H\,ds = i \int E\,dl$$
$$= Vi = Z_0 i^2.$$

This is the expression for W that we would naturally assume from circuit theory.

Evidently we may use either the language of the electromagnetic field to study the propagation of principal waves on transmission lines or the language of circuits (V and i), since when confined to the discussion of perfectly conducting cylinders they lead to identical conclusions.

The customary approach to transmission-line studies has been from circuit theory, but it is valuable, and at microwave lengths essential, to introduce field concepts if a faithful description of the phenomena is the aim.

Circuit concepts are unsuitable in introductory studies of wave guides, but when the facts have been firmly established and described in terms of field theory they may often conveniently be accurately restated in terms of circuit theory.

The principal or TEM-wave is not the only form of wave that can be propagated along transmission lines, and it will appear later that other forms of propagation are possible when the separation of the conductors is of the order of magnitude of the wave-length. Normally this condition does not obtain and these other waves are suppressed.

REFERENCES

Jackson, Willis. *High Frequency Transmission Lines*. Methuen Monograph.
Slater, J. C. *Microwave Transmission*. McGraw Hill Book Co.
Schelkunoff, S. A. *Electromagnetic Waves*. Van Nostrand Co. Inc.
Ramo, S. and Whinnery, J. R. *Fields and Waves in Modern Radio*. John Wiley.
M.I.T. Radar School. *Principles of Radar*. McGraw Hill Book Co.
Brainerd *et al*. *Ultra-High Frequency Techniques*. Van Nostrand Co. Inc.

Chapter 2

PROGRESSIVE ELECTROMAGNETIC WAVES IN WAVE GUIDES

2·1. General features of electromagnetic waves in metal wave guides

In an account of practical wave guides we are less concerned with general studies of all possible field configurations in a variety of wave guides than with an intimate appreciation of the properties of those few types of electromagnetic waves that are most commonly used in practice.

From a pedagogic standpoint it is most fortunate that the form of wave most frequently used, the H_{01}-wave in a rectangular wave guide, is also the one that is susceptible to elementary analysis. In this chapter not only shall we study the H_{01}-wave for its intrinsic importance, but we shall also employ it to demonstrate some properties common to all waves in wave guides.

Consider the parallel-strip transmission line of fig. 1·4 (a). It can evidently be converted into a rectangular wave guide of breadth a and depth b, by adding a pair of metal walls parallel to the dimension b. We inquire whether the TEM-wave of the strip transmission line is also found in the wave guide. That this type of wave cannot exist within the rectangular tube is easily seen, for its electric field would be required to exist tangential to one pair of walls and the magnetic field to stand perpendicular to these walls, a behaviour which is inconsistent with the boundary conditions 1·4 (3) and (4). Alternatively, since an electrostatic field cannot exist within a hollow conductor unless free charges are present in the cavity, we see that a principal wave cannot exist when there is only one conducting boundary present because its electric field pattern is that of a two-dimensional electrostatic field. We conclude that TEM-*waves are not propagated within wave guides.*

However, the waves that are actually propagated in wave guides differ markedly in many respects from the more familiar TEM-waves on transmission lines, and it is convenient first to summarize their more salient features. It will later appear that the particular type of wave that is selected for detailed discussion does in fact possess these properties.

Waves propagated in wave guides have the following characteristics:

(*a*) The oscillating fields **E** and **H** are not both transverse with respect to the direction of propagation (tube axis). Further, it is found that the waves in almost all instances are divisible into two classes:

(1) TE- (*transverse electric*) *or H-waves*. In these waves the electric field is entirely transverse, but the magnetic field possesses both a transverse component H_t and a longitudinal component H_l in the direction of propagation. The electric field **E** and the transverse component of the magnetic field vibrate in phase, but the two components of the magnetic field oscillate in quadrature (that is,

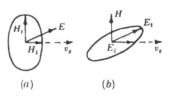

(*a*)   (*b*)

Fig. 2·1.

with a phase difference of 90°). The magnetic field therefore executes elliptic vibrations (except at exceptional positions where one or other of its components is permanently zero). The relation of **E** to **H** is shown in fig. 2·1 (*a*).

(2) TM- (*transverse magnetic*) *or E-waves*. In these, **H** is entirely transverse, but **E** has both a transverse component E_t and a longitudinal component E_l. The components of **E** oscillate in quadrature and with E_t in phase with H. These facts are illustrated in fig. 2·1 (*b*). (The TE and TM nomenclature is American and the E and H nomenclature is British.) There are many different kinds, or modes, of both E and H-vibrations that may occur in a wave guide.

(*b*) The *velocity of propagation* (phase velocity) of waves in guides exceeds the velocity of the TEM-wave of the same frequency in free space or on a transmission line, where ϵ and μ are the same.

Thus, if v_g is the speed of the wave-guide wave and v that of the TEM-wave, then

$$v_g > v = \frac{1}{\sqrt{(\mu\epsilon)}}.$$

Since $v_g = f\lambda_g \quad \text{and} \quad v = f\lambda,$

where λ_g and λ are the wave-lengths of the wave-guide wave and the TEM-wave respectively, it follows that

$$\lambda_g > \lambda.$$

The group velocity v_G (signalling velocity) is less than v_g. In fact

$$v_g v_G = v^2. \tag{1}$$

(c) *Cut-off.* Unless the TEM wave-length $\lambda = f/v$ is less than a certain value λ_c, called the cut-off wave-length, the disturbance cannot exist in the wave guide *as a wave.*

Each mode (or type) of wave has its own cut-off wave-length which depends on the geometry and the dimensions of the cross-section of the tube. The frequency $f_c = v/\lambda$ is called the cut-off frequency of the particular cut-off wave-length λ_c.

In a given wave guide there is one mode of propagation, called the *dominant mode*, whose cut-off wave-length λ_c exceeds that of any other mode. This greatest value of λ_c is of the order of magnitude of the smallest dimensions of the cross-section of the guide. For instance, if the guide is rectangular with dimensions a and b, with $a > b$, then

$$(\lambda_c)_{\mathrm{max.}} = 2a.$$

(d) The following relation obtains between the TEM wave-length λ, the wave-guide wave-length λ_g and the cut-off wave-length λ_c of any propagated mode:

$$\frac{1}{\lambda_g^2} = \frac{1}{\lambda^2} - \frac{1}{\lambda_c^2}. \tag{2}$$

Multiply each side of (2) by $1/f^2$ to obtain

$$\frac{1}{v_g^2} = \frac{1}{v^2} - \frac{1}{f^2 \lambda_c^2}. \tag{3}$$

Since v and λ_c are independent of f it follows that the velocity v_g depends on the frequency. In other words, the propagation is accompanied by dispersion.

2·2. Derivation of the electromagnetic field patterns of H_{0n} (TE$_{0n}$) modes

It has been remarked in the previous section that TEM- (or principal) waves are not propagated in metal tubes, but it remains to discover the form of the electromagnetic fields that satisfy the basic field equations of Maxwell and also the boundary conditions 1·4(3) and (4). We choose rectangular wave guides for immediate study, first, because they are of the greatest practical importance, and secondly, because a relatively elementary mathematical analysis leads to a satisfactory understanding of the basic physical processes both in these and in wave guides of other geometrical cross-sections.

It will appear that by simple superposition of a pair of elementary plane TEM free-space waves of the type described in § 1·2, a new field pattern results which is of such a form that it can be fitted correctly into a rectangular wave guide. Further, since the fields of the constituent plane waves are consistent with Maxwell's equations any field pattern which results from their simple superposition is also consistent with them.

It should not be thought that an elementary analysis of the type we shall employ of necessity gives a less fundamental description of the basic phenomena than a more sophisticated mathematical approach; on the contrary, the present method leads more rapidly to a sound physical grasp of the subject than a formal mathematical treatment. For this reason the mathematical treatment of wave propagation is relegated to the final chapter.

It is of interest to note that the method of resolving waves in rectangular tubes into elementary plane waves appears to have been used first by Lord Rayleigh* in connexion with the propagation of sound in pipes, and it has been mentioned by many subsequent writers of books and papers on wave guides.

Fig. 2·2 (a) represents a section made by the plane of the paper with a free-space TEM-wave whose wave fronts are perpendicular to the plane of the paper. The wave is travelling at velocity $v = 1/\sqrt{(\mu\epsilon)}$ in the direction indicated by the arrow. The section of the wave is so chosen that the magnetic field $\mathbf{H}$ is parallel to the plane of the paper. The heavy lines represent the magnetic lines and the arrows on them denote the directions of the magnetic vectors. The electric field $\mathbf{E}$ is represented by circles carrying a cross or a dot to indicate respectively that $\mathbf{E}$ is directed into or out of the plane of the section at right angles to $\mathbf{H}$ and to $\mathbf{v}$.

The full lines are wave fronts over which $\mathbf{E}$ and $\mathbf{H}$ are chosen to have maximum values alternately in one sense and the opposite sense. Consequently, the distance between adjacent full lines is half

---

* *Theory of Sound*, vol. 2, chapter XIII.

a wave-length $\frac{1}{2}\lambda$. Over the dotted lines, midway between, **E** and **H** are everywhere zero. The figure illustrates, therefore, the distribution of field in space at a chosen instant. Fig. 2·2 (*b*) shows a similar wave train of the same wave-length λ and field amplitude,

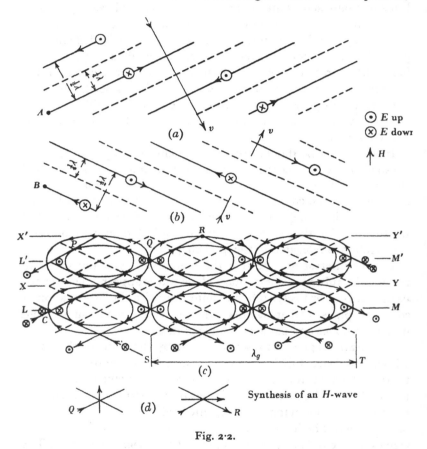

Fig. 2·2.

but moving in a different direction. In each case the direction of travel is perpendicular to the wave fronts. In fig. 2·2 (*c*) the wave trains are shown passing simultaneously through the same region by superimposing the individual patterns so that the points *A* and *B* coincide at *C*. The arrows which indicate the directions of the electric fields in the individual waves have, however, been placed at the edges of the diagram in order not to confuse the picture.

The instantaneous magnetic field at each position is the resultant of the two vector magnetic fields of the original plane waves. At certain points in the composite field of fig. 2·2 (c) the direction of the resultant may be seen by inspection. For instance, at all points on a dotted line the magnetic and electric fields of the one wave train are zero; consequently, the resultant field is identical with the field in the other wave train. Thus along any dotted line the direction of the field is parallel to the wave fronts of the other wave train, as, for instance, at P. Where two dotted lines intersect there is no field whatsoever. Since the original wave trains are assumed to possess equal field amplitudes, it follows that where two full lines cross the resultant is directed along the bisector of either the acute or the obtuse angle between them, that is, either parallel to the bisector of the acute angle between the propagation vectors $\mathbf{v}$ or at right angles to it. These possibilities are indicated at Q and R in fig. 2·2 (c) and are shown separately in fig. 2·2 (d).

When the direction of the resultant magnetic field is indicated at a sufficient number of points it is possible to sketch in the lines of magnetic force of the composite field pattern. These are evidently closed curves whose precise form is determined by the angle of intersection of the wave fronts of the original wave trains. The centre of each loop is also the point of intersection of two dotted lines and is therefore a point where both the electric and magnetic fields are zero.

Since the electric vectors of the original wave trains are directed either into or out from the plane of the diagram, the resultant electric field in the composite pattern is also normal to the plane of the paper and attains a maximum strength twice that in the separate wave trains. In fig. 2·2 (c) the circles with a cross or dot within indicate the sense of the resultant electric field. It is to be understood that the pattern extends in depth into and out of the plane of the paper.

Over planes such as XY and $X'Y'$ between adjacent layers of magnetic loops the magnetic field is *everywhere* tangential to these planes and the electric field is zero. Such planes are nodal planes of E, and antinodal planes of $H_{\text{longitudinal}}$. Conversely, over the intermediate planes LM, $L'M'$, etc., the magnetic field is everywhere perpendicular to them and the electric field reaches its maximum value on them. The distance ST (fig. 2·2 (c)) which comprises two loops is the wave-length λ_g of the composite pattern.

The pattern in fig. 2·2 (c) relates to a specific instant of time, but as the time increases the constituent wave trains (shown separately in figs. 2·2 (a) and (b)) move obliquely at velocity v and the composite pattern travels without distortion along XY at another velocity v_g. We proceed to find v_g.

In order to discover the velocity v_g of the pattern, all that is necessary is to find the velocity of some distinguishable feature of it. For instance, the magnetic field at C in fig. 2·2 (c) is associated with the point of intersection of two full lines (wave fronts) of the constituent waves shown separately in figs. 2·2 (a) and (b). These wave fronts are isolated in fig. 2·3, as CA and CD.

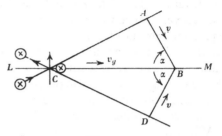

Fig. 2·3.

As the wave fronts progress at velocity v in the direction of the wave normals AB and DB respectively, their point of intersection C travels at a velocity, which we call v_g, along the bisector LM of the acute angle between the wave fronts. Since the points A, D and C arrive at D simultaneously it follows that

$$\frac{v}{v_g} = \frac{f\lambda}{f\lambda_g} = \frac{AB}{CB} = \cos\alpha,$$

where 2α is the angle ABD between the wave normals. Whence

$$v_g = v/\cos\alpha, \quad \lambda_g = \lambda/\cos\alpha, \tag{1}$$

where
$$v = \frac{1}{\sqrt{(\mu\epsilon)}}.$$

We may summarize what has been done as follows: By suitably superimposing two plane electromagnetic waves travelling obliquely across each other we obtain a new field pattern which has a completely different appearance from that of the original plane waves. The direction of propagation is along the bisector of the angle between the two wave normals (ray directions) of the elementary waves and the speed of propagation is

$$= v/\cos\alpha, \quad \text{which exceeds } v.$$

The magnetic field in the composite pattern comprises a system of closed loops, and it is evident that at most points in the pattern there is a component H_l of the magnetic field along or against the direction of propagation. The electric field is entirely transverse but is amplitude modulated in the plane of the magnetic field along directions at right angles to the direction of propagation.

From what was stated in § 2·1 we conclude that we have synthesized an H- (or TE) wave in free space.

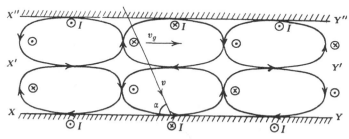

Fig. 2·4.

We may therefore regard the system either as a single H-wave or as a pair of crossing TEM-waves, according to convenience.

The H-wave is of unlimited extent, and the next task is to find how a portion of it can be fitted into a rectangular wave guide. We proceed in stages, first of all fitting a portion of it between the plates of a parallel-strip transmission line.

It has been remarked that in fig. 2·2 (c) the lines XY and $X'Y'$ represent planes over which **E** is everywhere zero and **H** is tangential. The behaviour of the electromagnetic field at these planes is precisely what we require at a plane metal boundary, and it is evident that the pattern can be fitted between a parallel pair of metal plates of infinite extent provided the plates coincide with a pair of nodal E-planes of the pattern such as XY and $X'Y'$ which contain one or more rows of closed loops between them. It follows that a composite wave pattern such as that indicated in fig. 2·4 is an example of a possible electromagnetic wave that could be propagated between a pair of parallel plates.

Surface currents flow in the plates at right angles to the direction of propagation at those places where the tangential magnetic field at the plates is not zero. The pattern may still be resolved into a pair of waves reflected obliquely back and forth between the plates, if so desired.

Fig. 2·5 represents the propagation of a three-layer *H*-wave between a pair of parallel plates of unlimited lateral extent. It is clear that the distance *b* between the plates is an exact multiple of the width, at right angles to the direction propagation, of the largest magnetic loops; or, more precisely, the distance *b* is an integral multiple of the distance between adjacent antinodal *E*-planes.

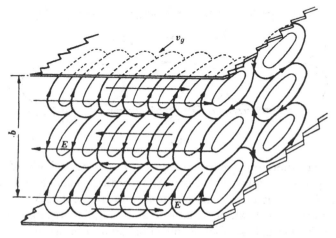

Fig. 2·5.

When a second pair of parallel metal plates is inserted at right angles to the other pair and at a separation *a*, the resulting tube is a rectangular wave guide. If the second pair of walls is inserted parallel to the plane of the magnetic loops, then it can be seen that the portion of the originally unlimited field which is now isolated by the four walls is one that can persist in the wave guide because the magnetic field is tangential at each wall and the electric field is perpendicular to one pair of walls and zero on the other pair. The pattern is consistent with Maxwell's field equations and also satisfies the necessary boundary conditions. The wave guide and wave pattern are shown in fig. 2·6.

Since the *H*- or TE-wave shown in fig. 2·6 is only one of many possible field patterns that can exist within the wave guide, it is convenient at this point to introduce the standard nomenclatures by which the individual waves are designated. Refer the wave pattern of fig. 2·6 to a set of Cartesian coordinate axes $OXYZ$ in which O coincides with the point A, OX with the edge AD, OY with the edge AB and OZ with the direction of propagation. We note that in travelling across the face of the wave-guide section $ABCD$ parallel to the edge AD (coordinate y remains constant) there is no change in the instantaneous value of either $\mathbf{E}$ or $\mathbf{H}$ due to the variation in the coordinate x. However, in a displacement across the

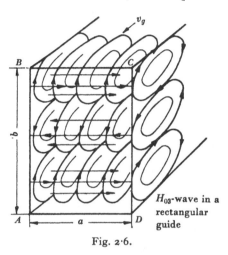

H_{03}-wave in a rectangular guide

Fig. 2·6.

face, parallel to OY (y changes but x remains fixed), the field components alter cyclically because the dimension b spans three complete loops. Because the wave is an H-wave (TE-wave) in which no change of pattern is associated with the coordinate x, and a three-layer change is associated with the coordinate y, it is designated as an H_{03}- or TE_{03}-wave. More generally, when the b dimension parallel to OY comprises n layers, and the dimension a parallel to OX no loops, the wave is an H_{0n}-wave. Suppose the wave guide and its pattern to be turned through 90° so that the dimension b lies along OX and a along OY. We now have n layers along OX and none along OY. The wave, although the same wave, would be called an H_{n0}-wave; that is, the first subscript is associated with OX and the second with OY.

One member of the H_{0n} family of waves is of outstanding practical importance; it is the H_{01}-wave and we discuss it in detail below. The wave-length λ_g of these waves in the guide is twice the length of a large loop in the direction of propagation.

2·3. The H_{01}- (TE$_{01}$) wave in a rectangular wave guide—dominant modes

This wave, as the subscripts indicate, comprises a single layer of loops between the walls parallel to the XOZ plane.

Fig. 2·7 represents a section of the field parallel to the ZOY plane, that is, the section contains the direction of propagation and the

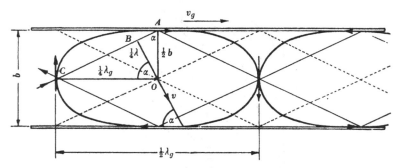

Fig. 2·7. Section of H_{01}-wave in a rectangular guide.

dimension b. Not only are two of the larger loops represented but also the wave fronts of the two obliquely reflected elementary plane waves. The ray direction (wave normal) of the wave front AC is BO, and to be consistent with figs. 2·2, 2·3 and 2·4 we must suppose that BO produced intersects the lower wall at an angle of elevation α. As indicated on the figure, we may recognize the following dimensions:

$$CO = \tfrac{1}{4}\lambda_g, \quad BO = \tfrac{1}{4}\lambda, \quad AO = \tfrac{1}{2}b,$$

where λ_g is the wave-length of the H_{01}-wave which is equal to the length of two loops because the fields are periodic in this distance along the direction of propagation.

λ is the wave-length of the elementary TEM-waves in the tube and b the y-dimension of the tube.

We find, directly from the diagram,

$$\sin \alpha = \lambda/2b, \tag{1}$$

$$\cos \alpha = \lambda/\lambda_g = v/v_g, \tag{2}$$

whence
$$\left(\frac{\lambda}{2b}\right)^2 + \left(\frac{\lambda}{\lambda_g}\right)^2 = 1$$

or
$$\frac{1}{\lambda_g^2} = \frac{1}{\lambda^2} - \frac{1}{(2b)^2}. \tag{3}$$

(3) is evidently a special case of 2·1 (2) with the cut-off wave-length $\lambda_c = 2b$.

We conclude that the cut-off wave-length of an H_{01}-wave is $2b$; that is, twice the y-dimension of the cross-section.

Let the TEM wave-length $\lambda = 2b = \lambda_c$ and the frequency $f = v/\lambda_c = f_c$, then according to equation (1) the angle of incidence of the elementary waves reflected to and fro between the walls is 90°. That is, the wave fronts of the elementary waves are parallel to the walls, the loops of the composite H_{0n}-waves are infinitely long and the composite wave pattern does not progress along the axis. Thus at the cut-off wave-length there is no component of the Poynting flux along the axis of the wave guide and no progressive wave is propagated.

It is easy to see that the cut-off wave-length of the H_{0n}-wave is b/n, for b now spans n layers and in fig. 2·7 OA becomes $b/2n$, and equation (3) becomes
$$\frac{1}{\lambda_g^2} = \frac{1}{\lambda^2} - \left(\frac{n}{2b}\right)^2, \tag{4}$$

or
$$\lambda_c = 2b/n. \tag{5}$$

The different field patterns corresponding to the various values of n are known as *wave modes*. Formula (5) shows that the higher the order of the mode (i.e. the larger n) the smaller the cut-off wave-length in a guide with fixed dimension b. In other words, the TEM wave-length of the disturbance must be further diminished in order to propagate a high-order mode down a tube than to propagate a low-order mode. Also, when the frequency f is given ($\lambda = v/f$) it is possible so to choose the dimension b such that the H_{01}-wave is propagated, but the H_{02}-, H_{03}-, etc., waves cannot be propagated.

In addition to the H_{0n}-wave patterns in a guide there are also more complicated H-wave patterns in which the field components are dependent upon the coordinates x as well as y. The general designation for these is H_{mn}. There is also a set of E-wave patterns (TM-waves) designated in general by E_{mn} (TM$_{mn}$).

Of all these possible wave types in rectangular wave guides the H_{01}-wave (where $b > a$, and is parallel to OY) is the *dominant mode*, that is, it has the greatest cut-off wave-length. It is therefore possible to propagate the H_{01}-wave (or H_{10} according as b lies along OY or OX) alone and to suppress all other types of progressive wave by appropriate choice of b. This feature of the H_{01}-wave gives it a practical importance for several reasons. First, it is of advantage to carry the power to and from the aerial in a single type of wave because the field configuration is then determinate and it is possible to radiate into the aerial from the end of the wave guide in a predictable fashion. When several propagated modes are permitted it is impossible to know what are their relative amplitudes and what are the field distributions at any section of the guide. Secondly, when a single wave type alone is permitted, all others being suppressed, the wave guide resembles a transmission line in which the power is carried by the principal wave and other possible modes of transmission are suppressed. In such wave guides many transmission-line techniques for eliminating reflexions along the course of the guide and at the termination may be adopted with very little modification.

These matters are further discussed in the sequel, but to illustrate what was said about the choice of wave-guide dimension it is pertinent to give the dimensions of a standard wave guide designed for use at a frequency of 3300 Mc./sec. ($\lambda = 9 \cdot 1$ cm.). This is a rectangular wave guide, whose internal dimensions are $1 \times 2\frac{1}{2}$ in., that is, $2 \cdot 54 \times 6 \cdot 35$ cm. The cut-off wave-length corresponding to the dimension $b = 6 \cdot 35$ cm. is $(\lambda_c)_b = 12 \cdot 7$ cm., which exceeds $\lambda = 9$ cm. of the wave to be propagated. The cut-off wave-length corresponding to the dimension $a = 2 \cdot 54$ cm. is $(\lambda_c)_a = 5 \cdot 08$ cm., which is much less than 9 cm. Let the edges of length b and of length a be respectively parallel to OY and OX, then it is evident that the guide can carry an H_{01}-wave but not an H_{10}-wave. The cut-off wave-length of the H_{02}-mode is $b = 6 \cdot 35$ cm., and this is the greatest TEM wave-length which can be propagated in this form down the tube. Clearly the 9 cm. wave cannot be so propagated. Therefore, in this wave guide the only wave that can be propagated at 3300 Mc./sec. is an H_{01}-wave with the electric field perpendicular to the long edges of the cross-section and with the magnetic loops in planes parallel to the broader pair of faces.

If we choose to regard the H_{01}-wave as a pair of wave trains reflected to and fro between the narrow faces of the guide, then the angle α (complement of angle of incidence) is, according to (1),

$$\alpha = \sin^{-1}\frac{\lambda}{2b} = \sin^{-1}\left(\frac{9 \cdot 1}{12 \cdot 7}\right) = 45°.$$

The wave-length λ_g of the H_{01}-wave is, from equation (3), $\lambda_g = 13$ cm.

2·4. Method of launching an H_{01}-wave in a rectangular wave guide

A common method of launching an H_{10}-wave is shown in fig. 2·8. A probe, whose length is $\frac{1}{4}\lambda$ (not $\frac{1}{4}\lambda_g$) protrudes into the wave guide

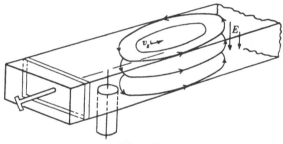

Fig. 2·8.

through a hole on the centre line of a broad face. The probe, in practice, would be an extension of the inner conductor of a coaxial feeder line. It radiates as a $\frac{1}{4}\lambda$ aerial, and at a distance of some 6 in. away the field, which near the probe is highly complex, has simplified to that of the single H_{10}-wave travelling down tube as shown. The movable plunger at the left of the figure is adjusted to act as a matching device for the coaxial line input, and the best position, found by trial, is that for which maximum power proceeds down the guide. This best position is roughly at a distance an odd multiple of $\frac{1}{4}\lambda_g$ from the probe.

2·5. System of wall currents of the H_{10}-wave

Currents flow in the surfaces of the walls where there exists a tangential component of the magnetic field, the direction of flow being at right angles to the local magnetic field, and it is important

to understand how the flow of current is related to the electromagnetic field of the wave.

There are several special applications of lengths of wave guide
with slots cut in the walls; for instance, in testing for the presence of
standing waves in a guide it is necessary to insert a probe through a
longitudinal slot; further, slots one-half wave-length long are used
for radiating directly from the wave guide into space. In all these
cases it is important to have a clear understanding of the 'arterial'
flow of current in the walls before making a 'surgical incision'.

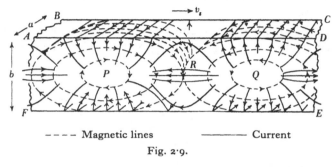

- - - - Magnetic lines   ———— Current

Fig. 2·9.

Fig. 2·9 shows the instantaneous directions of flow of the currents
in the walls of a rectangular wave guide carrying an H_{01}-wave. The
direction of flow is everywhere at right angles to the local magnetic
field. Thus, on the face $ADEF$ to which the magnetic loops are
parallel, the currents converge towards the region P, and away from
the region Q. The instantaneous surface charge densities at P and
Q are zero, but their rate of change is a maximum and P is about to
be charged positively and Q negatively with the inverse charging
processes at the regions opposite them on the other broad face.
At the instant to which the picture refers, no transverse electric field
exists at P or Q. The region R and its opposite on the other face are
fully charged and there is a transverse electric field between the
broad faces of the guide at R. This is consistent with figs. 2·5, 2·6
and 2·7, where the electric field is transverse and situated near the
ends of the magnetic loops but is zero at the centres of the loops.
The whole pattern of current flow is propagated in the direction of
the wave-guide axis at the speed v_g.

A narrow slot may be cut in a wall without affecting the propagation of the wave provided it does not cut across the lines of

flow of the current. For instance, a narrow longitudinal slot in the centre of the broad face $ADEF$ cut in the direction PRQ nowhere interrupts the flow of current in the wall and therefore has negligible influence on the propagation. Such a slot is used in standing wave measurements. Were a similar slot to be cut along the centre line of the narrow face $ABCD$ it would be at right angles to the current and the large reflected component of the H_{01}-wave would be generated at the slot unless it were very narrow.

2·6. Other wave modes in a rectangular wave guide

E-waves (TM-*waves*). In these, as already explained in §2·1, the magnetic field is entirely transverse, but the electric field possesses both a transverse and a longitudinal component. In order to discover the general form of the field pattern of the *E*-waves in a rectangular wave guide we again use the technique of superimposing the fields of a pair of elementary waves to obtain an *E*-wave for the resultant, instead of an *H*-wave. We then attempt to enclose the *E*-wave in a wave guide as before. To obtain an *E*-wave it is necessary to change the polarization of the elementary waves, so that the electric field **E** and not the magnetic field **H** as shown in fig. 2·2 lies in the plane of the paper. The full lines in fig. 2·2 now represent **E** and the circles the direction of **H** except that the sense of the latter should be the reverse of that indicated. In the composite pattern 2·2 (*c*) the closed loops now represent lines of electric force **E** and the circles (with the senses reversed) the magnetic field **H**. The field is therefore that of an *E*-wave (TM-wave) in space.

A portion of this field may also be isolated between a pair of parallel plane metal plates if the pattern is fitted as shown in fig. 2·10 (*a*).

It is evident that the boundary conditions 1·4 (3) and (4) are fulfilled. An oblique view of the two-layer *E*-wave is given in fig. 2·10 (*b*). It is easy to see that the formulae 2·3 (1), (2) and (3) are also applicable here in the case of a single-layer *E*-wave, where b is the distance between the plates. The cut-off wave-length is also $\lambda_c = 2b$ for a single-layer wave and $(2b/n)$ for an n-layer wave. When, however, we attempt to enclose the *E*-wave of fig. 2·10 in a rectangular wave guide to form an E_{0n}-wave analogous to the H_{0n}-wave, the addition of the pair of side walls destroys the field because the planes of the *E*-loops are parallel to these walls and the *H*-lines are perpendicular to them, so that the boundary conditions are violated. A *rectangular* wave guide cannot, therefore, carry an E_{0n}-wave.

The simplest E-wave in a rectangular wave guide is the E_{11}-wave (TM_{11}-wave), whose electromagnetic field pattern is shown in fig. 2·11, but is not derived.

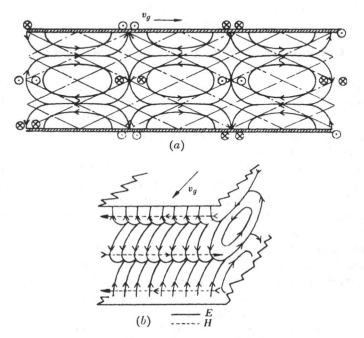

Fig. 2·10. E_{02}-wave between parallel plates.

Fig. 2·11 (*a*) represents a central section of the wave guide containing the axis and parallel to one pair of walls. It will be noted that the electric lines of force are half-loops that stand perpendicular to the walls. Along the axis of the wave guide the electric lines of force form an axial bundle. Fig. 2·11 (*b*) is a transverse section of the wave at P in fig. 2·11 (*a*) with the wave approaching the observer. The centre of the section is a source of electric lines which diverge from the axial bundle and terminate on the walls. The magnetic lines are closed loops lying in the cross-section. At Q, the pattern is similar but with the directions of the fields reversed.

The theory of the propagation of this wave leads to the following formula for the cut-off wave-length of the E_{11}-wave in a wave guide whose cross-sectional dimensions are a and b:

$$\frac{1}{\lambda_c^2} = \frac{1}{(2a)^2} + \frac{1}{(2b)^2},\qquad(1)$$

and the wave-length λ_g is

$$\frac{1}{\lambda_g^2} = \frac{1}{\lambda^2} - \frac{1}{\lambda_c^2}$$

as before.

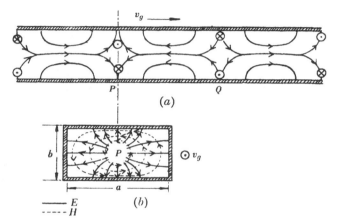

Fig. 2·11. E_{11}-wave in a rectangular tube.

The general E-wave (TM_{mn}). The wave pattern of an E_{mn}-wave may be regarded as a number of $m \times n$ of E_{11}-wave patterns fitted into a single tube and integrated into a continuous pattern. For instance, fig. 2·12 is the cross-section of the pattern of an E_{32}-(TM_{32}) wave at a position where the magnetic field is at maximum intensity. The pattern resembles a tiled floor with the E_{11} pattern as unit. In adjacent tiles the centre of one is a source of E-lines and the centre of the other a sink. We could, in fact, suppose metallic partitions inserted between the unit patterns so that the E_{mn}-wave resolves into $m \times n$ independent E_{11}-waves.

The subscripts m and n refer to the number of unit tiles along the horizontal and vertical dimensions a and b respectively.

We infer from what has been said, and from formula (1), that the cut-off wave-length of the E_{mn}-wave in a wave guide with dimensions a and b is the same as that of an E_{11}-wave in a wave guide of dimensions (a/m) and (b/n) or

$$\frac{1}{\lambda_c^2} = \left(\frac{m}{2a}\right)^2 + \left(\frac{n}{2b}\right)^2.$$ (2)

This formula is, in fact, correct.

The H_{mn}-(TE$_{mn}$) wave. Like the E_{mn}-wave pattern, that of the H_{mn}-wave can be regarded as an $m \times n$ manifold of a simple basic

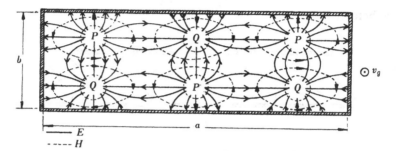

Fig. 2·12. E_{32}-wave in a rectangular guide.

pattern, that of the H_{11}-wave, which is illustrated in fig. 2·13. The corners of the section are sources and sinks, alternately, of magnetic lines of force which run near the corners of the wave guide in bundles. Each magnetic line is a closed loop. The electric field is transverse with the electric lines of force spanning the corners of the cross-section as shown. Formula (2) also gives the cut-off wavelength of the H_{mn}-wave.

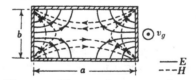

Fig. 2·13. H_{11}-wave in a rectangular guide.

Of the E_{mn} and H_{mn} types (or modes) only the H_{01}-mode is used in practice to carry power in a wave guide, for the reasons mentioned in §2·3. We shall therefore discuss the E_{mn} or H_{mn} no further at this stage.

2·7. Circular wave guides and higher modes in coaxial transmission lines

2·7·1. Although a rigorous account of the propagation of electromagnetic waves in circular tubes and of the higher modes in coaxial transmission lines requires a more advanced mathematical treatment than it is appropriate to introduce at this point, yet the physical approach employed in the previous section can be adapted with advantage to our present discussion.

2·7·2. *Supplementary waves in coaxial transmission lines*

Consider a single-layer H-wave between a pair of parallel plates as shown in fig. 2·14(a).

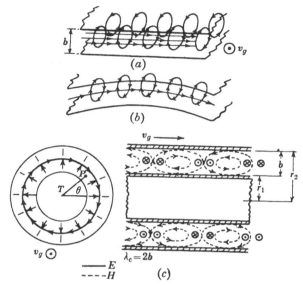

Fig. 2·14. H_{01}-wave on a coaxial line.

Next, suppose the plates to be curved as indicated in fig. 2·14(b) until finally the system is transformed into the coaxial transmission line of fig. 2·14(c). The H-wave between the plates transforms into a wave propagated between the inner and outer conductors of the coaxial line of the form shown.

When the spacing b between the conductors of the coaxial is small in comparison with the radius r the wave closely resembles, both in pattern and behaviour, the original H-wave between the flat plates. In particular, it would be expected that when $r \gg b$ the cut-off wave-length of this coaxial mode is $\lambda_c \doteqdot 2b$, as obtains with the parallel plate H-wave. A rigorous analysis confirms this surmise but affords very little additional information.*

* Stratton, *Electromagnetic Theory*, p. 548.

A more general H-wave in a coaxial system is obtained by 'wrapping up' an n-layer H-wave between parallel plates to give an n-layer wave between coaxial cylinders, whose cut-off wavelength is $\lambda_c \doteqdot 2b/n$, when $b \ll r$, where r is the mean radius of the coaxial system. This more general n-layer wave in the coaxial line is called an H_{0n}-wave, for there is here no variation with the polar angle coordinate θ (which replaces the Cartesian coordinate x) but an n-layer variation with the radial coordinate r.

We conclude that in addition to the usual principal wave on a coaxial line, other waves may be propagated whose properties are similar to those of waves in wave guides and very different from those of the principal wave. However, unless the wave-length of the principal wave is reduced until it is of the order of the distance between the inner and outer conductor these additional or *supplementary modes* are not excited as progressive waves.

2·7·3. H_{01}-wave (TE$_{01}$) in a circular wave guide

Suppose the inner cylinder of the coaxial line carrying the H_{01}-wave, of fig. 2·14 (c), to shrink until it becomes an axial wire. The

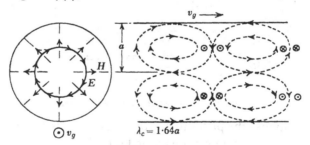

Fig. 2·15. H_{01}-wave in a circular guide.

magnetic loops now run parallel where they touch the surface of the wire and form an axial bundle. Since a parallel bundle of lines of force is self-supporting, the wire is superfluous and may be withdrawn leaving the wave pattern unchanged. We thus arrive at the pattern of an H_{01}-wave in a circular wave guide. The cut-off wavelength is no longer given accurately by the formula $\lambda_c \doteqdot 2b$, and the correct formula is $\lambda_c = 1\cdot64a$, where a is the internal radius of the tube. Fig. 2·15 indicates the general form of the electromagnetic field of an H_{01}-wave. The H_{0n}-wave in a circular wave guide may also be regarded as a limiting form of the H_{0n}-wave in a coaxial transmission line as the radius of the inner conductor is reduced. These transmission modes have, as yet, found no extensive application.

2·7·4. E_{0n}-waves (TM$_{0n}$)

The field configuration of the E_{0n}-mode on a coaxial may be derived from that of the n-layer E-wave between parallel plates (fig. 2·10) by bending as before, and the cut-off wave-length is again

$$\lambda_c \doteqdot 2b/n,$$

where b is the separation of the conductors. The inner conductor is surrounded by half-loops of electric force whose ends stand perpendicular to it and terminate on surface charges. Consequently, it is not permissible to reduce the inner conductor and to remove it, since its presence is required to support the pattern.

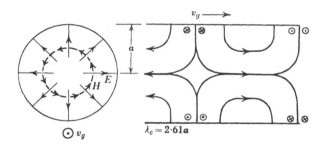

Fig. 2·16. E_{01}-wave in a circular guide.

To obtain the E_{01}-wave pattern in circular wave guide, we start with a square-section wave guide that carries an E_{11}-wave (fig. 2·11) and suppose it to be distorted into a circular wave guide. The E_{11}-wave pattern then transforms into the E_{01}-wave pattern in a circular wave guide indicated in fig. 2·16.

The magnetic lines of force are circles lying in the plane of the cross-section and the electric lines are half-loops standing on the surface of the tube. The cut-off wave-length of this wave is $\lambda_c = 2·61a$, where a is the radius of the tube.

The E_{01}-wave is of considerable practical importance because it is used in rotatable joints which are required to connect a fixed wave-guide feeder system to a rotatable aerial. This application will be described later.

An important feature common to all H_{0n}- and E_{0n}-waves in circular wave guides is that their fields are *completely symmetrical about the axis of the wave guide.*

2·7·5. *Wave modes without axial symmetry*

It remains to give examples of waves in circular wave guides and coaxial transmission lines whose field patterns do not possess axial symmetry. We again derive our final pattern by transforming a known pattern into the one that we seek.

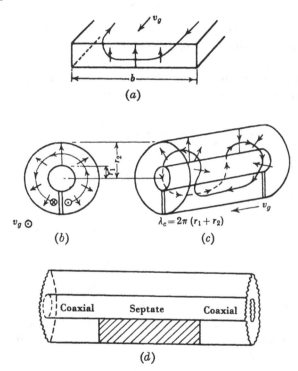

Fig. 2·17. Septate guide.

Consider an H_{10}-wave in a flat wave guide of the form indicated in fig. 2·17. Let the wave guide be bent into a coaxial system but one in which the original narrow faces of the wave guide meet and unite to form a metal septum between the inner and outer conductors as shown in fig. 2·17 (b) and (c). This structure is called a **septate** coaxial system. The original electromagnetic field of the

H_{01}-wave transforms to that shown in figs. (b) and (c) where the magnetic loops are bent over so that their longitudinal portions run in opposite senses, one on each side of the septum. The discontinuity in **H** across the septum is supported by currents in its surface. The electric field is radial with its intensity zero at the septum and a maximum opposite the septum.

A feature of interest about the septate-coaxial system is its cut-off wave-length compared with the longest cut-off wave-length of a circular wave guide (or even a coaxial transmission line with respect to supplementary modes). The cut-off wave-length of the prototype wave guide of fig. 2·17 (a) is $\lambda_c = 2b$. In the septate system the dimension b has become the mean circumference $\pi(r_1 + r_2)$, where r_1 and r_2 are the radii of the surfaces of the coaxial system.

The cut-off wave-length, when the separation of cylinders is small compared with r_1 is, therefore,

$$\lambda_c \doteqdot 2\pi(r_1 + r_2).$$

This is approximately twice as great as the greatest cut-off wave-length in a circular wave guide of the same outer radius r_2.

The field configuration in the septate coaxial system is very similar, in the region where the electric field is greatest, to the pattern of the principal wave on a coaxial transmission line and is consequently readily excited by the principal wave on a coaxial line whose inner conductor forms an extension of the inner cylinder of the septate-coaxial combination, as shown in fig. 2·17 (d), provided the wave-length of the principal wave is less than the cut-off wave-length of the septate guide. When the radius of the inner cylinder is reduced to zero the septate coaxial becomes a *septate wave guide*.

We consider next the field configuration of the H_{11}-wave in a circular wave guide. In fig. 2·18 (a) a pair of identical wave guides carrying identical H_{01}-waves is shown. In fig. 2·18 (b) these are bent and finally clamped together at (c) to form a double septate coaxial with the fields transformed as shown. Since the magnetic fields run parallel on opposite sides of each septum, the septa are superfluous and may be removed, leaving the coaxial system in which the field pattern is identical with that in fig. 2·18 (c). The cut-off wave-length is approximately that of either of the prototype rectangular wave guides, consequently

$$\lambda_c \doteqdot \pi(r_1 + r_2).$$

This mode possesses the greatest cut-off wave-length of all supplementary coaxial modes. When the inner conductor is shrunk to a wire and then removed the field transforms to that depicted in fig. 2·18 (d) which is that of an H_{11}-wave in a circular wave guide. Since r_1 is zero we have approximately

$$\lambda_c \doteqdot \pi r_2 = \pi a,$$

whereas a rigorous analysis gives

$$\lambda_c = 3 \cdot 42a,$$

where a is the inner radius of the wave guide.

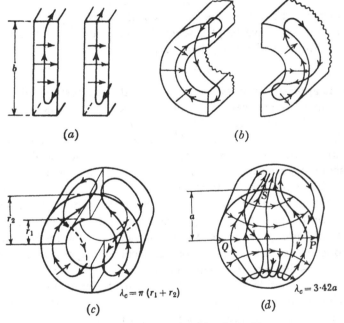

$$\lambda_c = \pi (r_1 + r_2)$$

$$\lambda_c = 3 \cdot 42a$$

Fig. 2·18. H_{11}-wave in a circular guide.

The cut-off wave-length of the H_{11}-wave is greater than that of any other mode in a circular wave guide and this mode is therefore the dominant mode.

The field pattern of an E_{11}-mode in a circular wave guide is sketched in fig. 2·19. Its cut-off wave-length $\lambda_c = 1 \cdot 64a$ is the same as that of the H_{01}-mode.

It is sometimes possible to recognize within the wave pattern of a wave in a given wave guide, the pattern of a wave in another wave guide with a different geometrical cross-section. For instance, consider the pattern of the H_{11}-wave in the square-section wave guide of fig. 2·20 (a).

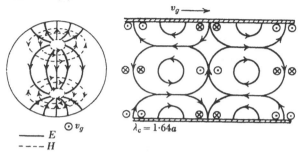

Fig. 2·19. E_{11}-wave in a circular guide.

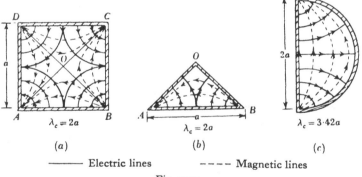

(a)   (b)   (c)

————— Electric lines   – – – – Magnetic lines

Fig. 2·20.

It is evident that thin diagonal metallic partitions AC and BD can be introduced without violating the electromagnetic boundary conditions, and that the field pattern in the triangular sections remain unaffected. Thus, a wave can be propagated in the quarter wave guide whose section is an isosceles right-angled triangle, whose field pattern is portion of the field pattern of the E_{11}-wave in a square wave guide. The cut-off wave-length is $\lambda_c = 2a$ in each case, where a is the edge of the square. Fig. 2·20 (c) shows one-half of the H_{11}-wave of fig. 2·18 (c) in a semicircular wave guide.

We conclude this subsection with a summary of the cut-off wave-lengths of those few wave types of practical importance:

(1) The principal or TEM-wave on a transmission line—no cut-off phenomenon.

(2) H_{01}-wave (TE$_{01}$) in a rectangular wave guide with magnetic loops parallel to the edge of the cross-section of length b. Cut-off wave-length $\lambda_c = 2b$. This is the dominant mode in a rectangular guide when b is the larger dimension.

(3) E_{01}-wave (TM$_{01}$) in a circular wave guide of radius a, $\lambda_c = 2 \cdot 61 a$.

(4) Dominant mode in septate-coaxial system—$\lambda_c \doteq 2\pi(r_1 + r_2)$.

(5) H_{11}-wave (TE$_{11}$) in a circular guide of radius a, $\lambda_c = 3 \cdot 42 a$—dominant mode in a circular wave guide.

2·8. Methods of launching those wave modes of practical significance

We have already indicated in fig. 2·8, and §2·4, one method of launching an H_{01}-wave in a rectangular wave guide, and this is the method that is commonly used. If the power radiated is large then a thick probe with a spherical end is used to avoid electrical break-down of the air. The wall opposite the probe may also be recessed into a spherical dome.

It should be noted that the system is reversible and that power may be abstracted from the wave guide by the probe and led away by the coaxial line, the piston position being adjusted to give maximum power at the output of the coaxial line. The same method can be used to launch an H_{11}-wave into a circular wave guide and to abstract power from the wave guide. The electric field in the H_{11}-wave is parallel to the probe.

A modification of this method of launching these H-waves and abstracting power from them is shown in fig. 2·21, which represents the launching section of a $2\frac{1}{2} \times 1$ in. wave guide for operation at a frequency of 3300 Mc./sec. ($\lambda = 9$ cm.).

It will be observed that the radiating wire extends the whole way across the narrow dimension of the wave guide and continues on as the central conductor of a coaxial stub short-circuited by a movable plunger. The diagram represents a central section. The wave guide is also terminated on one side of the input feeder by a shorting plunger which makes good contact with the walls by means of a

fringe of small phosphor-bronze spring contacts. This plunger can
be moved smoothly by means of a screw and control knob as shown.
A double adjustment of coaxial plunger and wave-guide plunger is

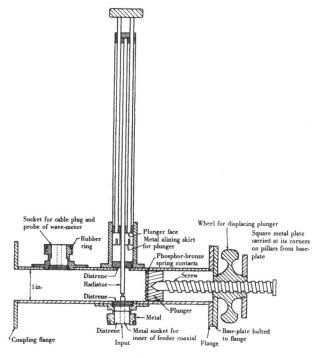

Fig. 2·21.

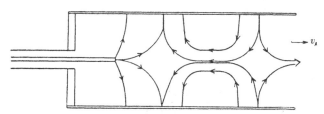

Fig. 2·22.

made until a satisfactory match between the input cable and the
wave guide is achieved.

A method of launching an E_{11}-wave in a rectangular wave guide
or an E_{01}-wave in a circular wave guide is shown in fig. 2·22.

The wave is excited by a $\frac{1}{4}\lambda$ probe that protrudes into the wave guide along its axis and through a hole in an end-plate. The electric lines of force can be crudely envisaged as being shed from the probe as indicated.

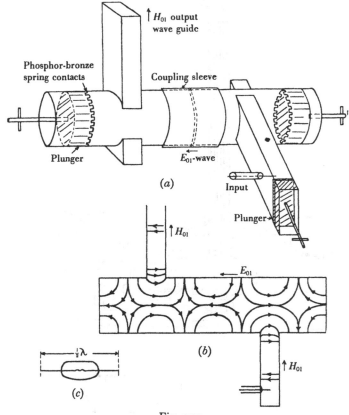

Fig. 2·23.

An interesting and important method of exciting an E_{01}-wave in a circular wave guide by means of an H_{01}-wave in a rectangular wave guide, and conversely, is illustrated in fig. 2·23.

The device is known as an H-to-E transformer, and the demonstration apparatus represented in fig. 2·23 (a) is one in which an H_{01}-wave excited at the input probe is first converted into an H_{01}-wave and then back again into an H_{01}-wave in the output rectangular wave guide. The relationship of the electric lines of force in the circular and rectangular wave guides is indicated in fig. 2·23 (b).

It can be seen that the transverse electric fields of the H_{01}-waves become longitudinal fields in the circular guides and are able to excite an E_{01}-wave in it. An important feature of the E_{01}-wave is its axial symmetry which enables it to excite an H_{01}-wave in the output rectangular wave guide on any azimuth. Consequently, if the two transformers are separate, as indicated in fig. 2·23 (a), but held together by a metal sheath over their circular portions, it is found that the power output is independent of the rotation of the output rectangular wave guide relative to the input. This important result receives an application in a rotatable wave-guide joint, but we postpone discussion of this to a later section. Fig. 2·23 (c) represents a convenient low-power lamp (10 mW.) whose leads are cut to be half a wave-length (4·5 cm. at 3300 Mc./sec.) from tip to tip. Such a lamp can be used to indicate the direction of the electric field in the wave, since it behaves as a resonant half-wave aerial. Even when a low-power reflector klystron with power output of 200 mW. at 3300 Mc./sec. is used as a source of power, the lamp should light brightly when placed parallel to the short edge of the mouth of the output wave guide and at the centre of the mouth. By removing the output half of the system the lamp may be introduced longitudinally into the circular wave guide when it will show the existence on the axis of the longitudinal electric field of the E_{01}-wave. Visual demonstrations of this type are valuable teaching aids.

2·9. Group velocity

According to equation 2·2 (1), the phase or pattern velocity is

$$v_g = v/\cos \alpha,$$

where α is the angle between the wave normals of the component wave fronts and the walls. This velocity refers to a steady state in which a continuous H_{10}-wave of constant amplitude completely fills the wave guide. The speed at which a limited wave train or a modulation of a long wave train travels along the guide is not the phase velocity v_g but the group velocity v_G. A signal may be supposed to be transported by the constituent waves reflected to and fro between the faces so that its velocity resolved along the wave guide axes is

$$v_G = \frac{v}{\cos \alpha} = \frac{v^2}{v_g},$$

whence $$v_g v_G = v^2,$$

as stated in 2·1 (1). This is an oversimplified account because a limited wave train possesses a spectrum of frequencies.

According to a standard formula, the group velocity of a limited wave train is given by

$$\frac{1}{v_G} = \left(\frac{dN}{df}\right),$$

where

$$N = \frac{1}{\text{wave-length}} = \frac{1}{\lambda_g}.$$

In a wave guide

$$\frac{1}{\lambda_g} = \sqrt{\left(\frac{1}{\lambda^2} - \frac{1}{(\lambda_c)^2}\right)} = \sqrt{\left(\frac{f^2}{v^2} - \frac{1}{(\lambda_c)^2}\right)},$$

it follows that

$$\frac{1}{v_G} = \frac{f}{v^2} \frac{1}{\sqrt{\left(\frac{1}{\lambda^2} - \frac{1}{(\lambda_c)^2}\right)}} = \frac{f\lambda_g}{v^2} = \frac{v_g}{v^2}$$

or

$$v_g v_G = v^2.$$

Chapter 3

FORMULAE FOR FIELD COMPONENTS— EVANESCENT MODES—ATTENUATORS— ATTENUATION DUE TO WALLS

3·1. Introduction

We begin this chapter with a recapitulation in mathematical language of the procedure used in Chapter 2, to find the field patterns of H_{0n}-waves, and thus arrive at formulae for the field components of these waves as functions of position in the wave guide.

Subsequently, we consider the nature of the electromagnetic field in a wave guide when the frequency is less than the cut-off frequency so that there is no propagated wave, and also the attenuation of a progressive wave due to the finite conductivity of the walls.

3·2. Field components of the H_{0n}-wave (TE_{0n}) in a rectangular wave guide

Let CD, in fig. 3·1, represent a wave front and OQ the direction of propagation of a plane electromagnetic wave polarized with **H** in the plane of the paper and **E** perpendicular to it.

Since we shall require to consider a second wave, the electric and magnetic fields of the first wave are written $\mathbf{E}_1$ and $\mathbf{H}_1$ and those of the second wave $\mathbf{E}_2$ and $\mathbf{H}_2$.

We refer the wave E_1-H_1 to a system of Cartesian coordinates chosen with the ZOY plane as the plane of the paper, and with the positive direction of OX directed into the paper. If, therefore, $\mathbf{H}_1$ and OQ (the wave normal) lie in the plane ZOY as shown, then $\mathbf{E}_1$ is parallel to $+OX$. If $\mathbf{H}_1$ is reversed then $\mathbf{E}_1$ is parallel to $-OX$.

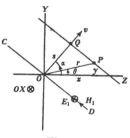

Fig. 3·1.

Over the equiphase plane CD through O suppose $\mathbf{E}_1$ and $\mathbf{H}_1$ to be

$$E_1 = \sqrt{\left(\frac{\mu}{\epsilon}\right)} H_1 = A_1 \cos{(\omega t + g)}, \tag{1}$$

where $\omega = 2\pi f$; t is the time and g is a phase constant whose value may be chosen to suit subsequent requirements.

Let QP be the wave front through the point $P(O, y, z)$. If the length of the normal OQ from the origin on to QP is s and the wavelength of the wave train is λ, then, over the plane QP,

$$E_1 = \sqrt{\left(\frac{\mu}{\epsilon}\right)} H_1 = A_1 \cos(\omega t - ks + g), \qquad (2)$$

where $k = 2\pi/\lambda$.

Let the angle QOZ be α; then we may express s in terms of α and the coordinates y and z of P.

For, if $OP = r$ and the angle $POZ = \theta$, then

$$s = r\cos(\alpha - \theta) = r(\cos\alpha\cos\theta + \sin\alpha\sin\theta)$$
$$= (y\sin\alpha + z\cos\alpha).$$

Since the plane QP is an equiphase surface, whose phase, according to (2), is determined only by s, the fields $\mathbf{E}_1$ and $\mathbf{H}_1$ at all points (x, y, z) of this plane are

$$E_1 = \sqrt{\left(\frac{\mu}{\epsilon}\right)} H_1 = A_1 \cos(\omega t - ky\sin\alpha - kz\cos\alpha + g). \qquad (3)$$

In fig. 3·2 we have the wave train $(E_2\text{-}H_2)$ with the same wavelength λ and with $\mathbf{H}_2$ and the ray direction OQ' again in the plane ZOY. However, the angle ZOQ' is here equal to $-\alpha$.

Over the plane $D'OC'$ through the origin let

$$E_2 = \sqrt{\left(\frac{\mu}{\epsilon}\right)} H_2 = A_2 \cos(\omega t + h),$$

where h is again an adjustable phase constant.

Similarly, over the wave front $Q'P'$ through the point $P(x'y'z')$ at a perpendicular distance s' from the origin

$$E_2 = \sqrt{\left(\frac{\mu}{\epsilon}\right)} H_2 = A_2 \cos[\omega t + h - k(z'\cos\alpha - y'\sin\alpha)]. \qquad (4)$$

Let the two waves run simultaneously across the region represented in the diagram, and choose the amplitude A_2 and the phase constant h so that at all times t, the resultant electric field shall be zero everywhere over the plane ZOX (fig. 3·3).

We require, therefore, $\mathbf{E} = \mathbf{E}_1 + \mathbf{E}_2 = 0$ for all values of z along OZ, since $\mathbf{E}_1$ and $\mathbf{E}_2$ are both perpendicular to the plane of the paper.

The resultant field E at an arbitrary point (x, y, z) is

$$E = E_1 + E_2$$
$$= A_1 \cos(\omega t - kz \cos\alpha - ky \sin\alpha + g)$$
$$+ A_2 \cos(\omega t - kz \cos\alpha - ky \sin\alpha + h). \tag{5}$$

Since $y = 0$ over the plane XOZ, we put $y = 0$ in this expression for E and equate it to zero. We require, therefore, at all times t and for all values of z,

$$A_1 \cos(\omega t - kz + g) + A_2 \cos(\omega t - kz + h) = 0.$$

That is, $A_1 = A_2 = A, \quad h = (g \pm \pi)$ radians.

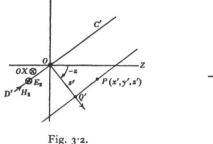

Fig. 3·2.

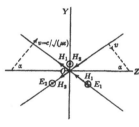

Fig. 3·3.

With these values of A_2 and h the expression (5) for the field at an arbitrary point becomes

$$E = 2A \sin(ky \sin\alpha) \cos(\omega t - kz \cos\alpha + g - \tfrac{1}{2}\pi), \tag{6}$$

where $k = 2\pi/\lambda$. This is the equation of a plane wave propagated in the direction $+OZ$ and amplitude modulated in the direction OY by the factor $\sin\left(\dfrac{2\pi}{\lambda} y \sin\alpha\right)$.

To find the resultant magnetic field $\mathbf{H}$ we resolve the constituent fields $\mathbf{H}_1$ and $\mathbf{H}_2$ into their components along OZ and OY and add corresponding components. From equations (3) and (4) and fig. 3·3 we deduce the following expressions for the components of the resultant field $\mathbf{H}$:

$$H_y = 2\sqrt{\left(\frac{\epsilon}{\mu}\right)} A \cos\alpha \sin(ky \sin\alpha) \cos(\omega t - kz \cos\alpha + g - \tfrac{1}{2}\pi),$$

$$H_z = 2\sqrt{\left(\frac{\epsilon}{\mu}\right)} A \sin\alpha \cos(ky \sin\alpha) \cos(\omega t - kz \cos\alpha + g + \pi), \quad \left.\right\} \tag{7}$$

to which we add (6) for completeness,

$$E = E_x = 2A \sin(ky \sin\alpha) \cos(\omega t - kz \cos\alpha + g - \tfrac{1}{2}\pi).$$

The two elementary waves have evidently combined to form an H-wave.

E_x and H_y are zero when $\sin(ky\sin\alpha)$ is zero, that is, over the planes $y = $ constant, where

$$y = \pm \frac{n\pi}{k\sin\alpha} \quad (n = 0, 1, 2, ..., \text{etc.}).$$

We may, as before, suppose any pair of these planes to be occupied by a metal sheet. Choose the pair $y = 0$ and $y = \dfrac{n\pi}{k\sin\alpha} = \dfrac{n\lambda}{2\sin\alpha} = b$. We then have an n-layer H-wave between a pair of parallel metal plates at a separation

$$b = \frac{n\lambda}{2\sin\alpha}. \tag{8}$$

When side walls at any distance a apart are inserted, we obtain an H_{0n}-wave in a rectangular wave guide.

Equation (8) may be written

$$\sin\alpha = \frac{n\lambda}{2b}, \quad \cos\alpha = \sqrt{\left(1 - \frac{n^2\lambda^2}{2b}\right)}, \tag{9}$$

and from expressions (7) the wave-guide wave-length λ_g is seen to be

$$\lambda_g = \lambda/\cos\alpha. \tag{10}$$

Thus, from (9) and (10) we obtain the standard formula 2·1 (2)

$$\frac{1}{\lambda_g^2} = \frac{1}{\lambda^2} - \frac{1}{\lambda_c^2}, \tag{11}$$

where the cut-off wave-length

$$\lambda_c = (2b/n). \tag{12}$$

Expression (7) may be simplified by making use of (9) and by putting $g = \frac{1}{2}\pi$. There results, when $2A$ is written as E_0,

$$\left.\begin{array}{l} E_x = E_0 \sin\left(\dfrac{n\pi y}{b}\right)\cos\left(\omega t - \dfrac{2\pi z}{\lambda_g}\right), \\[2ex] H_y = E_0 \sqrt{\left(\dfrac{\epsilon}{\mu}\right)}\sqrt{\left(1 - \dfrac{\lambda^2}{\lambda_c^2}\right)}\sin\left(\dfrac{n\pi y}{b}\right)\cos\left(\omega t - \dfrac{2\pi z}{\lambda_g}\right), \\[2ex] H_z = E_0 \sqrt{\left(\dfrac{\epsilon}{\mu}\right)}\dfrac{\lambda}{\lambda_c}\cos\left(\dfrac{n\pi y}{b}\right)\cos\left(\omega t - \dfrac{2\pi z}{\lambda_g} - \tfrac{1}{2}\pi\right). \end{array}\right\} \tag{13}$$

This is the H_{0n}-wave that will fit into the wave guide whose section is shown in fig. 3·4.

This electromagnetic field satisfies the boundary conditions at the metal surface and Maxwell's equations within the wave guide.

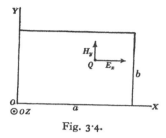

Fig. 3·4.

If, instead of using the trigonometrical forms (3) and (4) for the elementary wave, we use the equivalent exponential representations

$$E_1 = \sqrt{\left(\frac{\mu}{\epsilon}\right)} H_1 = A_1 e^{j[\omega t + g - k(y \sin \alpha + z \cos \alpha)]},$$
$$E_2 = \sqrt{\left(\frac{\mu}{\epsilon}\right)} H_2 = A_2 e^{j[\omega t + h + k(y \sin \alpha - z \cos \alpha)]}, \qquad (14)$$

we obtain on superimposing the fields, with $A_2 = A_1 = A$ and $h + \pi = g = \frac{1}{2}\pi$ in place of equations (13), the equivalent expressions

$$E_x = E_0 \sin\left(\frac{n\pi y}{b}\right) e^{j(\omega t - \gamma z)},$$
$$H_y = E_0 \sqrt{\left(\frac{\epsilon}{\mu}\right)} \sqrt{\left(1 - \frac{\lambda^2}{\lambda_c^2}\right)} \sin\left(\frac{n\pi y}{b}\right) e^{j(\omega t - \gamma z)},$$
$$H_z = -jE_0 \sqrt{\left(\frac{\epsilon}{\mu}\right)} \frac{\lambda}{\lambda_c} \cos\left(\frac{n\pi y}{b}\right) e^{j(\omega t - \gamma z)}, \qquad (15)$$

where $j \times j = -1$ and we have used the relation

$$e^{\pm j \frac{1}{2}\pi} = (\cos \tfrac{1}{2}\pi \pm j \sin \tfrac{1}{2}\pi) = \pm j.$$

The *propagation constant* $\gamma = 2\pi/\lambda_g$. The field expressions in (15) formally satisfy Maxwell's equations and the boundary conditions provided, in accordance with equation (11),

$$\gamma^2 = k^2 - \left(\frac{2\pi}{\lambda_c}\right)^2 = k^2 - \left(\frac{\pi n}{b}\right)^2.$$

3·3. Field components of H_{mn}- and E_{mn}-waves

For completeness, the expressions for the field components of the H_{mn}- and E_{mn}-modes are given below, but are not derived (see § 7·10·2). The field components of an H_{mn}-wave are

$$
\left.
\begin{aligned}
E_x &= \left(\frac{n\pi}{b}\right)\cos\left(\frac{m\pi x}{a}\right)\sin\left(\frac{n\pi y}{b}\right) e^{j(\omega t - \gamma z)}, \\[4pt]
E_y &= -\left(\frac{m\pi}{a}\right)\sin\left(\frac{m\pi x}{a}\right)\cos\left(\frac{n\pi y}{b}\right) e^{j(\omega t - \gamma z)}, \\[4pt]
H_x &= \left(\frac{m\pi}{a}\right)\sqrt{\left(\frac{\epsilon}{\mu}\right)}\sqrt{\left(1 - \frac{\lambda^2}{\lambda_c^2}\right)}\sin\left(\frac{m\pi x}{a}\right)\cos\left(\frac{n\pi y}{b}\right) e^{j(\omega t - \gamma z)}, \\[4pt]
H_y &= \left(\frac{n\pi}{b}\right)\sqrt{\left(\frac{\epsilon}{\mu}\right)}\sqrt{\left(1 - \frac{\lambda^2}{\lambda_c^2}\right)}\cos\left(\frac{m\pi x}{a}\right)\sin\left(\frac{n\pi y}{b}\right) e^{j(\omega t - \gamma z)}, \\[4pt]
H_z &= -j\left(\frac{2\pi}{\lambda_c}\right)\left(\frac{\lambda}{\lambda_c}\right)\sqrt{\left(\frac{\epsilon}{\mu}\right)}\cos\left(\frac{m\pi x}{a}\right)\cos\left(\frac{n\pi y}{b}\right) e^{j(\omega t - \gamma z)},
\end{aligned}
\right\} \quad (1)
$$

$$
E_z = 0,
$$

where

$$
\frac{1}{\lambda_c^2} = \left(\frac{m}{2a}\right)^2 + \left(\frac{n}{2b}\right)^2 \quad \text{and} \quad \gamma^2 = \left(\frac{2\pi}{\lambda_g}\right)^2 = (2\pi)^2\left[\frac{1}{\lambda^2} - \frac{1}{\lambda_c^2}\right]. \quad (2)
$$

It is easy to verify that expressions 3·2 (15) for the H_{0n}-wave, except for the arbitrary constant E_0, are obtained by putting $m = 0$ in (1) and (2).

The corresponding field expressions for the E_{mn}-wave in a rectangular wave guide are

$$
\left.
\begin{aligned}
H_x &= \left(\frac{n\pi}{b}\right)\sin\left(\frac{m\pi x}{a}\right)\cos\left(\frac{n\pi y}{b}\right) e^{j(\omega t - \gamma z)}, \\[4pt]
H_y &= -\left(\frac{m\pi}{a}\right)\cos\left(\frac{m\pi x}{a}\right)\sin\left(\frac{n\pi y}{b}\right) e^{j(\omega t - \gamma z)}, \\[4pt]
E_x &= -\left(\frac{m\pi}{a}\right)\sqrt{\left(\frac{\mu}{\epsilon}\right)}\sqrt{\left(1 - \frac{\lambda^2}{\lambda_c^2}\right)}\cos\left(\frac{m\pi x}{a}\right)\sin\left(\frac{n\pi y}{b}\right) e^{j(\omega t - \gamma z)}, \\[4pt]
E_y &= -\left(\frac{n\pi}{b}\right)\sqrt{\left(\frac{\mu}{\epsilon}\right)}\sqrt{\left(1 - \frac{\lambda^2}{\lambda_c^2}\right)}\sin\left(\frac{m\pi x}{a}\right)\cos\left(\frac{n\pi y}{b}\right) e^{j(\omega t - \gamma z)}, \\[4pt]
E_z &= j\left(\frac{2\pi}{\lambda_c}\right)\left(\frac{\lambda}{\lambda_c}\right)\sqrt{\left(\frac{\mu}{\epsilon}\right)}\sin\left(\frac{m\pi x}{a}\right)\sin\left(\frac{n\pi y}{b}\right) e^{j(\omega t - \gamma z)},
\end{aligned}
\right\} \quad (3)
$$

$$
H_z = 0,
$$

where
$$\frac{1}{\lambda_c^2} = \left(\frac{m}{2a}\right)^2 + \left(\frac{n}{2b}\right)^2.$$

It is easy to see that if either m or n is zero then every component vanishes and therefore that an E_{0n}- or an E_{m0}-wave is not a possible type of wave in a rectangular wave guide.

It follows from (1) and (3) that the components both of H_{mn}- and E_{mn}-waves satisfy the following relationship:

$$E_x H_x + E_y H_y + E_z H_z = 0,$$

which is the condition that the fields **E** and **H** should be at right angles. The fields are such that at all places and times the resultant electric and magnetic fields are at right angles.

Wave fields in circular guides and in coaxial lines also possess this property.

3·4. Wave impedance and Poynting flux

3·4·1. *Wave impedance*

We require, for subsequent discussions, an expression for the Poynting flux associated with each wave-guide mode.

For simplicity, consider first the H_{0n}-wave and refer to fig. 3·4. The transverse components of **E** and **H** at the point $Q(x,y)$ are E_x and H_y.

It follows from equations (15) that

$$\frac{E_x}{H_y} = \sqrt{\left(\frac{\mu}{\epsilon}\right)} \sqrt{\left(\frac{1}{1 - \lambda^2/\lambda_c^2}\right)} = \sqrt{\left(\frac{\mu}{\epsilon}\right)} \sqrt{\left(\frac{1}{1/\lambda^2 - 1/\lambda_c^2}\right)} \frac{1}{\lambda}$$
$$= \left(\frac{\lambda_g}{\lambda}\right) \sqrt{\left(\frac{\mu}{\epsilon}\right)} = 120\pi \sqrt{\left(\frac{K_e}{K_m}\right)} \frac{\lambda_g}{\lambda} \text{ ohms.} \quad (1)$$

This may be compared with the corresponding value of the transverse field components in a plane wave in free space which according to equation 1·2 (6), is $\sqrt{(\mu/\epsilon)} = 120\pi \sqrt{(K_e/K_m)}$, and was called the wave impedance of this wave.

Since the ratio E_x/H_y in the H_{0n}-wave is independent of position and of time it is also called the *wave impedance* or the *intrinsic impedance* of the wave guide for this wave. According to 3·2 (15), E_x and H_y vibrate with no phase difference; consequently the wave impedance is purely resistive. We shall denote the wave impedance of an H-wave by Z_H, therefore, in this instance,

$$Z_H = \sqrt{\left(\frac{\mu}{\epsilon}\right)} \frac{\lambda_g}{\lambda}. \tag{2}$$

For the H_{mn}-wave we find from (1) that

$$Z_H = \frac{E_x}{H_y} = -\frac{E_y}{H_x} = \sqrt{\left(\frac{\mu}{\epsilon}\right)} \frac{\lambda_g}{\lambda}. \tag{3}$$

We may similarly define the wave impedance of the E_{mn}-wave as Z_E and find from (3)

$$Z_E = \frac{E_x}{H_y} = -\frac{E_y}{H_x} = \sqrt{\left(\frac{\mu}{\epsilon}\right)} \frac{\lambda}{\lambda_g}. \tag{4}$$

The expressions (3) and (4) for Z_H and Z_E are also valid for waves in circular wave guides.

3·4·2. *Poynting flux*

Let the directions of the transverse components of **E** and **H** in either an E_{mn}- or an H_{mn}-wave at an arbitrary point $Q(x,y)$ of the cross-section be those shown in fig. 3·5. Let the resultant transverse field E make an angle θ with the horizontal edge a. From the figure

$$\left.\begin{array}{ll} E_x = E\cos\theta, & E_y = E\sin\theta, \\ H_x = -H\sin\theta, & H_y = H\cos\theta. \end{array}\right\} \tag{1}$$

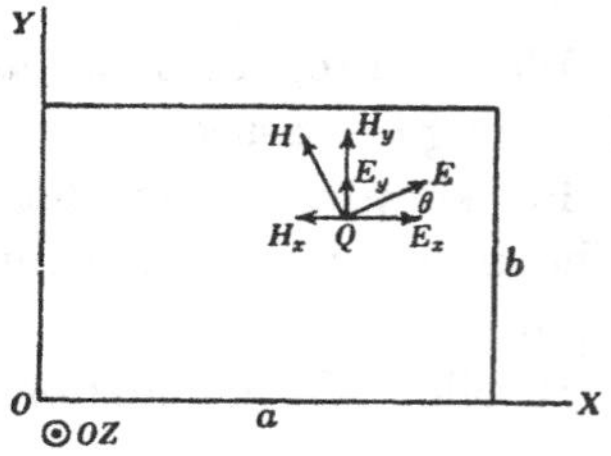

Fig. 3·5.

The Poynting flux density of power along the axis at Q is $P = EH$ W./sq.m. (§ 1·3 (1)), that is, from (1)

$$P = (E_x H_y - E_y H_x).$$

But, from 3·4·1 (3) and (4), $H_y = E_x/Z$, $H_x = -E_y/Z$, where Z represents Z_E or Z_H according as we are discussing an E- or an H-wave.

The expression for P becomes

$$P = \left(\frac{E_x^2 + E_y^2}{Z}\right) = \frac{E^2}{Z}. \tag{2}$$

This is the instantaneous flux density of power at Q. The quantity Z therefore behaves as circuit impedance provided it refers to the field components and flux density of power at a point, and that we suppose the transverse electric field E to be the voltage and the transverse field H the current, all referred to the same point Q on the cross-section.

The mean flux density of power is

$$\overline{P} = \frac{1}{2}\frac{E_0^2}{Z} = \frac{ZH_0^2}{2}, \qquad (3)$$

where E_0 and H_0 are the amplitudes of the resultant transverse fields. To obtain the average power crossing the whole section of the wave guide it is necessary to integrate (3) over the section, writing $E_0^2 = (E_{0x}^2 + E_{0y}^2)$ and $H_0^2 = H_{0x}^2 + H_{0y}^2$.

The total mean power carried by the H_{mn}-wave of equation 3·3 (1) is

$$\overline{W} = \frac{ab}{8Z_H}\left[\left(\frac{m\pi}{a}\right)^2 + \left(\frac{n\pi}{b}\right)^2\right]\mathrm{W}. \qquad (4)$$

With an H_{0n}- or H_{m0}-wave the expression for $\overline{W}$ must be doubled since in the integration there is only one square sinusoidal factor. Thus for an H_{0n}-wave, $\overline{W} = \frac{ab}{4Z_H}\left(\frac{n\pi}{b}\right)^2$. The expression for an E_{mn}-wave is, from 3·3·1 (3),

$$\overline{W} = \frac{ab}{8}Z_E\left[\left(\frac{m\pi}{a}\right)^2 + \left(\frac{n\pi}{b}\right)^2\right]. \qquad (5)$$

Since equations 3·3(1) and (3) give the relative amplitude of the components they are often given in other forms which, however, on examination are the same as equations 3·3(1) and (3) with every component multiplied by a common factor which will be usually different in the two cases. If A is this factor, then expressions (4) and (5) for $\overline{W}$ will contain an additional factor A^2.

The component of the Poynting flux perpendicular to the axis is found by associating H_z and E_z with the transverse E and H-components respectively in the H- and E-waves. Since, however, the z-components oscillate with a $90°$ phase difference from the associated transverse components in a wave guide with perfectly conducting walls, the mean flux of power outwards is zero.

3·5. Evanescent modes

We have hitherto refrained from discussing the nature of the electromagnetic field of a mode when the frequency f is less than the cut-off frequency f_c so that the mode cannot exist as a progressive wave, but it is now necessary to do so.

The two electromagnetic fields whose components are proportional respectively to the expressions given in equations 3·3 (1) and (3) represent formal solutions of Maxwell's equations of the electromagnetic field provided the conditions 3·3 (2) are satisfied. Every field component in 3·3 (1) and (3) contains the factor $e^{j(\omega t - \gamma z)}$, and when the propagation constant γ is a real quantity the electromagnetic fields are those of progressive H_{mn}- and E_{mn}-waves respectively.

Since, from 3·3 (2),

$$\gamma = 2\pi \sqrt{\left(\frac{1}{\lambda^2} - \frac{1}{\lambda_c^2}\right)} = \frac{2\pi}{\lambda} \sqrt{\left(1 - \frac{\lambda^2}{\lambda_c^2}\right)}, \tag{1}$$

the condition for a progressive wave is that the quantity under the square root should be positive; that is the TEM wave-length λ must be less than the cut-off wave-length λ_c, or in other words the frequency f must exceed f_c for the particular mn mode.

Suppose the frequency f to be reduced to a value less than f_c so that λ exceeds λ_c, then the quantity under the square root becomes negative and γ becomes a mathematically imaginary quantity. We may rewrite the expression for γ

$$\left. \begin{array}{l} \gamma = \pm j 2\pi \sqrt{\left(\frac{1}{\lambda_c^2} - \frac{1}{\lambda^2}\right)} = \pm j\alpha, \\[2mm] \text{where} \qquad j \times j = -1 \quad \text{and} \quad \alpha = 2\pi \sqrt{\left(\frac{1}{\lambda_c^2} - \frac{1}{\lambda^2}\right)}. \end{array} \right\} \tag{2}$$

The exponential factor now becomes

$$e^{j(\omega t \pm j\alpha z)} = e^{j\omega t} \times e^{\pm \alpha z}.$$

Let us choose, for convenience, the negative sign before αz (i.e. put $\gamma = -j\alpha$) and write the exponential factors

$$e^{j\omega t} \times e^{-\alpha z}.$$

These factors appear in the expressions for the field components. The electromagnetic field is now one of a totally different character from that of the progressive wave. First, the amplitudes of all components decay exponentially with distance along the axis, and at a sufficient distance from the source the disturbance becomes relatively negligible; secondly, there is no phase dependence on the distance z as there is in a progressive wave as may be seen from 3·3 (1) and (3), remembering that λ/λ_c is now greater than unity. A given component oscillates everywhere in phase with itself although its amplitude diminishes as z increases. A mode whose frequency f exceeds the cut-off frequency f_c so that it behaves as described above, is said to be *evanescent* (vanishing rapidly).

The properties of evanescent modes are of theoretical importance in studies of the effect of obstacles and geometrical discontinuities in wave guides and of practical importance in the piston attenuator, an example of which is described in the next section.

3·6. The piston attenuator

We first give the theory of this device. Consider an arbitrary electromagnetic field oscillating sinusoidally with frequency f within a wave guide, and suppose f to be less than f_c, the cut-off frequency of the dominant mode in the guide. The arbitrary electromagnetic field can be represented as a series of coexistent evanescent modes whose amplitudes and phases must be chosen appropriately. Since it has been assumed that f is less than the cut-off frequency of the dominant mode, it follows that the field amplitudes in each mode (m, n) diminish exponentially with distance z from the source of the disturbance according to a factor $\exp(-\alpha_{mn}z)$, where, from 3·5 (2),

$$\alpha_{mn} = 2\pi \sqrt{\left(\frac{1}{(\lambda_c)_{mn}^2} - \frac{1}{\lambda^2}\right)}. \tag{1}$$

If we further postulate that $f \ll f_c$, that is $\lambda \gg \lambda_c$, then

$$\alpha_{mn} \doteqdot 2\pi/(\lambda_c)_{mn}.$$

Since the higher the order of the mode (mn) the smaller the cut-off wave-length, it is evident that the higher order modes are attenuated more strongly than the lower order modes; consequently, by proceeding a sufficient distance z from the source of disturbance the residual field is effectively that of the dominant mode and exhibits a simple exponential dependence on distance.

The attenuation coefficient is $\alpha = 2\pi/\lambda_c$, where λ_c is the cut-off wave-length of the dominant mode. We note, therefore, that provided $\lambda \gg \lambda_c$, the attenuation coefficient α of the residual dominant mode (and of all other modes) is virtually independent of frequency and depends only on the geometry and dimensions of the wave guide.

An instructional wave guide for showing the exponential reduction in the amplitude of an evanescent mode is illustrated in fig. 3·6. It comprises an ordinary piece of $2\frac{1}{2} \times 1$ in. wave guide which leads smoothly into a length of narrow wave guide $1\frac{25}{32} \times 1$ in. The narrow portion carries a longitudinal slot on the centre line of its broad face. The dimension of the narrow wave guide is so chosen that at the lower wave-length range ($\lambda = 8·6$ cm.) of a reflector klystron (Sutton tube) the wave-length is almost equal to the cut-off wave-length λ_c, but at the other end of the tuning range ($\lambda = 9·3$ cm.) the H_{10}-wave is fully evanescent. The H_{10}-mode is launched into the broad wave guide and becomes evanescent in the narrow guide and no other modes are excited. The variation of field strength with axial distance can be found by means of a standing wave indicator (§ 4·4) whose probe is introduced into the field through the slot.

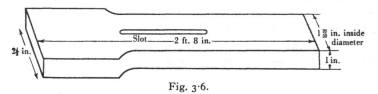

Fig. 3·6.

The d.c. from the crystal of the standing wave indicator is proportional to the square of the amplitude of the field within the wave guide when the rectified current from the crystal is less than $20 \mu A$. The narrow wave guide is terminated in a reflectionless wooden load which excites no disturbing field.

Fig. 3·7 is a plot of crystal current against distance on semi-logarithmic paper for three oscillator wave-lengths. The linearity of the curves establishes the exponential dependence of amplitude (here it is power) on axial distance z. The wave-lengths λ are sufficiently near to λ_c to require the full formula (1) to be used to check the measured and calculated values of $2\alpha_{10}$.

The exponential dependence of field amplitude on distance is employed in the piston attenuator which is a device employed in special forms of signal generators at microwave-lengths for reducing the amplitude of the output e.m.f. by any number of decibels within a given range. Fig. 3·8 is a diagram illustrating how a piston attenuator can be incorporated in a signal generator.

The equipment is T.R.E. Signal Generator Type 47.

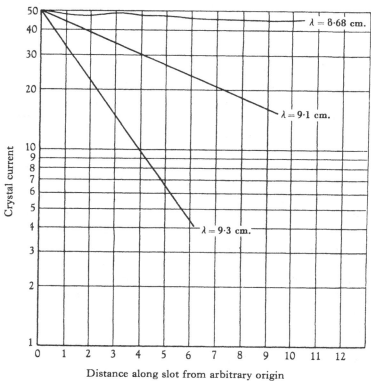

Distance along slot from arbitrary origin

Fig. 3·7.

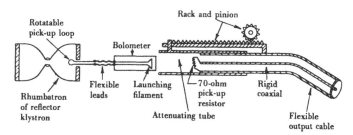

Fig. 3·8.

The source of power is a reflector klystron ($\lambda = 9 \cdot 1$ cm.) shown on the left. The e.m.f. abstracted can be altered at will by rotating the coupling loop by means of the knob shown at the left of fig. 3·8. The output from the loop is fed through flexible leads to a bolometer that comprises a straight filament inside an evacuated glass tube. The bolometer filament has a double function; first, it launches the electromagnetic disturbance into the attenuating tube (piston attenuator) to the right of the diagrams; secondly, its brightness is used to indicate when the disturbance reaches an arbitrary standard strength, by turning the coupling loop to the klystron until the bolometer is observed through an eyepiece just to glow. At a sufficient distance z within the circular tube of the piston attenuator, whose diameter is small enough to ensure that the dominant H_{11}-mode is strongly evanescent, the disturbance is a pure H_{11} evanescent mode. A 70-ohm resistor acts as a pick-up loop for the output coaxial which delivers the e.m.f. to an output plug on the front panel of the signal generator. The position of the pick-up resistor in the tube can be adjusted by a rack and pinion movement as shown. There is a position of closest approach of the resistor to the source within which the output reacts on the input, and it is arranged that the resistor cannot approach the bolometer filament within this distance.

Let V_1 be the signal output when the resistance is at distance z_1 from the filament and V_2 at distance z_2. From what has been said it follows that

$$\frac{V_2}{V_1} = e^{-\alpha(z_2 - z_1)} \quad \text{or} \quad \log_e\left(\frac{V_1}{V_2}\right) = \alpha(z_2 - z_1) \text{ nepers,}$$

i.e.
$$20 \log_{10}\left(\frac{V_1}{V_2}\right) = 8 \cdot 686\alpha(z_2 - z_1) \text{ db.}$$

The reduction of the signal strength measured in decibels is therefore proportional to the displacement of the pick-up loop. The attenuation dial which controls the movement of the pick-up resistor can therefore be calibrated directly in decibels, the zero on the dial corresponding to the position of closest permissible approach. The maximum reduction of output of this arbitrary standard is 100 db., and the decibel scale on the attenuation dial is linear.

The principal use of signal generators of this type is to check receiver performance at wave-lengths of 9 cm. Their chief defect is the small output which in this case is 60 mV. at the o db. setting, from an internal impedance of 75 ohms. The Signal Generator Type 47 (T.R.E.) is a useful field equipment for use on S-band. Signal generators that incorporate piston attenuators have also been developed for operation in the X-band ($\lambda = 3\cdot3$ cm.); they are the R.R.D.E. Signal Generator Type 8 and the T.R.E. Signal Generator Type 102. Instead of using a visual method of adjusting the bolometer current, they employ a more sensitive bridge method.

3·7. Attenuators

It is convenient to mention here a wave-guide attenuator of a type different from the piston attenuator. This comprises a length of wave guide into which is introduced a resistive lamina with its plane parallel to the electric lines of force of the H_{10}-wave, and to the narrow face of the wave guide. Power is dissipated in the resistive sheet, and the amplitude of the wave is diminished as it is propagated past it. There are two principal forms of this attenuator, the 'flap' type and the 'push across' type. In the former the lamina is introduced into the wave guide through a longitudinal slot in a broad face. It is hinged to the wave guide at one end and can be lowered into the wave guide to any desired degree by motion of a dial operating a cam. This dial can be calibrated directly in decibels of attenuation. A defect of this design is the fact that currents flowing in the resistive sheet also flow in the portion of it outside the wave guide and radiate into the surrounding region and cause trouble.

As in fig. 4·29 (a) this defect is avoided in the 'push across' form of attenuator in which the resistive sheet lies entirely within the wave guide but can be moved normal to itself and to the narrow faces from a position where it is lying flat against the wall and produced little attenuation ($\frac{1}{4}$ db.) to the central section of the wave guide where it is in the maximum electric field and produces maximum attenuation. A pair of thin rods run across the wave guide and through the resistive sheet which is rigidly attached to them as in fig. 4·29 (b).

These rods allow the resistive film to be moved across the wave guide by the operation of a cam and dial. The dial can be calibrated in decibels.

Common forms of resistive sheet are composed of carbon deposited on bakelite, but these do not retain their calibration well.

More recently, platinum-on-glass sheet has been developed in the United States and is very stable. A typical resistive value is 200–300 ohms per square. 'Cracked' carbon films are being developed for this purpose in this country and appear to be sufficiently stable.

These attenuators can be calibrated against a signal generator incorporating a piston attenuator which, as we have seen, gives attenuation directly in decibels.

In X-band (3·3 cm.) wave-guide test equipment it is desirable to decouple the oscillation by including an attenuating section of wave guide between it and the main run of wave guide. This section is an ordinary length of guide with a resistive strip of bakelite loaded with graphite which fits against one of the narrow walls. In this way a fixed attenuation of 10 db. can be introduced between oscillator and load.

3·8. Poynting flux in an evanescent mode

In § 3·4·2 it was shown that the component of the Poynting vector along the axis of the wave guide was a real quantity and that therefore there is a flow of power along the axis when the electromagnetic field is that of a progressive wave. When, however, the electromagnetic field is that of an evanescent mode the transverse components of E oscillate in quadrature (90° phase difference) with those of H, and there is no mean flow of power along the wave guide. This may be appreciated on referring to equations 3·3 (1) and (3) for the components of H_{mn}- and E_{mn}-waves in a rectangular wave guide. It will be noted that in the third and fourth expressions of each group the factor $\sqrt{(1 - \lambda^2/\lambda_c^2)}$ appears. When the electromagnetic field is that of a progressive wave $\lambda < \lambda_c$ and the expression is real but with an evanescent mode, λ exceeds λ_c and the factor should be written $-j\sqrt{[(\lambda/\lambda_c)^2 - 1]}$. Consequently, in equations 3·3 (1) and (3) the components of H oscillate together but in quadrature with those of E. When the wave impedance and axial Poynting flux are derived, as in § 3·4, they appear as imaginary quantities; that is, the wave impedance is a reactance and not a resistance. This reactance is inductive for evanescent H-modes and capacitive for E-modes.

Since energy flows into and out of the field every quarter cycle, but none is carried away, the fields of evanescent modes are called *storage fields*. They resemble the fields of inductances and condensers.

A paradox. In the piston attenuator power is transferred from the source to the output along a wave guide in which only evanescent modes exist. This fact appears to be inconsistent with the statement that an evanescent mode does not convey power along the wave guide. To resolve this paradox, refer to equations 3·3 (1) and (3). We have seen that when $\sqrt{(1 - \lambda^2/\lambda_c^2)}$ is written $-j\sqrt{(\lambda^2/\lambda_c^2 - 1)} = -j\alpha$ the amplitude of each field component diminishes with increasing z through the factor $e^{-\alpha z}$. In the H_{mn}-wave of equations 3·3 (1) the field components H_x and H_y therefore oscillate with a phase retardation of 90° on E_x when all components vary with increasing z through the factor $e^{-\alpha z}$.

If, however, we use the alternative sign for the square root and write $+j\sqrt{(\lambda^2/\lambda_c^2 - 1)} = +j\alpha$, the dependence on z is now through

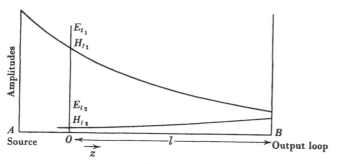

Fig. 3·9.

the factor $e^{+\alpha z}$, and H_x and H_y vibrate with a phase advance of 90° on E_x. When the output loop of the piston attenuator is introduced into the field of the persisting mode it distorts it, and further evanescent modes are excited which are superimposed on that from the source. When the output loop and source are widely separated, at a sufficient distance from either, and between them, the persistent mode from each is the dominant mode, all others being unimportant.

In fig. 3·9 the point A marks the position of the source (bolometer filament) and B the position of the pick-up loop. At a point O between A and B the field is that of the two persisting modes from A and B superimposed. The exponential decay of the amplitudes is indicated by the curves. Distance z along the wave guide is referred to O as an arbitrary origin. Let the loop lie at a distance $z = OB = l$, and suppose that the transverse components at the fields in the two modes at position z are (E_{1t}, H_{1t}) and (E_{2t}, H_{2t}). We may write

$$\left.\begin{array}{ll} H_{1t} = Ae^{j\omega t}e^{-\alpha z}, & E_{1t} = Be^{j(\omega t + \frac{1}{2}\pi)} \times e^{-\alpha z}, \\ H_{2t} = Ce^{+\alpha z}e^{j(\omega t - \pi - \beta)}, & E_{2t} = De^{+\alpha z} \times e^{j(\omega t + \frac{1}{2}\pi - \beta)}, \end{array}\right\} \quad (1)$$

where A, B, C and D are the amplitudes at $z = 0$ and β is a phase constant of the second mode relative to the first. Since each persisting mode is the dominant mode it follows that

$$\frac{A}{B} = \frac{C}{D}. \quad (2)$$

We may assume that the amplitude of the mode excited at the output loop is proportional to the field E_{1t} at this loop, and it follows from equation (1) that

$$\left.\begin{array}{ll} Ce^{\alpha l} \propto Ae^{-\alpha l}, & De^{\alpha l} \propto Be^{-\alpha l}, \\ C \propto Ae^{-2\alpha l}, & D \propto Be^{-2\alpha l}. \end{array}\right\} \quad (3)$$

or

Equation (3) is consistent with (2).

Fig. 3·10 (a) and (b) show the representative vectors of the oscillations of the fields (H_{1t}, E_{1t}) and (H_{2t}, E_{2t}) respectively at the position

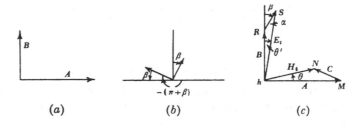

(a)   (b)   (c)

Fig. 3·10.

$z = 0$. The resultant electric and magnetic fields E_t and H_t at this position are obtained by vectorial addition of the component fields of the two evanescent fields. This addition is carried out in fig. 3·10 (c). The amplitude and phase of the resultant field H_t at $z = 0$ are respectively LN and θ, and those of E_t, LS and $(\frac{1}{2}\pi - \theta')$. Clearly, the resultant fields H_t and E_t do not oscillate in quadrature. There is therefore a component of the Poynting vector along Oz, which is real and represents a mean flow of power from the source of the loop.

The mean flow of power is therefore proportional to

$$LS.LN\cos(\tfrac{1}{2}\pi-\theta-\theta') = LS.LN\sin(\theta+\theta').$$

When $C \ll A$ and $D \ll B$, this expression simplifies because θ and θ' are then small angles and $\sin(\theta+\theta') \doteqdot (\theta+\theta')$.

We have also $\qquad \theta \doteqdot \dfrac{C}{A}\sin\beta, \quad \theta' \doteqdot \dfrac{D}{B}\sin\beta,$

so that, from (2), $\qquad\qquad \theta = \theta'.$

Consequently the mean flow of power at $z = 0$ from the source to the loop is proportional to

$$LS.LN\sin(\theta+\theta') \doteqdot 2AB\theta = 2AD\sin\beta = 2A\sin\beta e^{-2\alpha l}.$$

Thus the mean power delivered to the loop at position l relative to an arbitrary fixed origin is proportional to $e^{-2\alpha l}$. This is the law of the piston attenuator. From what has preceded, it is evident that exponential dependence on distance is only obtained when the amplitudes C and D are much smaller than A and B, and the attenuator should be constructed so that the evanescent mode excited at the output loop is always relatively feeble at the source. The position of the loop corresponding to zero decibels on the dial is chosen to satisfy this requirement.

3·9. Attenuation of progressive waves due to finite conductivity of the walls

It has been supposed hitherto that the walls of the wave guide were perfectly conducting and consequently that progressive waves were propagated in it with no loss of power. In practice, however, the metal walls possess a large, but finite, electrical conductivity, and the currents that flow in them when an electromagnetic wave is propagated along the guide are accompanied by the generation of heat. The energy transformed into heat is abstracted from the power carried by the wave whose field components are therefore attenuated with axial distance z away from the source. In the expressions for the amplitude of each component of the electric field $\mathbf{E}$ and of the magnetic field $\mathbf{H}$ there must now be included a factor $\epsilon^{-\alpha_l z}$. The ratio of the field amplitudes at two corresponding positions on the cross-section at distances z and $(z+d)$ is therefore $\epsilon^{\alpha_l d}$, and the logarithm of this ratio to base ϵ is $\alpha_l d$; this is the attenuation measured in nepers suffered by the wave. The loss per unit length is therefore α_l nepers or $8 \cdot 686\alpha_l$ db. The loss coefficient depends on the electric conductivity of the material of the walls, the frequency, the mode of wave being transmitted and the dimensions and the geometry of the wave guide. When the attenuation $\alpha_l \lambda_g$ nepers suffered in a distance equal to the wave-length λ_g, is very much less than unity, as occurs in practice in wave guides with silver, copper or brass walls, the appropriate formulae for α_l may be derived by the following method.

It is assumed that in wave guides with highly conducting metal walls, the field components in the wave are the same as those in a hypothetical wave guide with perfectly conducting walls, whereas, in fact, there is a small component of the electric intensity **E** tangential to the metal surface. The components of **E** and **H** each contain the factor $\epsilon^{-\alpha_l z}$, consequently the power w transmitted over a cross-section of the wave guide, which is obtained by integrating the product of the transverse components of **E** and **H** over the cross-section, contains the factor $\epsilon^{-2\alpha_l z}$. Thus

$$w = w_0 \epsilon^{-2\alpha_l z},$$

where w_0 is the power transmitted at $z = 0$, and

$$\frac{dw}{dz} = -2\alpha_l w.$$

But the loss of power per unit length $-dw/dz$ is the energy dissipated in the walls of the wave guide in ohmic heating.

This energy loss per unit length can be separately calculated from the currents in the walls which are directly proportional to the tangential components of the magnetic field at the wall, and in terms of the conductivity of the metal and the frequency. Call this energy loss per unit length A. Then

$$A = -\frac{dw}{dz} = 2\alpha_l w,$$

or $\qquad\qquad \alpha_l = \dfrac{A}{2w}$ nepers/unit length.

Thus, the attenuation coefficient is found theoretically by calculating the energy lost per unit length in ohmic heating, and the flux of power w over the corresponding cross-section. This procedure leads to the following formulae for attenuation coefficients of progressive waves in rectangular wave guides (see § 7·15).

The symbols which appear in these formulae have the following interpretations:

f, *the frequency* of the wave in cycles per second.

f_c, *the cut-off frequency* for the wave mode concerned.

a and b are the dimensions of the wave-guide cross-section measured in metres, the dimension a is associated with the mode integer m and b with n.

R_s *is the surface resistance* in ohms per unit square of the metal surface (the quantity R_s was discussed in § 1·4; according to that section $R_s = 1/\sigma\delta$), that is, power is dissipated in unit area of the surface at the rate $\frac{1}{2}R_s I^2$ W., where I is the total surface current in amperes per unit length. R_s is found from the following formula (§ 7·14):

$$R_s = 2\pi\sqrt{(10^{-7}\rho f)} \text{ ohms/unit square}, \tag{1}$$

where ρ is the specific resistance of the wall metal in ohms per metre cube.

For copper, $\rho = 1 \cdot 6 \times 10^{-8}$ ohms-m., consequently with copper walls
$$R_s = 8\pi \times 10^{-8}\sqrt{f}.$$

The surface resistance of any other metal may be obtained by multiplying this value by the factor

$$N = \sqrt{\frac{\rho_{\text{metal}}}{\rho_{\text{copper}}}}.$$

α_l is the loss coefficient in nepers per metre. To obtain the loss in decibels per metre it is necessary to multiply the value in nepers by 8·686.

Formulae for attenuation coefficients in rectangular air-filled wave guides

The H_{10}-*wave* (electric field parallel to the edge b of the cross-section):

$$\alpha_l = \frac{R_s}{120\pi b}\frac{1}{\sqrt{\{1-(f_c/f)^2\}}}\left[1+2\frac{b}{a}\left(\frac{f_c}{f}\right)^2\right] \text{ nepers/m.} \tag{2}$$

To obtain α for the H_{01}-mode, interchange a and b in this formula.

The E_{mn}-*wave:*

$$\alpha_l = \frac{2R_s}{120\pi b}\frac{1}{\sqrt{\{1-(f_c/f)^2\}}}\left[\frac{m^2(b/a)^3+n^2}{m^2(b/a)^2+n^2}\right] \text{ nepers/m.} \tag{3}$$

The H_{mn}-wave:

$$\alpha_l = \frac{2R_s}{120\pi b}\left[\frac{(b/a)\{m^2(b/a)+n^2\}}{\{m^2(b/a)^2+n^2\}}\left\{1-\left(\frac{f_c}{f}\right)^2\right\}+\left(\frac{b}{a}+1\right)\left(\frac{f_c}{f}\right)^2\right]$$

$$\times \frac{1}{\sqrt{\{1-(f_c/f)^2\}}} \text{ nepers/m.} \quad (4)$$

According to these formulae the attenuation in a rectangular guide depends on (a) the size of the tube, (b) the ratio of the cross-sectional dimensions, (c) the resistivity of the walls, (d) the frequency.

We shall not, however, discuss the influence of these factors in detail because in radar practice at a frequency of 3000 Mc./sec. the only wave that is employed to carry power through a long run of wave guide is the H_{10}-wave in a rectangular guide, and it will suffice to discuss the attenuation of this wave in the actual $2\frac{1}{2} \times 1$ in. and 3×1 in. wave guides used in service equipments.

The losses in copper wave guides with these standard internal dimensions are calculated as a function of frequency, from formula (2). The results are exhibited in the curves of fig. 3·11 (a) and (b) which give the loss in decibels per metre suffered by an H_{10}-wave, as a function of frequency f.

According to formula (2) the loss becomes infinite at the frequency $f = f_c$ due to the factor $1/\sqrt{\{1-(f_c/f)^2\}}$ and also at $f = \infty$ due to the factor $\sqrt{f}$. We may therefore expect a_l to reach a minimum value, in a given wave guide at some frequency greater than f_c. These properties of the curves are shown in fig. 3·11 (a) and (b). The asymptotic approach of a_l to infinity as f approaches f_c, and the minimum on each curve, are indicated.

The values for α_l relate to wave guides with copper walls, but to obtain the attenuations when another metal is used it is necessary only to multiply the ordinates of the curve by a factor $N = \sqrt{\dfrac{\rho_{\text{metal}}}{\rho_{\text{copper}}}}$. Thus

Metal	Brass	Silver	Aluminium
N	2·1	0·97	1·27

With iron, the factor N is

$$N = \sqrt{\left(\frac{\rho_{\text{iron}}}{\rho_{\text{copper}}}\right)} \times \sqrt{(\text{permeability of iron})}.$$

Because of the high permeability of iron and steel the attenuation in tubes of these metals is large.

The position of the S-band of frequencies is indicated on each curve.

The smaller $2\frac{1}{2} \times 1$ in. wave guide, to which fig. 3·11 (a) relates, is used in R.A.F. airborne equipments at wave-length of about 9 cm.,

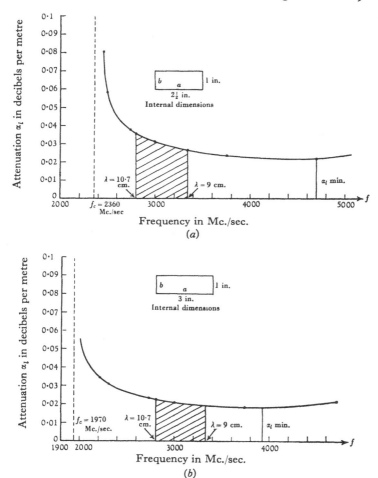

Fig. 3·11. Attenuation coefficient α in standard copper wave guides as a function of frequency.

that is, at the right-hand end of the band shown in the figure, where the attenuation is about 0·027N db./m. Ground equipments, on the other hand, use wave-lengths at the other end of the S-band of

10 and 10·7 cm. respectively in Naval and R.A.F. ground equipments and Army equipments. This brings the representative points on to the rising portion of the curve, consequently it is desirable to displace the operating band along the curve towards the minimum. This is achieved by employing the larger 3 × 1 in. wave guide for which the attenuation curve is shown in fig. 3·11 (b). This size of wave guide is less convenient for airborne use because of its greater size and weight than the 2½ × 1 in. guide. It can be seen that any further reduction in the dimensions of the airborne wave guide would move the operating frequency nearer to f_c with a resulting increase in attenuation. The attenuation at the operating frequencies in both wave guides is therefore approximately 0·025N db./m., and in a brass wave guide this is 0·025 × 2·1 = 0·0525 db./m. The corresponding loss in the standard polythene cable, Uniradio 21, is about 0·6 db./m., that is, about ten times as great.

Since the attenuation is proportional to the surface resistance it is important to avoid corrosion of the interior surface of the guide due to weather or salt spray. For this reason wave guides are often silver or cadmium plated internally, or hermetically sealed at their free ends by cellophane diaphragms, the air within being kept dry with silica gel cells which communicate with the interior through small holes in the walls.

REFERENCES

§ 3·9. Kuhn, S. Calculation of attenuation in wave guides. *J. Instn Elect. Engrs*, vol. 93, part III A, no. 4, p. 663, 1946.

GENERAL

Ramo, S. and Whinnery, J. R. *Fields and Waves in Modern Radio*. John Wiley.

Chapter 4

WAVE-GUIDE TECHNIQUES

4·1. Introduction

The two outstanding properties of wave guides that recommend their use in preference to cables in transmission systems at microwave-lengths are large-power handling capacity and small attenuation of the transmitted power. Although the general features of wave propagation in wave guides were well known for many years before the war, yet a great many subsidiary technical problems required solution before workable transmission systems for incorporation in radar equipments could be achieved.

In this and the following chapter we are concerned with those practical techniques which, almost without exception, have been developed during the war to meet the requirements of microwave radar.

4·2. Choice of wave-guide geometry and dimensions

When a wave-guide run is used to carry power from a transmitter to a load it is important for several reasons, that the dominant mode only is propagated and that all others are evanescent. It is not otherwise possible to ensure that the field distribution at the end of the wave guide, or along it, is known with certainty. If modes higher than the dominant can be propagated, then they may be excited by the dominant mode at irregularities and geometrical discontinuities in the guide, and appear at the output end. Further, matching procedures analogous to those employed in transmission line practice are used in wave guides to eliminate reflected components of the dominant mode, and for their successful application it is essential that all modes except the dominant shall be evanescent.

It is evident, therefore, that in the two obvious forms of wave guide that might be employed, the rectangular and the circular, the dimensions of the cross-sections must be such that the dominant mode only is propagated. In the rectangular wave guide the dimensions of the cross-section, a and b, are chosen so that only the H_{10}-mode with the electric vector parallel to the shorter dimension is propagated, and in the circular wave guide the dominant H_{11} would be the transmitted mode. It is essential that the polarization of the transmitted wave should not vary along the guide and in this requirement the circular wave guide fails in practice. If the cross-section departs from the truly circular form then in general the H_{11}-wave resolves into a pair of modes with different velocities, which recombine to give an H_{11} mode where the tube again has a circular cross-section. The field pattern, however, of the reformed H_{11}-wave will be elliptically polarized, in general, with respect to the original. Thus, the polarization is not stable with respect to small deformations in the wave guide. For this reason, circular wave guides are used in short lengths only and in special devices such as rotatable joints.

The rectangular wave guide is therefore adopted as the standard. It was explained in § 2·3 that in a standard $2\frac{1}{2} \times 1$ in. wave guide operated at 3300 Mc./sec. the H_{10}-wave only is propagated. In order to reduce the weight of the wave guide it would be possible to reduce the long dimension from 6·35 cm. ($2\frac{1}{2}$ in.) to any value greater than 4·5 cm. at which the 9 cm. wave would be cut off and to obtain a propagated component. There are two reasons why it is undesirable to operate the wave guide near cut-off. First, as may be seen from formula 3·9 (2), the ohmic attenuation increases rapidly near cut-off and the guide would therefore be wasteful of power. It is apparent from the curve in fig. 3·11 that in the $2\frac{1}{2} \times 1$ in. wave guide a 3300 Mc./sec. wave has an attenuation coefficient which is not excessive and not much greater than the minimum. Secondly, near cut-off the dispersion of the wave is excessive. According to equation 2·1 (3) the speed of propagation depends on the frequency f. The rate of change of speed v_g with frequency is, from 2·1 (3),

$$\frac{dv_g}{df} = -\left(\frac{\lambda_g^3}{\lambda_c^2}\right).$$

Since λ_g tends to infinity as f approaches f_c it is clear that the rate of change of v_g with frequency becomes very large near cut-off. Consequently, the different components in the spectrum of a transmitted pulse travel at different speeds, and if the wave guide is long the shape of the pulse can be impaired. The size $2\frac{1}{2} \times 1$ in. is a suitable compromise between these competing mechanical and electrical requirements. For ground radar equipments that operate on a frequency of 3000 Mc./sec. ($\lambda = 10$ cm.) it is found desirable to enlarge the cross-section to 3×1 in.

Similar considerations obtain at short wave-lengths where the standard X-band ($\lambda = 3 \cdot 3$ cm.) wave guides have internal dimensions of $1 \times \frac{1}{2}$ in. (British) and $0 \cdot 9 \times 0 \cdot 4$ in. (American).

4·3. Avoidance of reflected waves within a wave guide

It is undesirable in any high-frequency transmission systems to permit a reflected wave to return to the generator, but at centimetre wave-lengths it is especially important to eliminate the reflected wave.

The reason is the following: Whereas at wave-lengths greater than a few metres the master oscillator of the transmitter is followed by buffer amplifiers which isolate it from the load, at centimetre wave-lengths the oscillator is a magnetron which is coupled directly into its load. When no reflected wave returns to the input the load presented to the magnetron can be made resistive, but if a reflected wave returns it adds in general a reactive component to the load. This is transformed by the coupling to a reactance in parallel with the equivalent oscillatory circuit of the magnetron. Both the frequency and the power output of the magnetron change with this reactive component. It is only possible therefore to cause the magnetron to behave according to specification provided the reflected wave is eliminated. The chief sources of reflected waves are, first, the termination of the wave guide where the power is fed to the aerial, and secondly, joints, bends and other discontinuities along the course of the wave guide.

As already mentioned, one effect of a reflected wave is to shift the operating frequency of the magnetron. This effect is called *frequency pulling*. In addition to this there is a further phenomenon known as *frequency splitting* which occurs with long wave-guide runs and is caused by a wave reflected at the end of the wave guide. This wave introduces a reactive component at the input to the wave guide, whose magnitude in a long wave guide is very frequency sensitive. A full analysis leads to the conclusion, justified by experience, that an unstable state results in which the magnetron is able to oscillate on any one of a set of discrete frequencies, and in practice tends to operate at random on one or other of them. This phenomenon is only found when the wave guide is several wave-lengths long. The ideal wave-guide installation, therefore, is short and straight with as few joints as possible. This ideal, however, is seldom achieved, and it is a matter of importance so to design joints and bends that reflexions from them are sufficiently weak as to be innocuous. To test whether or not a given design of joint or bend is free from reflexions a standing wave indicator is used together with a reflexionless termination at the end of the wave guide to eliminate any reflected wave from the end.

Because of the importance of these devices they will be described next.

4·4. Standing wave indicators

For work on the S-band of wave-lengths (9–10·7 cm.) a standing wave indicator comprises the following components:

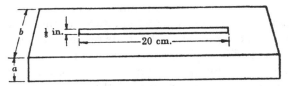

Fig. 4·1. Longitudinal slot.

(a) A short piece of standard S-band wave guide ($2\frac{1}{2} \times 1$ in. at 9 cm. and 3×1 in. at 10–10·7 cm.) with a longitudinal slot $\frac{1}{8}$ in. wide and about 8 in. long cut on the centre line of a broad face as shown in fig. 4·1. The outer surface of this face is made as truly flat as possible.

This piece of wave guide carries standard flanges at its ends which permit it to be incorporated in the wave-guide run to be tested. The slot must be cut accurately central and of uniform width if the standing wave indicator is to function reliably. In an X-band (3·3 cm.) standing wave indicator the greatest care must be taken to make a precision cut, since all dimensions are reduced. The slot in the $1 \times \frac{1}{2}$ in. wave guide is now only $\frac{3}{32}$ in. wide and 3 in. long. The piece of wave guide is made especially uniform in dimensions throughout, and the outer surface of the face carrying the slot is made accurately flat. With these precautions the probe which is inserted through the slot can be made to preserve the same clearance from its edges and to protrude into the wave guide to the same depth at all positions along the slot. As explained in §2·5, a longitudinal slot on the centre line of a broad face does not affect the propagation of an H_{01}-wave.

(b) *A detector unit*, comprising a probe for insertion into the guide in order to abstract an e.m.f. from the electromagnetic field within, a crystal detector and a coaxial matching stub and piston, and the microammeter to register the rectified current from the crystal. The whole unit is mounted on a brass carrier with a plane under-surface except for a rectangular ridge that fits accurately into the slot in the wave guide mentioned above. The depth of the ridge

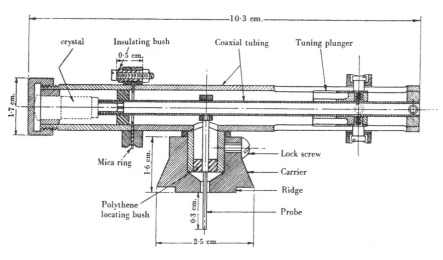

Fig. 4·2. Longitudinal section of indicator unit.

is the same as the wall thickness of the wave guide. The probe runs through a central hole in the carrier and the ridge. The construction of the detector unit can be understood from fig. 4·2.

The probe of the detector projects through the carrier and penetrates into the guide a small distance parallel to the electric field of the H_{01}-wave which induces an e.m.f. in the probe. This e.m.f. drives a current along the coaxial system of the standing wave indicator and through the crystal. The crystal develops a d.c. e.m.f. across the mica insulating ring that insulates the crystal holder from the remainder of the standing wave indicator, and this e.m.f. is applied to the terminals of the microammeter through the leads. A lock screw that fixes the carrier to the standing wave indicator allows the carrier to be adjusted in position so that the penetration of the probe into the wave guide can be altered until a convenient microammeter deflexion is obtained with the particular electromagnetic field under examination. The terminating plunger of the coaxial stub is adjusted until the microammeter registers a maximum deflexion when the indicator is stationary on the wave-guide slot. The extension of the probe into the guide is in practice about $\frac{1}{16}$ in. only, and it disturbs the field to a negligible extent. For readings less than about 20 μA. (corresponding to an r.f. input to the crystal of about 30 μW.) the meter response is proportional to the square of the amplitude of the field. Standing wave indicators at X-band are similar, but with smaller overall dimensions.

An alternative form of indicator for use at the short wave-lengths (X-band) is illustrated in fig. 4·3. A straight piece of $1 \times \frac{1}{2}$ in. X-band

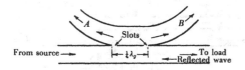

Fig. 4·3. Directive feed.

wave guide has a pair of transverse slots cut in the broad face. The slots are each $\frac{3}{32}$ in. wide and $\frac{1}{4}\lambda_g$ apart. A piece of curved wave guide is fixed to the straight wave guide so that the slots are common to the walls of both wave guides on the bend of the curved guide, as shown.

Since the slots are transverse they interrupt longitudinal current in the straight wave guide, and equal and opposite line charges accumulate on opposite edges of each slot giving rise to a transverse oscillating electric field across the slots. The pair of slots radiates into the upper wave guide into the arms A and B. If the wave in the main wave guide is travelling from left to right then it can be seen that the radiations from the slots, because of their $\frac{1}{4}\lambda_g$ separation, add in arm B but cancel in arm A. Conversely, if the wave in the main guide is travelling from right to left then the pair in slots radiates with arm A and not into B. When a complete or partial standing wave exists in the main wave guide, formed by two wave trains propagated in opposite senses, the amplitudes of the waves that travel up the arms A and B are proportional to those of the progressive waves in the main wave guide. If, therefore, it were possible to terminate these arms in matched crystal detectors of equal sensitivity, the standing wave ratio in the main wave guide would be proportional to the square root of the ratio of their d.c. outputs. It is difficult to obtain a pair of crystals whose sensitivities remain equal for long periods, consequently it is better to match the arm B which is excited by the wave from the generator, and to use the response in arm A to indicate whether or not a reflected component exists in the wave guide. It is also possible to calibrate the microammeter response against standard mismatches, so that an indication of the standing wave ratio is given when a standard amount of power is fed into the wave guide.

The pair of slots that radiate into one arm but not the other when a single travelling wave exists in the main guide is an example of a *directive feed*. A single pair of slots is frequency sensitive and an alternative directive feed which employs four or more slots or holes is less frequency dependent. A directive feed can also employ a single central circular hole of a special diameter, which depends on the wave-length. Directive feeds were first used on transmission lines to isolate the two travelling waves in a partial standing wave.

4·5. Crystal detector

At microwave-lengths, thermionic detectors are not used because a greater signal to noise ratio is obtained from a special form of

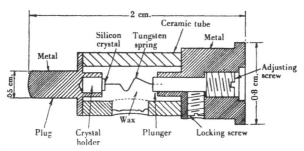

Fig. 4·4. Section of crystal capsule.

crystal detector, which we describe but briefly, since our chief concern is with wave guides and not valves.

Fig. 4·4 is an enlarged representation of a crystal detector. Detection occurs at the point of contact of a tungsten wire and a fragment of crystal silicon. The wire which is 0·2 mm. in diameter is formed into a spring, and the adjusting screw is turned until a contact is obtained with the silicon and a force of the order of 50 g. weight is applied. The detector is tapped until a back to front resistance ratio of about 10:1 is obtained, the forward resistance being of the order of 200–300 ohms. The ceramic tube is then filled with wax to hold the tungsten spring in position. These crystals deteriorate if peak powers exceeding about $\frac{1}{10}$ W. are passed through them. A typical d.c. characteristic curve is given in fig. 4·5. A typical

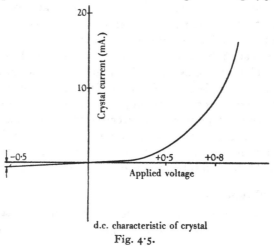

d.c. characteristic of crystal
Fig. 4·5.

mounting of a crystal in a detector is that already shown in fig. 4·2. A crystal holder and pick-up loop for use with a cavity resonator (§ 6·8) is shown in fig. 4·6. The pick-up loop is the output loop of the cavity and abstracts a small amount of power sufficient to give a deflexion on the microammeter fed through the coaxial output line. The coupling into the cavity can be adjusted either by variation of the size of the loop, by the distance it extends into the cavity, or by rotation of the plane of the loop. This arrangement is used on the S-band of wave-lengths, and it is important to limit the power to the crystal to a value below the 'burn out' value of the order of $\frac{1}{10}$ W.

In an *X-band crystal mixer*, the crystal is placed directly in the wave guide and is mounted in it as shown in fig. 4·7. This mixer is used to terminate the wave guide in laboratory test experiments in which the power level is low and derived from klystron oscillators.

It can be seen that the crystal is mounted across the line with its axis parallel to the electric vector. The wave guide is circular and is connected to the rectangular wave guide run by a transformer of

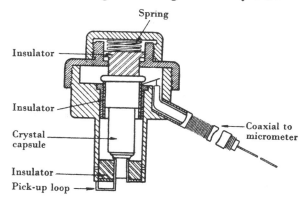

Fig. 4·6. Crystal holder and loop.

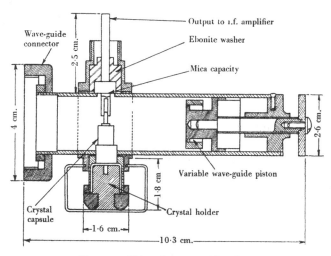

Fig. 4·7. X-band wave-guide mixer.

the type shown in fig. 4·22 (c). This converts the H_{01}-wave in the rectangular guide into an H_{11} in the mixer. The resistive component of the crystal impedance, regarded as a generator, matches the output transmission line. The tuning plunger is used to eliminate the wave reflected by the mixer back to the generator, and the dimensions of the mixer are chosen so that this matching can be achieved by means of a single movable piston. In such a mixer the whole power carried by the incident wave is absorbed in the crystal. Since the d.c. current output is roughly proportional to the square of the field amplitude in the wave-guide wave, the microammeter reading is proportional to the power reaching the end of the wave guide.

A matched mixer can be used in this way to find the attenuation suffered by a wave in a chosen section of a wave guide, from the

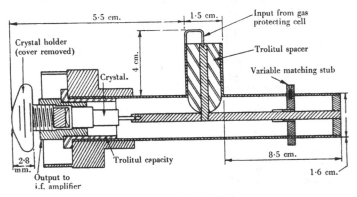

Fig. 4·8. *S*-band mixer.

ratio of the powers reaching the mixer before and after the insertion into the wave-guide run of the section under test.

When it is required to amplify the output from the crystal, as, for instance, in a radar receiver, then it is necessary to change the frequency to an intermediate frequency, of say 45 Mc./sec., in order that the signal may be amplified by normal radio methods.

The mixer of fig. 4·7 therefore carries a probe, not shown in the figure, for injecting a signal from a local oscillator into the mixer cavity, and the difference frequency is abstracted from the crystal output and led away to the amplifier. A mica disk, placed as shown, forms a capacity which by-passes the currents at microwave frequencies but offers sufficient impedance to e.m.f.'s at the difference frequency for almost the full e.m.f. of the difference frequency to appear at the output.

Fig. 4·8 shows a typical *S*-band (9 cm.) mixer used in airborne radar equipments. The signal from the aerial is abstracted from the wave guide by a combination of probe and plunger, such as that indicated in fig. 2·8, and is carried through a coaxial cable to the input loop of a gas-filled resonant cavity which protects the crystal from damage by the transmitter pulse. The signal is abstracted from this cell by the loop indicated and reaches the crystal. This mixer also carries a side probe for the local oscillator input, but this is not shown. Crystal capsules are marked with coloured spots to indicate two important characteristics, the wave band on which they are best operated and their resistance to 'burn out'.

A *yellow spot* is marked on *S*-band (9·1–10·7 cm.) crystals, but a *green spot* on *X*-band (3·3 cm.) crystals.

A *red spot* indicates high resistance to burn out, and an *orange spot*, medium resistance to burn out.

The red and orange spot crystals also carry the coloured spot indicating the wave band.

Crystals with the wave band spot only have small resistance to burn out.

4·6. *X*-band wave-meter

A convenient form of wave-meter for use with an *X*-band test set is shown in fig. 4·9. It is merely a coaxial transmission line with

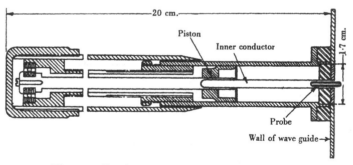

Fig. 4·9. Coaxial wave-meter fitted on wave guide.

a movable shorting plunger. The inner conductor projects a small distance into the wave guide as a probe parallel to the electric field of the wave, and the whole wave-meter is rigidly mounted on the wall of the wave guide. A micrometer screw gives the displacement of the plunger. At resonance the probe reflects a wave to the generator and the wave-guide input impedance becomes reactive and mismatches the generator. The level of power delivered by the klystron oscillator falls, and this is indicated by a change in the meter reading of the crystal mixer used to terminate the wave guide. Resonant positions of the wave-meter plunger occur at separations of $\frac{1}{2}\lambda$, λ, $\frac{3}{2}\lambda$, etc., and from them, using the micrometer scale, the TEM wave-length λ may be found with an accuracy of 0·01 cm.

4·7. Reflexionless terminations

A common requirement is so to terminate a wave guide that no reflected wave returns to the oscillator. The X-band mixer described in §4·5 is an example of a reflexionless termination which also indicates the power reaching the end of the wave guide. When it is

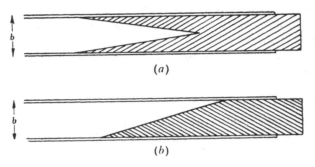

Fig. 4·10. Reflexionless loads.

desired merely to prevent reflexion and not to measure power, then it is convenient to use the simple 'wooden loads' illustrated in fig. 4·10, in which b is the long dimension. Wood strongly absorbs electromagnetic waves at microwave-lengths, consequently if a wooden 'clothes peg' (fig. 4·10 (a)) or wooden 'wedge' (fig. 4·10 (b)) is inserted into the rectangular wave guide which it fits closely, the pair of component plane waves into which the H_{01}-wave can be resolved enter the wood at the sloping surfaces and, if the slope is sufficiently small, with little backwards reflexion. These waves are then absorbed in the butt-end of the wooden load. To ensure that negligible power is reflected, the sloping faces should exceed $2\lambda_g$ in length, the standing wave ratio then being less than 1·05/1. A wooden cone with a butt-end is used with circular wave guides. As an alternative to wood, a composition of bakelite and graphite is used, and this is more suitable at X-band frequencies.

Wood is unsuitable for use with high powers since it chars, and a mixture of carbon and sand is used instead. The principle of a recent model of high-power reflexionless load is illustrated in fig. 4·11.

The wave guide ends in a steel section which fans out into a horn with end-flanges as extensions to the broad face. Between these flanges is placed a large slab of ceramic loaded with carbon. The purpose of the horn is to reduce the intrinsic impedance of the wave to that of a plane wave in the ceramic so that the power enters the ceramic slab with very little reflexion. The surfaces of the ceramic slab and of the wave-guide horn are large and serve to dissipate the heat at a rate sufficient to prevent destruction of the load.

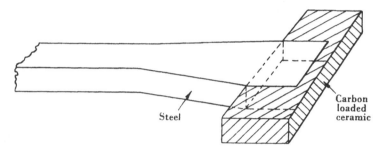

Fig. 4·11. Reflexionless load for high power.

At the greatest powers an all-steel load has been employed with success, the h.f. resistance of steel being high.

An ideal theoretical termination which is completely reflexionless is a thin resistive film whose resistance per square is equal to the intrinsic impedance of the H_{01}-wave, namely, $120\pi \sqrt{\left(\dfrac{\mu}{\epsilon}\right)} \dfrac{\lambda_g}{\lambda}$ ohms (§ 5·6). This film, placed in the normal cross-section of the wave guide and backed by a short-circuited $\frac{1}{4}\lambda_g$ extension of the wave guide, will completely absorb an incident H_{01}-wave. This device is not used in wave guides, but a resistive disk is used in a reflexionless termination for coaxial cables. Fig. 4·12 shows a non-reflecting crystal mounting for terminating a coaxial cable (Uniradio 21) and for use on the S-band of frequencies. This termination is virtually reflexionless, the driving e.m.f. for the crystal being developed across the resistive terminating disk. A lateral tapping probe (not shown) permits a small fraction of the power to be led away to a coaxial line wave-meter (§6·8, fig. 6·6) and the wave-length to be measured without 'pulling' of the oscillator as the wave-meter is tuned to resonance.

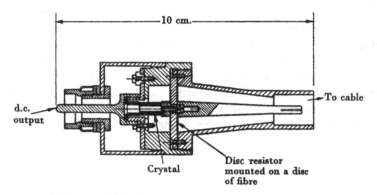

Fig. 4·12. Reflexionless termination for coaxial cable.

4·8. Wave-guide test set

A typical arrangement for the measurement of standing wave ratios is shown in fig. 4·13. The oscillator feeds into a section of wave guide carrying the wave-meter (at X-band, fig. 4·9) or feeding from a pick-up loop into a coaxial wave-meter (fig. 6·6). This is

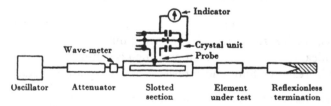

Fig. 4·13. Schematic of test equipment.

followed by an attenuating section, such as described in the final paragraph of §3·7. Then, follow in turn, the standing wave indicator, the element under test (bend, joint, etc.) and finally a section of wave guide terminated in a reflexionless load. Any standing wave is therefore due to the element under test and not to the termination. If the attenuation produced by the element is required, then the change in crystal output is observed with and without the element in the wave-guide run. At S-band wave-lengths the crystal would be fed from a loop which taps power from the wave guide in front of the wooden load, the crystal being supported as in figs. 4·6 or 4·8. At X-band wave-lengths, the reflexionless load would be replaced by the crystal mixer of fig. 4·7.

4·9. Wave-guide couplings and plungers

In practice, long runs of wave guide comprise several shorter sections held together by coupling devices. It is of the greatest importance to design couplings so that they do not excite a reflected wave, since small reflexions from a number of couplings may add to give an appreciable reflected wave at the input. The principal sources of reflexion at the junction are: (a) misalignment of the walls of the two sections of wave guide causing a step in the wall of the wave-guide run at the junction; (b) gaps between the walls across the junction which may interrupt the longitudinal wall currents.

Flange couplings. The standard coupling for S-band (9–11 cm.) wave guides is the plane flange coupling shown in fig. 4·14. These are soldered to be flush with the ends of the wave-guide sections, and the bolt-holes are drilled so that when the flanges are bolted together as shown in fig. 4·14 (b) the walls of the wave guides are in contact everywhere at the junction with no misalignment. If the flanges are not plane, gaps will result, as shown at G in fig. 4·15. The surfaces of the flanges between C and G may be considered as a pair of transmission lines of length l and short-circuited at C. If by mischance l is approximately equal to $\frac{1}{4}\lambda$ (λ = TEM wave-length), an antinode of electric field and a node of magnetic field will occur at G in the standing wave excited between the flanges. There is therefore no current flowing into or out of the gap, and if the gap occurs in the walls at a place where longitudinal current should flow then a serious reflexion will result.

At shorter wave-lengths (X and K-band) the tolerances are more strict and it is difficult to avoid reflexions with simple flange couplings of this type. Couplings for X-band wave guides are therefore more elaborate, and a typical design is shown in fig. 4·16. The flanges are deep and are clamped together by a ring nut which presses on the one (female) and screws on to the other (male). A step at the junction is avoided by accurately placed locating pins.

Choke coupling. As mentioned above, reflexions may arise from gaps in the wave-guide wall at a junction. In the choke coupler a

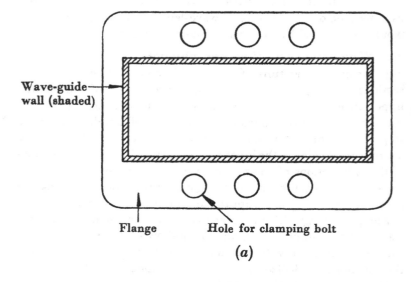

Wave-guide
wall (shaded)

Flange   Hole for clamping bolt

(a)

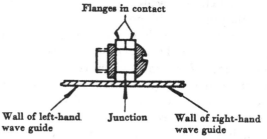

Flanges in contact

Wall of left-hand   Junction   Wall of right-hand
wave guide   wave guide

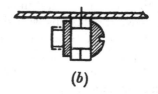

(b)

Fig. 4·14. *S*-band wave-guide coupling.

gap is deliberately introduced, but its dimensions and form are chosen so that it effectively behaves as a perfect junction. The principle of the 'choke' joint is illustrated in fig. 4·17.

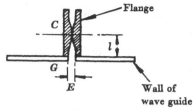

CBA is an L-shaped recess or 'ditch' cut into one of the flanges, the other flange remaining plane. The lengths GB and BA are each $\frac{1}{4}\lambda$, and when the flanges are in contact a

Fig. 4·15.

double quarter-wave transmission-line transformer is formed between them. The short circuit at A is therefore transferred to the gap at G, and in the standing wave that forms in the ditch an antinode of magnetic field lies in the gap and is continuous with the

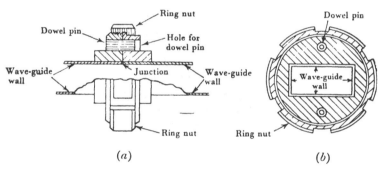

(a)   (b)

Fig. 4·16. X-band wave-guide coupling.

magnetic field of the wave in the wave guide, ór what amounts to the same, the full longitudinal current flows into the gap at one edge and out at the other in the same phase as though the gap were absent. The ideal ditch would be one that followed the contour of the wave-guide section, but this could not be readily mass-produced and a circular recess is the only practical form. However, the minimum size of the ditch is governed by the dimensions of the wave guide, and it is found possible to fit the circular ditch around the American (o·9 × o·4 in.) X-band wave guide, but less readily around the British (1 × $\frac{1}{2}$ in.) wave guide, since the total distance from the centre of the long dimension in the wall to the bottom of the ditch must be $\frac{1}{2}\lambda$, in a pair of quarter wave steps. The double-quarter wave transformer is designed so that the characteristic impedance of the portion *CB* is small (narrow gap) and that of *BA* large, in order to improve the frequency characteristic of the choke joint.

Choke plungers. The double-quarter wave choke is also used to obtain a perfectly reflecting piston or plunger to provide a movable short-circuiting termination for a wave guide. It is difficult to construct an ordinary plunger so that it both makes good contact

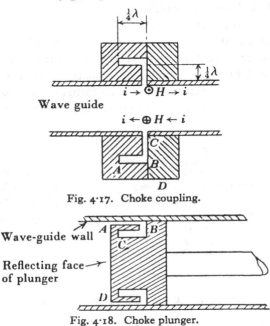

Fig. 4·17. Choke coupling.

Fig. 4·18. Choke plunger.

everywhere with the walls of the wave guide and at the same time does not fit so tightly that it cannot easily be moved. In *X*-band wave guides plungers take the form shown in fig. 4·18. From what was said above it is evident that when the distances *AB* and *BC* are each $\frac{1}{4}\lambda$, the face *AD* behaves as though the gaps at *A* and *D* were bridged by metal. In rectangular wave guides the recesses run across parallel to the broad face of the wave guide and there are no recesses at the narrow face. In circular wave guides the recesses are circular, and the diagram represents diametral section. At the longer wave-lengths (9–11 cm.) a suitable contact is obtained by using a ring of phosphor-bronze spring contacts as shown in fig. 2·21.

4·10. Bends and corners in wave guides

Bends. Wave-guide bends are of the two forms shown in fig. 4·19 (*a*) and (*b*). The bend of fig. 4·19 (*a*) is called an *H*-bend because the magnetic field lies in the plane of the bend, and that of fig. 4·19 (*b*) an *E*-bend because the electric field lies in the plane of the bend.

The effect of bending the tube is to change the wave impedance (§ 3·4·1) of the wave in the bend to a value different from that of the H_{10}-wave in the straight wave guide. It is found that discontinuous changes in wave impedance of this character generate a reflected wave whose relative amplitude depends on the precise circumstances, but in general increases with the magnitude of the discontinuity.

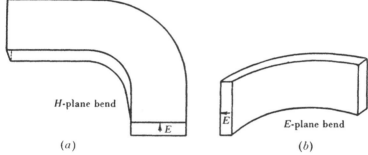

H-plane bend

(*a*)

E-plane bend

(*b*)

Fig. 4·19. Wave-guide bends.

When the radii of the bends are large compared with λ_g, the disparity in the wave impedances in the bent and straight wave guides is small and the reflexions are not serious. Consequently, when bends are used the radius of the inner face must exceed λ_g, and this requirement makes bends inconveniently bulky in *S*-band wave guides (9–11 cm.), but at *X*-band bends are entirely satisfactory and are in general use. A second reflexion occurs at the other end of the bend, and by making the mean length a multiple of $\frac{1}{2}\lambda_g$ the reflexions at the input and output of the bend can be caused to cancel. The total reflexion is then very small. Bends of length $\frac{1}{2}\lambda_g$ have been manufactured in the United States for use where space is limited and are very satisfactory. Care is taken that the section is not distorted, because the performance would then deteriorate.

Bends in circular wave guides are most unsatisfactory because it is found that the H_{11}-wave entering the bend resolves into two waves which travel at different speeds and recombine to give an H_{11}-wave in which the plane of polarization is rotated. This is an additional reason for using rectangular wave guides in preference to circular wave guides.

Corners. As mentioned above, bends are too bulky for general use with S-band wave guides, especially in airborne installations, and the *cut-off corners* illustrated in fig. 4·20 (*a*) and (*b*) replace them.

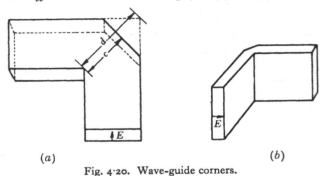

(*a*)   (*b*)

Fig. 4·20. Wave-guide corners.

Fig. 4·20 (*a*) and (*b*) represent respectively an H-bend and an E-bend. In each case the corner is removed and then closed by a flat plate. The best value for the ratio of the distances c to d to give least reflexion depends on the dimensions and form of the wave guide and the wave-length, and is found by experiment. In a standard $2\frac{1}{2} \times 1$ in. wave guide for use at a wave-length of 9 cm. the distance $c = 6$ cm. and $d = 9$ cm. for an H-corner; for the E-corner, $c = 2·2$ cm. and $d = 3·5$ cm. In general, H-corners are more satisfactory than E-corners, especially in X-band wave guides. In fact, the use of corners for X-band wave-lengths has now been abandoned. However, with a British size wave guide the optimum value of c is 1·1 cm. for E-corner and 2·5 cm. for an H-corner.

A satisfactory corner is obtained by using a double corner derived from a piece of straight wave guide by cutting it as shown in fig. 4·21 (*a*), and then reassembling the pieces to form the double corner of fig. 4·21 (*b*). It is easy to prove that the angle θ of the double corner is related to the angle ϕ of the cut, as follows:

$$\phi = (90° - \tfrac{1}{4}\theta).$$

Therefore, to obtain a 90° double corner we require an angle of cut $\phi = 67·5°$. In fig. 4·21, $\phi = 75°$ and $\theta = 58°$; $90 - \tfrac{1}{4}\theta = 75·5°$.

Fig. 4·21 (c) is a curve showing the dependence of the mean length of the bend L_{mean} divided by λ_g on the operating wave-length λ at which the corner gives minimum reflexion. The curve relates to the H-bend of fig. 4·21 (b) with $\theta = 90°$ and represents the results of experiments on a large-size 3×1 in. S-band wave guide. This curve may be used to design satisfactory double bends for X-band wave guides by scaling down λ and using the appropriate λ_g.

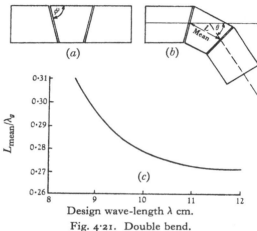

Fig. 4·21. Double bend.

The optimum mean distance L_{mean} for a double E-plane bend proves to be $\frac{1}{4}\lambda_g$, and the design is also less critical than for the H-plane bend.

4·11. Twists and tapers

A rectangular wave guide may be twisted in order to turn the plane of the electric field through some desired angle—usually 90° —as indicated in fig. 4·22 (a). Little reflexion results provided the length of the twisted portion per 90° twist exceeds λ_g.

It is occasionally necessary to employ a length of wave guide whose cross-section is less than that of the standard wave guide; for instance, when a paraboloid with a rectangular aperture is irradiated directly from the end of the wave guide placed at the focus, it is sometimes necessary to reduce the narrow dimension of the wave guide in order to radiate broadly into the wide dimension of the mirror. The transition from the standard wave guide to the narrow wave guide is made through a *tapered section* such as that shown in fig. 4·22 (b). Provided the taper is not abrupt, the wave passes across it with negligible reflexion. Tapers are also employed to connect a circular wave guide carrying an H_{11}-wave to a rectangular wave guide carrying an H_{01}-wave (fig. 4·22 (c)).

Rectangular wave guides with cross-sections of different dimensions, such that their intrinsic impedances (§ 3·4·1) are different, may be approximately matched by interposing a quarter-wave transformer between them. If Z_1 and Z_2 are the intrinsic impedances of the two wave guides, then the interposed section, in analogy with

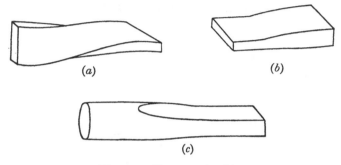

Fig. 4·22. Tapers and twists.

transmission-line procedure, is constructed to have an intrinsic impedance $Z_3 = \sqrt{(Z_1 Z_2)}$ and a length $\frac{1}{4}\lambda_g$, where λ_g is the wavelength of the H_{01}-wave in it (§ 5·15). This analogy with transmission-line practice neglects the excitation of evanescent modes at the discontinuities at the junctions, whose storage fields behave like those of shunting reactances. In practice tapered sections are more convenient.

4·12. E_{01}-H_{01} transformers and rotating joints

It was mentioned in § 2·8 (fig. 2·23) that an E_{01}-wave in a circular wave guide could be excited by the field of an H_{01}-wave in a rectangular guide and conversely. This device, the E-H transformer, finds an important application in the rotating wave-guide joint. The aerial systems of most centimetre wave radar equipments comprise a reflector, usually a paraboloid or a parabolic cylinder, which is required to rotate about a vertical or a horizontal axis, or both. Power is fed to the mirror from the end of a wave-guide run, and it is required to keep the main run of wave guide fixed while the end-section rotates with the mirror. Thus, one or more rotary joints must be introduced into the wave-guide run such that the amplitude and polarization of the waves fed to the mirror are independent of its orientation.

The account given in §2·8 of the E-H transformer is over-simplified; in addition to the E_{01}-wave an H_{11} is also excited in the circular wave guide and both waves are propagated along it. We may suppose that a portion of the H_{01}-wave turns the corner and proceeds down the circular wave guide as an H_{11}-wave as indicated in fig. 4·23 (a).

With respect to this H_{11}-wave the circular wave guide behaves as a series continuation of the rectangular wave guide. In addition, a portion of the wave is reflected from the face of the circular wave guide and superimposes on the entering wave to produce a con-centration of electric field near the axis of the circular wave guide, and thus excites the E_{01}-wave. It is evident that the transformer should be designed so that the excitation of the E_{01}-wave is en-couraged and that of the H_{11}-wave inhibited. Ancillary devices are introduced to accomplish this. In one design of transformer a 'matching step' is placed in the circular wave guide opposite the point of entry of the H_{01}-wave from the rectangular wave guide (fig. 4·23 (a) and (d)).

By making the distance from the face of the step to the axis of the wave guide equal to $\frac{1}{4}\lambda_g$ (where λ_g is the H_{01} wave-length in the rectangular guide) a high concentration of axial electric field is obtained in the circular wave, whereas the transverse electric field required for the excitation of the H_{11}-wave is reduced to zero. To eliminate the residual H_{11}-wave ring filters carried on distrene supports are placed in the circular wave guide, as shown in fig. 4·26 (d). The filter is best placed $\frac{1}{4}\lambda_g$ (λ_g is the wave-length of the H_{11}-wave) from the bottom of the circular wave guide, in order to avoid a resonant condition in the H_{11}-mode, with resulting large and damaging currents in the ring and a mismatch at the end of the rectangular wave guide. A pair of rings, as shown in fig. 4·23 (d), should not be placed $\frac{1}{2}\lambda_g(\lambda_g - H_{11})$, since a high Q resonant system results. The best spacing is an odd number of $\frac{1}{4}\lambda_g$.

In a complete rotating joint as shown in fig. 4·23 (d) and (e) the best overall length should be such as to obtain cancellation of the residual mismatches at each end. The need for all these ancillary requirements makes this rotating joint less convenient than a more recent alternative in which the E-H transformer is designed as indicated in fig. 4·23 (f). The matching step is here replaced by a

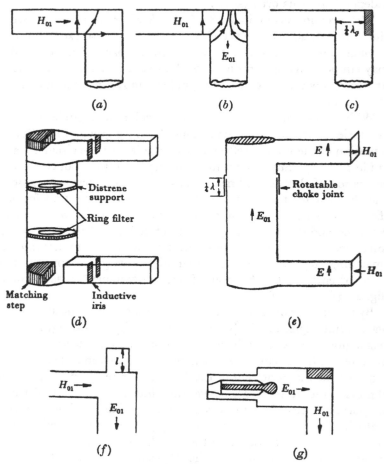

Fig. 4·23. E-H transformers.

circular stub on the end of the circular wave guide, and the diameter of the stub is so chosen that its length l is $\frac{1}{2}\lambda_g$ (E_{01}) but $\frac{3}{4}\lambda_g$ (H_{11}). If, in addition, the narrow dimension of the rectangular wave guide is appropriately chosen, the power carried by the H_{11}-wave in the circular wave guide represents only 0·4 % of the total, and there is no need for ring filters. A good match at the input can also be obtained by choosing a suitable diameter for the circular wave guide.

Fig. 4·23 (e) indicates how a $\frac{1}{4}\lambda$ choke joint can be incorporated to permit free rotation of the output *E-H* transformer to the aerial, relative to the input from the fixed wave-guide run.

The inductive irises in fig. 4·23 (d) serve to eliminate any residual mismatch at the input and output. Fig. 4·23 (g) shows the application of an *E-H* transformer to give a high-power output system for a magnetron feeding a wave guide.

4·13. Coaxial wave-guide transformers

Simple methods of transferring a wave from a coaxial cable to a wave guide have been mentioned in § 2·8 (figs. 2·8 and 2·21). These methods are those of the $\frac{1}{4}\lambda$ *probe* and the *straight through coaxial*, and they are illustrated again in fig. 4·24 (a) and (b).

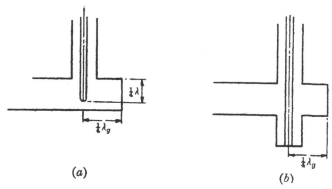

(a)   (b)

Fig. 4·24. Simple coaxial wave-guide transformers.

A convenient broad band transformer is the *door knob* shown in fig. 4·25 (a).

The coaxial stub of fig. 4·24 (b) is inverted into a pedestal and the sliding piston is given a semicircular face. From experiment it is found what are the best dimensions for a wide band transformer, and one is then constructed to be pre-tuned in the centre of the band of operating frequencies.

A modification of the door knob gives a very convenient form of crystal mixer. We refer to fig. 4·25 (b). The door knob here carries a 'choke' recess which prevents the r.f. signal from escaping from the wave guide but permits the i.f. or d.c. component to be abstracted as shown. Fig. 4·25 (*c*) shows an alternative form of piston attenuator (§ 3·6). The construction of the 'door knob' allows the pick-up loop to be moved within the wave guide carrying the evanescent mode.

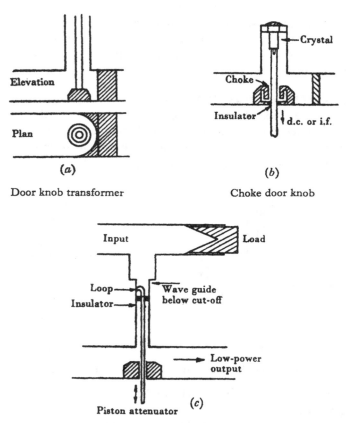

Fig. 4·25. Door knob transformers.

Another construction is the *bar and post* of fig. 4·26 (*a*) and (*b*). It is a wide band system. The inner of the coaxial is joined to a cross-bar spanning the wide dimension of the wave guide as shown in fig. 4·26 (*b*). Since the cross-bar is at right angles to the electric field in the H_{01}-wave it has no longitudinal currents induced in it, and it may be used to abstract the i.f. or d.c. output from the crystal as shown in fig. 4·26 (*c*).

The *bar and post* can also be modified to produce a balanced mixer as shown in fig. 4·26 (*d*) and (*e*). The electric field E of the signal

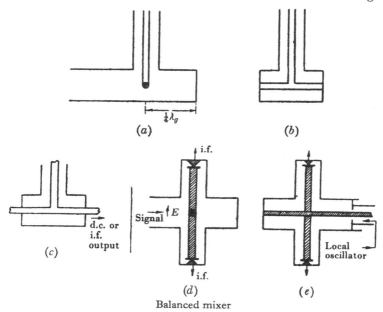

Fig. 4·26. Bar-and-post coaxial wave guide transformers.

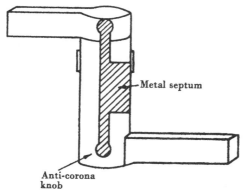

Fig. 4·27. *S*-band rotating joint.

from the wave guide drives currents in antiphase in the two crystals, but the e.m.f. from the local oscillator, introduced along the cross-bar, drives currents in the crystals in the same phase but radiates nothing into the wave guide. By recombining the outputs from the two crystals in the correct sense the noise from the local oscillator may balance out although the i.f. outputs reinforce. In this way the noise factor of a receiver may be greatly improved.

Fig. 4·27 shows a form of rotating joint suitable for S-band airborne equipments. It employs a septate-coaxial combination (§2·7·5) which permits a considerable reduction in the diameter of the circular tube in comparison with the E-H system of fig. 4·23, with the result that the circular tube at S-band in the septate-coaxial joint has a diameter of the same order of magnitude as that of the circular tube in the E-H transformer at X-band. With $\lambda = 9\cdot2$ cm. the diameter of the circular tube in the septate system is only $1\frac{1}{4}$ in. and is small enough to run through the axis of a scanner.

4·14. Flexible wave guides

Gauze wave guides. The first flexible wave guide of practical value was the gauze wave guide. It was made by wrapping the gauze round a rectangular mandrel of the correct dimensions, and by soldering the gauze to produce a rectangular tube. This was then given a thick rubber covering. With the gauze wires at $45°$ to the axis of the wave guide, the guide can be bent but not twisted, and with the wires parallel and perpendicular to the axis the guide twists but does not bend.

Tight-flex wave guide. A successful American flexible X-band wave guide is manufactured by winding silver plated metal tape about $\frac{1}{8}$ in. wide on to a rotating rectangular former and at the same time folding over the edges of adjacent turns so that they interlock. The guide is then provided with a rubber covering. This gives a durable and flexible wave guide which performs well. The loss per 18 in. length is about $\frac{1}{4}$ db., and the standing wave ratio with moderate bending is better than $1\cdot2/1$. At the expense of some flexibility the wave guide can be made airtight by winding eutectic wire into the seam and then fusing it.

Multiple choke wave guide is a flexible wave guide for use at the shorter wave-length (X-band and K-band). It comprises a stack of choke joints with some degree of movement between adjacent joints. The joints are choke-flange couplings mounted in a rubber holder which places corresponding points on adjacent joints $\frac{1}{4}\lambda_g$ apart.

4·15. Production of circularly polarized waves

A simple method of obtaining circularly polarized H_{11}-waves in a circular wave guide is illustrated in fig. 4·28 (*a*). It is based on the principle of the quarter-wave plate of optics. An H_{11}-wave, whose electric field is represented by E in fig. 4·28 (*a*) is launched in a circular wave guide and strikes a trolitul plate set diagonally at an

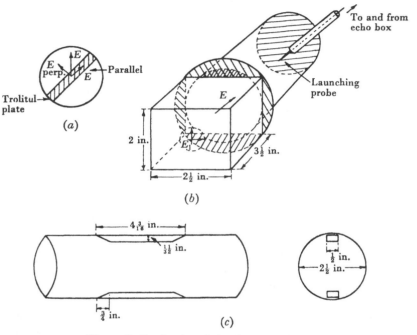

Fig. 4·28. Production of circular polarization.

angle of 45° to the direction of E. The H_{11}-wave may be regarded as a pair of H_{11}-waves polarized at right angles with equal electric fields E_{parallel} and $E_{\text{perp.}}$ at the centre of the wave guide, parallel and perpendicular to the diameter through the trolitul plate. When the trolitul plate is sufficiently thick the major portion of the electric field of the E_{parallel} wave lies within it. Consequently, this wave is propagated through the portion of the wave guide containing the plate, at a slower phase velocity than the $E_{\text{perp.}}$ wave, and if the length of the plate is correctly chosen, its phase can be retarded on that of the other component by 90° on emergence from the plate.

Since the two waves start with equal amplitudes, they recombine to produce a circularly polarized wave. This method has found no practical application.

Another method is shown in fig. 4·28 (b). The device comprises a piece of circular wave guide into which an H_{11}-wave is launched from a $\frac{1}{4}\lambda$ probe as shown. To the end of the circular wave guide is attached a rectangular wave guide whose dimensions are such that both an H_{10}- and an H_{01}-wave can be propagated but at different phase velocities. The electric field along the diameter of the circular wave guide is parallel to a diagonal of the rectangular wave guide, and the H_{11}-wave in the circular wave guide therefore becomes a pair of waves, an H_{10} and an H_{01}, of approximately equal amplitudes, in the rectangular wave guide. The length of the rectangular wave guide is such that the phase difference between these waves at the open end is 90°. It follows that a circularly polarized wave is radiated into space in the direction of the wave-guide axis, and that a dipole placed on the line of the axis with its length lying in the plane perpendicular to the wave-guide axis will have the same e.m.f. induced in it whatever its orientation in this plane. Conversely, by the principle of reciprocity, when the dipole transmits into the wave guide the output from the coaxial feeder is independent of the orientation of the dipole as it is spun in a plane perpendicular to the wave-guide axis.

This device finds application in a test equipment for a particular pulse radar equipment in which a spinning dipole radiates into a paraboloid, consequently the plane of polarization in the beam radiated by the mirror spins with the dipole. The output coaxial from the circular wave guide leads to an 'echo box' (§ 6·8), and the same test signal is returned to the radar set whatever the instantaneous orientation of the dipole.

It has been mentioned that elliptical polarization results in general when an H_{11}-wave is led into a length of wave guide with an elliptical cross-section. However, the mass production of an elliptical wave guide to specification is impracticable, and a simpler method of accomplishing the same result is to use the *ribbed wave guide* of fig. 4·28 (c).

This comprises a circular wave guide within which are fixed a pair of longitudinal bars diametrically opposite. When an H_{11}-wave is sent into the wave guide with its E-field along a diameter at 45°

to that joining the bars, then at the bars it divides into two waves travelling at different speeds. By choosing the correct length for the bars a circularly polarized wave results when these waves recombine. The dimensions given in the figure are correct for a 9 cm. wave.

4·16. Phase-shifting devices

It is convenient to have available a means of controlling the phase of the output at the end of a wave guide, and there are several methods of achieving this control. One simple method is to introduce a sheet of dielectric into the H_{01}-wave parallel to the electric field.

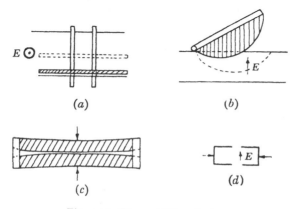

Fig. 4·29. Phase-shifting devices.

The two methods of introducing the sheet are exactly the same as already described for the resistive sheets in the attenuators (§ 3·7). They are illustrated in fig. 4·29 (a) and (b).

In fig. 4·29 (a) the phase shifter is of the 'push across' type of construction in which a plate of trolitul carried on a pair of cross-bars can be moved from the wall of the wave guide where the electric field is zero to the central section where it is a maximum, and it exerts most effect on the velocity of propagation of the wave past it. This design is used at wave-lengths of 9–11 cm. At X-band wave-lengths the 'flap' design of fig. 4·29 (b) is more convenient. Here a mica or ceramic sheet is lowered to any desired extent into the electric field, through a slot in the broad face of the wave guide. At still shorter wave-lengths (K-band) the split wave guide of fig. 4·29 (c) and (d) is convenient. It is a section of wave guide with a pair of extended longitudinal slots, one down the centre of each broad face. By applying lateral pressure as shown, by means of a screw, the broad dimension of the wave guide can be changed and the phase velocity of the H_{01}-wave altered with it. Thus the effective electrical length of the section can be controlled at will.

A more elaborate phase-shifting device is illustrated in fig. 4·30. It uses the device of fig. 4·28 (c) for obtaining circular polarization. This, as we saw, is equivalent to the quarter-wave plate of optics. The continuous phase shifter of fig. 4·30 (a), designed at the

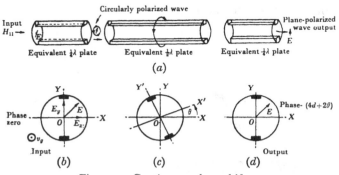

Fig. 4·30. Continuous phase shifter.

Admiralty Signal Establishment, comprises a quarter-wave plate, followed by a half-wave plate and finally a second quarter-wave plate. The three sections are placed together (not separated as shown) and coupled by choke joints. The quarter-wave plates remain fixed with their metal bars in line and the central half-wave plate is rotated. An H_{11}-wave is fed into the first quarter-wave plate with its E vector along a diameter at $45°$ to the diameter joining the ribs (fig. 4·30 (b)). Consequently, the wave emerging from the quarter-wave plate is circularly polarized. If the angle between the diameter joining the ribs in the half-wave plate and the corresponding diameters in the quarter-wave plates is θ as shown in fig. 4·30 (c), then the wave emerges from the second quarter-wave plate plane polarized and parallel to the entering wave. Its phase, however, can be changed by an amount proportional to the rotation of the half-wave section. When the latter rotates through an angle θ radians, then the phase of the electric field at the output changes by 2θ radians.

We proceed to establish this result. Choose Cartesian axes OX and OY with OY parallel to the diameter jointing the ribs of each quarter-wave plate (fig. 4·30 (b)). The electric vector of the H_{11}-wave at the centre of the input cross-section to the first quarter-wave plate is polarized at an angle of 45° to OX.

Its components at the input may therefore be written

$$E_x = E_y = Ae^{j\omega t}. \tag{1}$$

At the output of the quarter-wave section these components oscillate in quadrature and become

$$E_x = Ae^{j(\omega t-\alpha)}, \quad E_y = jAe^{j(\omega t-\alpha)}, \tag{2}$$

where α is the phase retardation in the absence of the ribs.

Suppose the half-wave plate to make an angle θ with respect to the quarter-wave plates and attach corresponding axes OX' and OY' to it, as shown in fig. 4·30 (c). The direction cosines of these axes with respect to the OXY system are

$$l_1 = \cos\theta, \quad l_2 = \sin\theta, \quad m_1 = -\sin\theta, \quad m_2 = \cos\theta.$$

The components (E'_x, E'_y) of the field (E_x, E_y), referred to the $OX'Y'$ system at the input to the half-wave plate are comprehended in the following scheme:

	E_x	E_y
E'_x	l_1	m_1
E'_y	l_2	m_2

$$\left.\begin{aligned} l_1 = m_2 = \cos\theta, \\ l_2 = -m_1 = \sin\theta, \end{aligned}\right\} \tag{3}$$

that is,

$$E'_x = l_1 E_x + m_1 E_y, \quad E'_y = l_2 E_x + m_2 E_y, \tag{4}$$

where E_x and E_y are given by (2).

At the output of the half-wave plate the phase of E'_x is retarded by -2α and that of E'_y by $-(2\alpha-\pi)$ relative to the phases of the input. Consequently, at the input to the second quarter-wave plate

$$E'_x = Ae^{j(\omega t-3\alpha)}(l_1+jm_1), \quad E'_y = -Ae^{j(\omega t-3\alpha)}(l_2+jm_2). \tag{5}$$

Refer this field, using the scheme (3), to the OXY coordinate system; then, at the input to the second quarter-wave plate the field is

$$E_x = (l_1 E'_x + l_2 E'_y), \quad E_y = (m_1 E'_x + m_2 E'_y). \tag{6}$$

To obtain the field components at the output multiply the E_x and E_y in (6) by the factors $e^{-j\alpha}$ and $je^{-j\alpha}$ respectively. The output field is therefore

$$
\begin{aligned}
E_x = (l_1 E_x' + l_2 E_y') e^{-j\alpha} &= A e^{j(\omega t - 4\alpha)} [l_1^2 - l_2^2 - j(l_1 m_1 + l_2 m_2)] \\
&= A e^{j(\omega t - 4\alpha)} [\cos^2\theta - \sin^2\theta - 2j\sin\theta\cos\theta] \\
&= A e^{j(\omega t - 4\alpha)} (\cos\theta - j\sin\theta)^2 \\
&= A e^{j(\omega t - 4\alpha - 2\theta)}.
\end{aligned}
$$

Similarly,
$$E_y = A e^{j(\omega t - 4\alpha - 2\theta)}.$$

The H_{11}-wave at the output is therefore plane polarized and parallel to that at input, but its phase is retarded on the input phase by $(4\alpha + 2\theta)$. The term 4α is a constant of the equipment and is the phase lag in the system without ribs, but the term 2θ is variable.

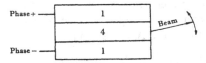

Fig. 4·31. Beam swinging.

Thus the output phase changes by 2θ when the central section is rotated through θ. Thus a 90° rotation produces a 180° phase change. This affords a method of sweeping the phase of the output signal at a great rate by rapid rotation of the central tube. The device has been used to swing a beam from an aerial (Naval 980/1 set) in the vertical plane. This array comprises three cheese aerials fed from horns, in the power ratio 1:4:1 as shown in fig. 4·31. The two outer cheese aerials are each fed through a phase shifter (fig. 4·30) whose central sections rotate in opposite directions, and the beam swings up and down in synchronism.

The further discussion of special techniques is postponed to the next chapter, which begins with a treatment of wave-guide impedance.

REFERENCES

General

Brown, L. W. Problems and practice in the production of wave guide transmission systems. *J. Instn Elect. Engrs*, 1946, vol. 93, part III A, no. 4, p. 689.

Directive Couplers

Surdin, M. *Ibid.*, p. 725.

Chapter 5

WAVE-GUIDE IMPEDANCE AND FURTHER TECHNIQUES

5·1. Introduction

Although it has been stated in § 4·3 that it is important to eliminate reflexions it has not yet been explained how to avoid the most serious of the reflected components—that from the end of the wave guide. To appreciate the technique for eliminating this component, and also the principles of related techniques, it is first necessary to consider the theory upon which they are based.

The practical problem of eliminating unwanted reflexions, that is, of removing standing waves, has long been familiar in transmission-line practice and a number of standard procedures employing devices such as quarter-wave transformers and stubs have been developed to solve it. The theory of these methods of 'matching' is based directly on the well-known theory of transmission lines which uses the concepts of characteristic impedance, line impedance and admittance and reflexion coefficient. It has fortunately proved possible to develop a rigorous theory of propagation in wave guides which is exactly parallel to the standard theory of transmission lines. Not only are the language of impedances and admittances, and the standard formulae preserved in this theory, but the practical methods of matching wave guides are entirely equivalent to those used with transmission lines.

Before discussing the theory of transmission in wave guides it is convenient to recapitulate the essential features of transmission-line theory in a form that can be readily adapted to the requirements of wave guides.

5·2. Transmission-line theory

5·2·1. *Voltage and current in a progressive wave*

In addition to the usual principal or TEM-waves, whose properties were described in § 1·5, other waves which are not TEM-waves may exist, as was explained in § 2·7. However, in practice, the dimensions and spacing of the conductors are such that the dominant or principal mode only is progressive and all supplementary modes are evanescent.

In the standard theory of transmission lines it is assumed that the currents and voltages that appear in the formulae are those of the principal wave, and the theory ignores the possible existence of supplementary modes except perhaps to represent their effects by shunting reactances at the discontinuities where they are excited. We consider first the relation of the current flow to the associated voltage in a single progressive principal wave on a transmission line.

Fig. 5·1 represents such a wave travelling from right to left on a loss-free transmission line with characteristic impedance Z_0.

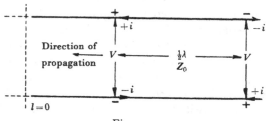

Fig. 5·1.

We note that at points on the line where the surface charge and voltage are positive the current i flows in the direction of propagation, and that where the charge and voltage are negative i flows against the direction of propagation. A distribution of voltage, the same as that shown in fig. 5·1, associated with a reversed distribution of current corresponds to a wave travelling in the opposite sense from left to right.

Distance along the transmission line measured from a reference section $l = 0$, will be denoted by l and is positive for points to the right of $l = 0$ and negative for points to the left. Since, according to § 1·5, the ratio V/i of voltage to current is the characteristic impedance Z_0 of the line, the mathematical description of the wave of fig. 5·1 propagated from right to left (in the sense of decreasing l) is the following:

$$V_1 = A_1 \cos(\omega t + kl) = Z_0 i_1, \qquad (1)$$

in which $k = 2\pi/\lambda$, $\omega = 2\pi f$, and the current i_1 is treated as positive when it flows from right to left in the upper of the conductors in the diagram.

Similarly, the voltages V_2 and currents i_2 in a progressive wave propagated from left to right are, from what was remarked above,

$$V_2 = A_2 \cos(\omega t - kl + \phi) = -Z_0 i_2, \tag{2}$$

where ϕ is an arbitrary phase constant.

In what follows, it is convenient to replace the trigonometrical functions in (1) and (2) by their exponential representations and to write (1) and (2) in the equivalent form

$$V_1 = A_1 e^{j(\omega t + kl)} = Z_0 i_1, \quad V_2 = A_2 e^{j(\omega t - kl + \phi)} = -Z_0 i_2. \tag{3}$$

5·2·2. *Reflexion coefficient of a terminating impedance*

We proceed to consider the distribution of voltage and current on the line when it is terminated, at $l = 0$, in an arbitrary impedance $Z_l = R_l + jX_l$.

Suppose a wave
$$V_1 = A_1 e^{j(\omega t + kl)} = Z_0 i_1$$

from the generator G (fig. 5·2) to impinge on the load Z_l. We call this wave the incident wave. Whereas in the progressive wave the

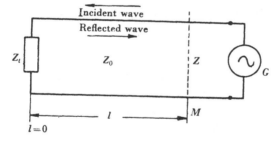

Fig. 5·2.

ratio $V_1/i_1 = Z_0$, the impedance Z_l requires the complex ratio V_l/i_l of the voltage V_l across it to the current i_l in it to be equal to Z_l; consequently the conditions at the termination cannot be represented in terms of a single progressive wave unless $Z_l = Z_0$. In the general case (in order to satisfy the conditions at the termination) it is necessary to introduce a reflected wave

$$V_2 = A_2 e^{j(\omega t - kl + \phi)} = -Z_0 i_2$$

that travels from Z_l to the generator.

The amplitude A_2 and phase constant ϕ are such that when the currents and voltages in the incident and reflected waves are superimposed the ratio of the resultant voltage V to the resultant current i assumes the value $V/i = Z_l$ at position of the load $l = 0$. The distribution of voltage and current along the line is therefore

$$V = (V_1 + V_2) = e^{j\omega t}[A_1 e^{jkl} + A_2 e^{-j(kl-\phi)}], \quad\Big\}$$
$$Z_0 i = Z_0(i_1 + i_2) = e^{j\omega t}[A_1 e^{jkl} - A_2 e^{-j(kl-\phi)}]. \quad\Big\} \qquad (1)$$

We define a quantity

$$\rho = \frac{A_2}{A_1} e^{j\phi} = |\rho| e^{j\phi} \qquad (2)$$

which we call the reflexion coefficient of the termination. It is a complex quantity whose magnitude $|\rho| = A_2/A_1$ is equal to the ratio of the amplitudes of the reflected and incident voltages at the termination and whose argument ϕ is the advance in phase of the voltage accompanying the reflexion.

Equations (1) for the resultant voltage and current may therefore be written

$$V = A_1 e^{j\omega t}[e^{jkl} + \rho e^{-jkl}], \quad Z_0 i = A_1 e^{j\omega t}[e^{jkl} - \rho e^{-jkl}]. \qquad (3)$$

We require the ratio V/i to equal Z_l when $l = 0$, whence, from (3),

$$Z_l/Z_0 = \left(\frac{1+\rho}{1-\rho}\right). \qquad (4)$$

The quantity Z_l/Z_0 is called the 'normalized' or 'specific' terminating impedance and is conveniently written

$$z_l = \frac{Z_l}{Z_0} = \frac{R_l + jX_l}{Z_0} = r_l + jx_l. \qquad (5)$$

Equation (4) becomes $\qquad z_l = \left(\frac{1+\rho}{1-\rho}\right). \qquad (6)$

This important formula relates the normalized impedance of the terminating load to its reflexion coefficient. The formula may also be transformed to

$$\rho = (z_l - 1)/(z_l + 1). \qquad (7)$$

If the characteristic impedance is a resistance, as obtains in practice at high frequencies, we conclude from equations (6) and (7) that, since z_l is then a real quantity, the reflexion coefficient is also real. This means that the phase change ϕ on reflexion from a resistive termination is, from (7), either zero ($z_l > 1$) or 180° ($z_l < 1$).

Since the reflexion coefficient $|\rho|$ is less than unity there is loss of amplitude on reflexion from a resistance.

In the special case where $Z_t = Z_0$ ($z_t = 1$), then ρ is zero and there is no reflected wave and the line is said to be 'matched'.

When the termination is a pure reactance $Z_t = jX_t$ ($z_t = jx_t$), then from (7)

$$|\rho|^2 = \rho\bar{\rho} = \frac{(jx_t-1)(-jx_t-1)}{(jx_t+1)(-jx_t+1)} = \frac{(1+x_t^2)}{(1+x_t^2)} = 1, \qquad (8)$$

in which $\bar{\rho}$ is the complex conjugate of ρ,

$$\rho = |\rho|\,e^{j\phi} = |\rho|\,[\cos\phi + j\sin\phi]$$

$$= -\frac{(jx_t-1)^2}{(x_t^2+1)} = \frac{(x_t^2-1)+2jx_t}{(x_t^2+1)}.$$

Thus
$$\tan\phi = \frac{2x_t}{(x_t^2-1)} = \frac{2/x_t}{(1-1/x_t^2)}$$

$$= \frac{2\tan\frac{1}{2}\phi}{(1-\tan^2\frac{1}{2}\phi)},$$

whence
$$\tan\left(\tfrac{1}{2}\phi\right) = 1/x_t. \qquad (9)$$

Thus in reflexion from a reactive load the amplitude is unchanged, but the phase is advanced when x_t is inductive (positive) and retarded when x_t is capacitive.

When Z_t is a general impedance $(R_t + jX_t)$ the reflected wave differs from the incident wave both in amplitude and phase at the termination.

From
$$\rho = (z_t-1)/(z_t+1)$$

it is easy to show that when $z_t = r_t + jx_t$

$$|\rho|^2 = \frac{(r_t-1)^2 + x_t^2}{(r_t+1)^2 + x_t^2}, \quad \tan\phi = \frac{2x_t}{r_t^2 + x_t^2 - 1}. \qquad (10)$$

Equations (10) include (8) and (9) as special cases.

5·2·3. *Line impedance and input impedance*

Consider a section M of the transmission line (fig. 5·2) at a distance l from the load.

The voltage V and the current i at this section are given by equations 5·2·2 (3), and their ratio V/i defines an impedance Z

which we call the *line impedance* at the distance l from the load. It follows that

$$Z/Z_0 = z = \left[\frac{e^{jkl} + \rho e^{-jkl}}{e^{jkl} - \rho e^{-jkl}}\right]$$

$$= \left[\frac{1 + \rho e^{-2jkl}}{1 - \rho e^{-2jkl}}\right], \tag{1}$$

where z is the normalized or specific impedance corresponding to Z. We note on comparing (1) with 5·2·2 (6) that the quantity

$$\rho_l = \rho e^{-2jkl} \quad (k = 2\pi/\lambda) \tag{2}$$

bears the same relation to z as ρ does to z_l. We therefore define ρ_l as the *reflexion coefficient of the line impedance* z.

It is, in fact, the ratio of the voltages in the reflected and incident waves at this position l. Equation (2) shows how the reflexion coefficient transforms with displacement l from the load.

To express z in terms of z_l we replace ρ in (1) by its value in terms of z_l as given by 5·2·2 (7), and obtain

$$z = \left[\frac{(z_l + 1)\,e^{jkl} + (z_l - 1)\,e^{-jkl}}{(z_l + 1)\,e^{jkl} - (z_l - 1)\,e^{-jkl}}\right]$$

$$= \left[\frac{z_l \cos kl + j \sin kl}{\cos kl + j z_l \sin kl}\right],$$

or

$$z = \left[\frac{z_l + j \tan kl}{1 + j z_l \tan kl}\right]. \tag{3}$$

This formula shows how the normalized line impedance z transforms along the line as l is increased.

When the distance l is the total length of the line, then z is the normalized input impedance. The load presented to the generator is then

$$Z = z Z_0$$

$$= Z_0 \left[\frac{Z_l + j Z_0 \tan kl}{Z_0 + j Z_l \tan kl}\right]. \tag{4}$$

This is a standard formula that relates the input impedance Z to the terminating impedance Z_l when the line is loss-free and of length l.

It is often more convenient to employ admittances instead of impedances in transmission-line theory and practice. The characteristic admittance of the line is $Y_0 = 1/Z_0$, and the normalized admittances corresponding to the admittances $Y = 1/Z$ and $Y_t = 1/Z_t$ are

$$y = Y/Y_0 = Z_0/Z = 1/z, \quad y_t = Y_t/Y_0 = Z_0/Z_t = 1/z_t.$$

In terms of admittances formulae (3) and (4) become

$$y = \left[\frac{y_t + j \tan kl}{1 + j y_t \tan kl}\right], \tag{5}$$

$$Y = Y_0\left[\frac{Y_t + j Y_0 \tan kl}{Y_0 + j Y_t \tan kl}\right]. \tag{6}$$

They are analytically identical with (3) and (4).

The input impedance of a line short-circuited at the termination is found, by putting $Z_t = 0$, in (3), to be

$$Z_t = j Z_0 \tan kl, \tag{7}$$

whereas that of a line open-circuited at the end ($Z_t = \infty$) is

$$Z_t = Z_0/j \tan kl. \tag{8}$$

In each case the impedance is a pure reactance which is inductive or capacitive according to the sign of $\tan kl$.

5·2·4. *Distribution of voltage and current in a standing wave*

The voltage and current distributions on the line terminated by the impedance Z_t are, according to 5·2·2 (1), (2) and (3),

$$\left.\begin{aligned}V &= A_1[\cos(\omega t + kl) + |\rho|\cos(\omega t - kl + \phi)], \\ Z_0 i &= A_1[\cos(\omega t + kl) - |\rho|\cos(\omega t - kl + \phi)].\end{aligned}\right\} \tag{1}$$

The voltage reaches its maximum amplitude $V_{max.}$ at those positions l where the vibrations of the incident and reflected waves differ in phase by $2\pi n$ ($n = 0, 1, 2,$ etc.) radians and its minimum amplitude where the phase difference is $(2n+1)\pi$. The voltage maxima therefore lie at the distances l such that

$$kl = -kl + \phi + 2n\pi \quad (k = 2\pi/\lambda),$$

or
$$l = \frac{\lambda\phi}{4\pi} + \tfrac{1}{2}n\lambda. \tag{2}$$

The value of the maximum amplitude of voltage is

$$V_{max.} = A_1(1 + |\rho|). \tag{3}$$

Similarly, the voltage minima $V_{\text{min.}}$ are given by

$$V_{\text{min.}} = A_1(1 - |\rho|), \tag{4}$$

and lie at the positions l_2, where

$$l_2 = \frac{\lambda\phi}{4\pi} + \left(\frac{n\lambda}{2} + \frac{\lambda}{4}\right). \tag{5}$$

Thus, adjacent maxima and minima are $\frac{1}{4}\lambda$ apart.

From the second of equations (1) it is easy to see that the current minima are

$$i_{\text{min.}} = A_1\frac{(1 - |\rho|)}{Z_0} = \frac{V_{\text{min.}}}{Z_0}, \tag{6}$$

and lie at the positions l obtained from (2). That is, the $i_{\text{min.}}$ coincide in position l with the $V_{\text{max.}}$.

Similarly, $$i_{\text{max.}} = A_1\left(\frac{1 + |\rho|}{Z_0}\right) = \frac{V_{\text{max.}}}{Z_0}, \tag{7}$$

and the $i_{\text{max.}}$ coincide in position with the $V_{\text{min.}}$.

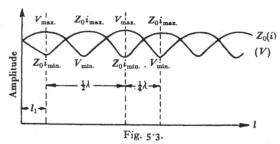

Fig. 5·3.

This distribution of current and voltage in a partial standing wave is shown in fig. 5·3. We note that the voltage maximum nearest to the termination is, from (2), at a distance from it

$$l_1 = \frac{\lambda\phi}{4\pi}. \tag{8}$$

Consequently, measurement of l_1 serves to determine the phase advance ϕ of the voltage on reflexion from the termination.

5·2·5. Standing wave ratio

Two quantities suffice to specify the partial standing wave. The first, already encountered in the previous section, is the distance $l_1 = \lambda\phi/4\pi$ of the first voltage maximum from the termination.

The second is the *standing wave ratio* S defined as follows:

$$S = \frac{V_{max.}}{V_{min.}} = \frac{i_{max.}}{i_{min.}}. \tag{1}$$

Frequently, the reciprocal $s = 1/S$ is called the standing wave ratio. It follows from equations 5·2·4 (3) and (4) that

$$S = \frac{(1+|\rho|)}{(1-|\rho|)} = \frac{1}{s}. \tag{2}$$

Both S and l_1 are measured by means of a standing wave indicator.

Measurements of S and l_1 are of the greatest practical importance, since they give immediately the value of the normalized terminating impedance z_t.

To see why this is so, consider first the nature of the normalized line impedances, as given by formula 5·2·3 (3), at positions of voltage maxima and voltage minima.

We deduce from equations 5·2·4 (1), (2) and (5) that at the positions of voltage maxima and minima the resultant voltage and current vibrate in phase and that the line impedance is therefore purely resistive. Since $V_{max.}$ coincides with $i_{min.}$ and $V_{min.}$ with $i_{max.}$ these resistances represent the maximum and minimum values attained by the magnitude of the line impedance anywhere on the line. Denote them by $R_{max.}$ and R_{min}, then

$$R_{max.} = \frac{V_{max.}}{i_{min.}} = Z_0\left(\frac{1+|\rho|}{1-|\rho|}\right), \tag{3}$$

whence, from equation (2),

$$r_{max.} = \left(\frac{1+|\rho|}{1-|\rho|}\right) = S = \frac{1}{s}, \tag{4}$$

and

$$R_{min.} = \frac{V_{min.}}{i_{max.}} = Z_0\left(\frac{1-|\rho|}{1+|\rho|}\right), \tag{5}$$

or

$$r_{min.} = \left(\frac{1-|\rho|}{1+|\rho|}\right) = s = \frac{1}{S}. \tag{6}$$

Thus

$$R_{max.}R_{min.} = Z_0^2, \quad r_{max.}r_{min.} = 1. \tag{7}$$

The magnitude $|\rho|$ of the reflexion coefficient of the termination is immediately derived from (4),

$$|\rho| = \left(\frac{S-1}{S+1}\right) = \left(\frac{1-s}{1+s}\right). \tag{8}$$

We remark that the quantities directly measured in standing wave measurements by means of a voltage standing wave indicator are the magnitude $|\rho|$ and the phase angle $\phi = 4\pi l_1/\lambda$ of the reflexion coefficient ρ of the termination.

The relation (5·2·2 (6)) between the terminating impedance z_t, which we require, and its reflexion coefficient ρ is

$$z_t = \frac{1}{y_t} = \left(\frac{1+\rho}{1-\rho}\right) = \left(\frac{1+|\rho|\,e^{j\phi}}{1-|\rho|\,e^{j\phi}}\right).$$

Replace $|\rho|$ by its value (8) in terms of S (or s) to obtain

$$z_t = \left(\frac{S-j\tan kl_1}{1-jS\tan kl_1}\right) = r_t + jx_t, \quad y_t = \left(\frac{s-j\tan kl_1}{1-js\tan kl_1}\right) = g_t + jb_t. \quad (9)$$

The following is an alternative derivation of this result. The general transmission-line formula 5·2·3 (3) shows how the terminating impedance z_t transforms to the value z of the line impedance on proceeding a distance l towards the generator. This formula may be inverted to express z_t in terms of z to show how an impedance z transforms to z_t in a displacement $-l$ (towards the load). We find

$$z_t = \left[\frac{z-j\tan kl}{1-jz\tan kl}\right], \quad y_t = \left[\frac{y-j\tan kl}{1-jy\tan kl}\right]. \quad (10)$$

But at the distance l_1 from the load the impedance z is equal to $r_{max.} = S$ ($y = g_{min.} = s$). Consequently the normalized load impedance (admittance) is

$$z_t = \left[\frac{S-j\tan kl_1}{1-jS\tan kl_1}\right], \quad y_t = \left[\frac{s-j\tan kl_1}{1-js\tan kl_1}\right]. \quad (11)$$

The formulae for r_t, x_t, g_t and b_t are derived directly from (11). They are

$$\left.\begin{array}{ll} r_t = \dfrac{S(1+\tan^2 kl_1)}{(1+S^2\tan^2 kl_1)}, & x_t = \dfrac{(S^2-1)\tan kl_1}{(1+S^2\tan^2 kl_1)}, \\[3mm] g_t = \dfrac{s(1+\tan^2 kl_1)}{(1+s^2\tan^2 kl_1)}, & b_t = -\dfrac{(1-s^2)\tan kl_1}{(1+s^2\tan^2 kl_1)}. \end{array}\right\} \quad (12)$$

In practice these components are more readily obtained from the circle diagram* as shown in fig. 5·4 (see fig. 5·59).

---

* For the theory and applications of the circle diagram see W. Jackson and L. G. H. Huxley, *J. Instn Elect. Engrs*, 1944, vol. 91, part III, p. 10; W. Jackson, *High-Frequency Transmission Lines*, chapter VI (Methuen).

The standing wave ratio S is plotted on the resistive axis at B and the u-circle through B is traversed counter-clockwise until the n-arc for which $n = (0\cdot25 - l_1/\lambda)$ is encountered at C. The point C in the complex plane is $z_l = r_l + jx_l$.

The value of the terminating impedance is then $Z_l = Z_0 z_l$.

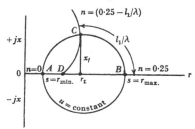

Fig. 5·4.

5·2·6. *Power carried to the terminating load*

The instantaneous flow of power across any section is $W = Vi$. This may be written

$$W = Vi = (V_1 + V_2)(i_1 + i_2)$$
$$= (V_1 + V_2)\frac{(V_1 - V_2)}{Z_0} = \frac{(V_1^2 - V_2^2)}{Z_0}. \tag{1}$$

The power travelling to the load is therefore the difference in the powers carried by the incident wave (V_1, i_1) and that carried by the reflected wave (V_2, i_2).

The mean power absorbed by the load is

$$\overline{W} = \frac{1}{2}\left(\frac{A_1^2 - A_2^2}{Z_0}\right). \tag{2}$$

5·2·7. *Formulae for a line with loss*

We here summarize, for completeness, the standard formulae for a transmission line with loss. The amplitude of a travelling wave here decays exponentially with distance from the point of excitation. The incident and reflected progressive waves are now represented by the following expressions in place of 5·2·2 (1) and (2):

$$\left.\begin{array}{c} V_1 = Z_0 i_1 = A_1 e^{+\alpha l} e^{j(\omega t + kl)} \\ = A_1 e^{j\omega t} e^{Pl}, \\ V_2 = -Z_0 i_2 = \rho A_1 e^{-\alpha l} e^{j(\omega t - kl)} \\ = \rho A_1 e^{j\omega t} e^{-Pl}, \end{array}\right\} \tag{1}$$

where the *propagation constant* $P = \alpha + j2\pi/l$ and

$$\rho = |\rho| e^{j\phi} = (z_l - 1)/(z_l + 1) = (Z_l - Z_0)/(Z_l + Z_0)$$
$$= (1 - y_l)/(1 + y_l). \tag{2}$$

The resultant voltage and current are

$$V = A_1 e^{j\omega t}[e^{Pl} + \rho e^{-Pl}], \quad Z_0 i = A_1 e^{j\omega t}[e^{Pl} - \rho e^{-Pl}]. \tag{3}$$

The line impedance Z at distance l from Z_l is from (3)

$$Z/Z_0 = \frac{V}{i} = \left[\frac{1 + \rho e^{-2Pl}}{1 - \rho e^{-2Pl}}\right] = z = \frac{1}{y}. \tag{4}$$

The reflexion coefficient at position l, corresponding to the impedance z, is

$$\rho_l = \rho e^{-2Pl}. \tag{5}$$

Replace ρ in (4) by its value in terms of z_l in (2) and reduce to the standard form

$$z = \left[\frac{z_l + \tanh Pl}{1 + z_l \tanh Pl}\right], \quad y = \left[\frac{y_l + \tanh Pl}{1 + y_l \tanh Pl}\right]. \tag{6}$$

5·2·8. *Stub matching*

One of the commonly used methods of eliminating the reflected wave from the major portion of the transmission line is 'stub matching' which we briefly describe.

It has been explained in § 5·2·5 that at positions of voltage maxima and minima the line impedance attains its maximum and minimum values respectively and that the impedance is then purely resistive.

Further, according to equation 5·2·5 (7),

$$r_{\max}.r_{\min.} = 1 = g_{\min.}g_{\max.}, \quad g_{\max.} = S, \quad g_{\min.} = s, \tag{1}$$

where $g_{\min.} = 1/r_{\max.}$ and $g_{\max.} = 1/r_{\min.}$ are the normalized line conductances at the positions of voltage maxima and minima.

Since at an intermediate position between a voltage maximum and a minimum the magnitude of the line admittance $y = g + jb$ is $y = \sqrt{(g^2 + b^2)}$, it follows from (1) that there is a position at which the conductance g is equal to unity, that is, the normalized line admittance is

$$y' = 1 \pm jb'.$$

This position is shown in fig. 5·5 at distance l' towards the generator measured from a voltage maximum. It follows that if a compensating susceptance $\mp jb'$ (shown dotted in the figure) is added in shunt across the line at the position l' then the portion of the line between this position and the generator is terminated in the admittance

$$1 \pm jb' \mp jb' = 1,$$

from which, according to 5·2·2 (7), there is no reflected wave. If the voltage maximum is chosen near the load, then it is possible in this way to remove the reflected component from the major portion of the transmission line. The line then presents to the generator an impedance equal to the characteristic impedance Z_0. Alternatively,

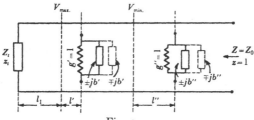

Fig. 5·5.

we may proceed from a voltage minimum a distance l'' towards the generator to arrive at a position where the line admittance is

$$y'' = 1 \mp jb''$$

and produce a 'match' by adding the shunt susceptance $\pm jb''$.

The shunt susceptance in each case is almost always provided by a short-circuited length of transmission line whose length l''' can be continuously varied. Such a 'stub' is convenient for connexion in shunt, with its length at right angles to the line, and its input susceptance can be given any value between $\pm j\infty$ according to the length l'''.

The values of the lengths l', l'' and l''' are most simply obtained from a circle diagram, and a description of the procedure will be found in the references given in § 5·2·5. A circle diagram is given in fig. 5·59 at the end of this chapter.

5·3. Generalization of theory to include propagation in wave guides

In attempting to apply the theory of transmission lines, as developed in §§ 5·2, to wave guides a difficulty is encountered at the outset. The theory as given depends on the fact that it is possible to define the voltage and current in a principal wave on a transmission line in an unambiguous manner (§ 1·5); further, the total characteristic impedance Z_0, when defined as the ratio of voltage to current in a progressive wave, also leads to the expressions

$$W = Vi = \frac{V^2}{Z_0} = Z_0 i^2$$

for the instantaneous power carried across a section of the line by the wave. The characteristic impedance therefore behaves consistently as the same resistance.

Suppose, however, that we are concerned with propagation in a rectangular wave guide. We first choose its dimensions so that the dominant mode $[H_{01}]$ only is propagated as a progressive wave (§§ 2·3 and 4·2) in analogy with the transmission line on which the principal wave is the only progressive mode. The H_{01}-wave may be described through its electric field which, according to equation 3·2 (13), is

$$E = E_x = E_0 \sin\left(\frac{\pi y}{b}\right) \cos\left[\omega t - \gamma z\right], \qquad (1)$$

where $\gamma = 2\pi/\lambda_g$ and b is the longer dimension of the cross-section, the shorter dimension being a.

The difficulty arises when an unambiguous definition of voltage, current and wave guide characteristic impedance are sought. We might suppose that the current is the total longitudinal current flowing across a transverse line in the broad face, and that the voltage amplitude is the maximum field E_0 multiplied by the short dimension a; that is, $V = E_0 a$. Alternatively, we could define the voltage to be the mean value of E over the section multiplied by a; that is, $2/\pi E_0 a$. Each of these divided by the current gives a different total characteristic impedance.

Further, the mean Poynting flux (§ 3·4·2 (3)) is

$$\overline{P} = \frac{1}{2} \frac{E_0^2}{Z_g} = \frac{Z_g H_0^2}{2},$$

where Z_g is the intrinsic impedance (§ 3·4·1) of the H_{01}-wave.

The total mean power carried by the wave is

$$\overline{W} = \frac{1}{4} \frac{E_0^2}{Z_g} ab = \frac{(E_0 a)^2}{4 Z_g} \left(\frac{b}{a}\right).$$

This, with the definition $V = E_0 a$, leads to an equivalent total wave-guide impedance $2Z_0(a/b)$, which is not the same as the ratio of $(E_0 a)$ to the total longitudinal current.

These and other examples show that an additional criterion is required to determine the most suitable definition of total impedance of a wave guide. It appears in § 7·16 that a convenient expression is $Z_{0g} = (a/b) Z_H$ (§ 3·4·1).

We proceed to restate the theory of §§ 5·2 in a form more acceptable for our present purpose. We first note a radical difference between the measuring techniques adopted respectively with transmission lines operated at low frequencies where the wave-length is enormously greater than the dimensions of voltmeters, ammeters or wattmeters that are used to measure voltage, current and power, and those operated at the shorter wave-lengths, in particular at wave-lengths of 10 cm. and less, where it is not permissible to determine voltage, current and power in this fashion since the storage fields at the evanescent modes generated at the discontinuities represented by the meters introduce unpredictable series and shunt reactances. The introduction of meters into the latter system produces large reflected components of the progressive waves, and the meter readings are unreliable or meaningless.

At microwave-lengths, both for transmission lines and wave guides, the two fundamental measuring devices are the standing wave indicator and the calorimeter. The former is used to explore the distribution of relative voltage, current or electromagnetic field amplitude in the dominant mode and to measure the standing wave ratio S and the location of the maxima or minima on the line relative to the termination (distance l_1 of 5·2·4(8)). From such measurements we deduce the reflexion coefficient $\rho_l = |\rho_l| e^{j\phi}$ of the termination with respect to voltage or field amplitude (§ 5·2·5) from

$$|\rho_l| = (S-1)/(S+1), \quad \phi = 4\pi l_1/\lambda. \tag{2}$$

Here ϕ is the phase advance of the reflected vibration on the incident vibration at the termination $l = 0$, and $|\rho_l|$ is the ratio of the amplitudes of the reflected and incident vibrations. At a distance l towards the generator from $l = 0$, the amplitudes in the reflected and incident waves are still in the ratio $|\rho_l|$, but the phase of the incident wave is advanced by $2\pi l/\lambda$ and that of the reflected wave is retarded by the same amount. At this distance l, the advance in phase of the reflected on the incident wave is reduced to

$$(\phi - 4\pi l/\lambda) = (\phi - 2kl).$$

The effective reflexion coefficient of the geometrical cross-section of the line at distance l_1, defined as the ratio of the voltage or field in the wave leaving the section (reflected wave) to that entering it (incident wave), is

$$\left. \begin{array}{l} \rho_l = |\rho_t|\, e^{j(\phi - 2kl)}, \\ \rho_l = \rho_t e^{-j2kl}. \end{array} \right\} \qquad (3)$$

that is

Thus, through equations (2) and (3), the standing wave indicator measures the reflexion coefficient ρ_t of the termination and the equivalent reflexion coefficient ρ_l at any distance l from it.

It should be remembered that with a square-law response it is necessary to take the square root of the standing wave-meter indications to obtain S. The standing wave indicator also determines the wave-length.

If the line or wave guide is dissipative, then relation (3) is replaced by

$$\begin{aligned} \rho_l &= |\rho_t|\, e^{-2\alpha l} e^{j(\phi - 2kl)} \\ &= \rho_t e^{-2Pl}, \end{aligned} \qquad (4)$$

where the propagation constant $P = (\alpha + jkl)$.

In order to correlate this theory with the earlier theory of transmission lines (§§ 5·2) we transform the reflected coefficients into the following quantities which we call the normalized impedances. The normalized terminating impedance z_t is defined to be

$$z_t = \left(\frac{1 + \rho_t}{1 - \rho_t}\right) = \frac{1}{y_t}, \qquad (5)$$

and the normalized line impedance z at distance l is correspondingly defined as

$$z = \left(\frac{1 + \rho_l}{1 - \rho_l}\right) = \frac{1}{y}, \qquad (6)$$

where y_t and y are the normalized admittances corresponding to z_t and z.

These normalized line impedances and admittances, it should be noted, are merely transformations of the measured reflexion coefficients and are not defined through voltage to current ratios.

When ρ_l in (6) is replaced by its value in terms of ρ_t in (4), we find

$$z = \frac{1}{y} = \left(\frac{1 + \rho_t e^{-2Pl}}{1 - \rho_t e^{-2Pl}}\right). \tag{7}$$

Eliminate ρ_t from (7) by means of (5) and reduce to the standard forms

$$z = \left(\frac{z_t + \tanh Pl}{1 + z_t \tanh Pl}\right), \quad y = \left(\frac{y_t + \tanh Pl}{1 + y_t \tanh Pl}\right), \tag{8}$$

where $P = \alpha + j2\pi/\lambda = \alpha + jkl$.

With a loss-free line $\alpha = 0$ and (8) reduces to the standard forms

$$z = \left(\frac{z_t + j\tan kl}{1 + jz_t \tan kl}\right), \quad y = \left(\frac{y_t + j\tan kl}{1 + jy_t \tan kl}\right). \tag{9}$$

These transformed reflexion coefficients that we have called normalized impedances and admittances clearly satisfy the standard transmission-line formulae 5·2·3 (3) and (5) and 5·2·7 (6); we have therefore obtained a complete correlation with the theory as developed in §§ 5·2.

On the other hand, the standard formulae involving the *normalized impedances* are now based on a theory which is applicable to any transmission system involving the propagation of a *single* quantity such as voltage, electric field strength or pressure, whose amplitude distribution with distance l can be found by means of a standing wave indicator. The theory does not require the introduction of a second quantity such as current and therefore is not dependent on the possibility of finding a suitable definition of total characteristic impedance of the system. If, however, the system does possess a total characteristic impedance Z_0, as does a transmission line, then the normalized impedances which are dimensionless quantities can be converted to true-line impedances by multiplying them by Z_0. The resulting impedances then satisfy the standard formula 5·2·3 (4).

The corresponding remarks apply to the conversion of the normalized admittances y to true admittances yY_0. With wave guides some writers appear to consider that the normalized impedances multiplied by the intrinsic (wave) impedance of the guide are the true equivalent of transmission-line impedances (considered as voltage to current ratios). According to our present standpoint these are merely converted reflexion coefficients multiplied by a quantity with the physical dimensions of an impedance.

The normalized impedances and admittances determined from reflexion coefficients are in general complex numbers of the form

$$z = r+jx, \quad y = g+jb. \tag{10}$$

r and x are respectively called the normalized resistance and susceptance and g and b the normalized conductance and susceptance.

We note that in § 5·2·8 the information required for stub matching was provided by the standing wave indicator, and that we were concerned not with absolute impedances but normalized impedances alone. The method of basing the definitions of impedance and admittance on reflexion coefficient and the experimental data afforded by the standing wave indicator permits us to introduce matching techniques for eliminating reflected waves in transmission systems such as wave guides and speaking tubes (where the concepts of voltage and current are not readily definable) that are entirely analogous to the stub-matching procedures of transmission-line practice. An important consequence of the formal identity of the final formulae for the normalized impedances and admittances obtained by the method of §§ 5·2 and that of the present section is that it is possible in many cases to represent a complicated field phenomenon in a wave guide by means of an equivalent transmission line shunted by circuit elements.

The second fundamental measuring device that we require at microwave-lengths, whether with transmission lines or wave guides, is a device for measuring power. At high powers this is a calorimeter and at low powers a bolometer or thermistor. From the power it is possible to estimate the amplitude A of the voltage on a transmission line and the amplitude E_0 of the maximum field in a wave guide from

$$\overline{W} = \frac{A^2}{2Z_0} \quad \text{and} \quad \overline{W} = \frac{E_0^2 ab}{4Z_g}$$

respectively, where Z_g is the intrinsic impedance of the wave guide.

It is then possible, if so desired, to calibrate standing wave indicators against power-measuring devices so that they indicate the power carried to the load as well as the standing wave ratio.

We note that matching and power measurement are distinct procedures.

Although standing wave measurements lead only to normalized impedances, on the other hand, in theoretical investigations of the propagation of electromagnetic waves the use of intrinsic or wave impedances is often of considerable convenience.

Normalized impedance in wave guides. We proceed to apply the theory of the previous section to propagation in wave guides. These we shall assume, unless otherwise stated, to be rectangular wave guides in which progressive waves are propagated in the H_{10}-mode only, other modes if present appearing only in an evanescent form.

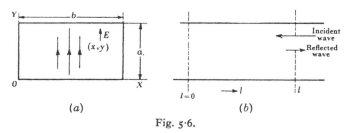

(a)   (b)

Fig. 5·6.

Consider a wave guide carrying both an incident and a reflected H_{10}-wave (fig. 5·6) whose fields can be expressed, respectively, by

$$E_1 = A \sin\left(\frac{\pi x}{b}\right) e^{j(\omega t + \gamma l)}$$

and

$$E_2 = |\rho_l| A \sin\left(\frac{\pi x}{b}\right) e^{j(\omega t - \gamma l + \phi)}$$

$$= \rho_l A \sin\left(\frac{\pi x}{b}\right) e^{j(\omega t - \gamma l)},$$

(11)

where $\gamma = 2\pi/\lambda_g$ and λ_g is the wave-length of the H_{10}-wave.

The resultant field across the section at distance l towards the source from that chosen as $l = 0$ (fig. 5·6(b)) is

$$E = A \sin\left(\frac{\pi x}{b}\right) e^{j\omega t} \left[e^{\gamma l} + \rho_l e^{-\gamma l}\right].$$

(12)

A wave-guide standing wave indicator of the type described in § 4·4 serves to determine the distribution of relative amplitude of the field E over a range of distances l, at the centre of the broad face where $\sin(\pi x/b)$ is unity.

When the standing wave ratio $S = E_{max.}/E_{min.}$ and the distance l_1 of the nearest maximum to $l = 0$ are found the reflexion coefficient appropriate to the section $l = 0$ is

$$\rho_t = |\rho_t| e^{j\phi},$$

where $|\rho_t| = (S-1)/(S+1)$ and $\phi = 4\pi l_1/\lambda_g$,   (13)

as explained in §§ 5·2 and 5·3.

The normalized impedance and admittance at the section $l = 0$ are

$$z_t = \frac{1}{y_t} = \left(\frac{1+\rho_t}{1-\rho_t}\right)$$

$$= (r_t+jx_t) = \frac{1}{(y_t+jb_t)}. \qquad (14)$$

The components r_t, x_t, g_t and b_t are obtained from formulae 5·2·5 (12) or by means of the circle diagram as explained at the end of § 5·2·5.

The reflexion coefficient ρ_l, normalized impedance z and admittance y at the section at distance l are

$$\left.\begin{aligned}
\rho_l &= \rho_t e^{-j2\gamma l} \, (\gamma = 2\pi/\lambda_g) \\
z &= \left(\frac{1+\rho_l}{1-\rho_l}\right) = \frac{1}{y} \\
&= \left(\frac{z_t+j\tan\gamma l}{1+jz_t\tan\gamma l}\right), \\
y &= \left(\frac{y_t+j\tan\gamma l}{1+jy_t\tan\gamma l}\right),
\end{aligned}\right\} \qquad (15)$$

in accordance with the discussion in § 5·3.

These equations give the variation of impedance and admittance along the wave guide. It is assumed that the standing wave measurements are taken in a section remote from discontinuities so that evanescent modes are unimportant.

Some examples are considered for illustration.

Wave guide with a reflexionless termination. Suppose the wave guide to be terminated in a reflexionless load as shown in fig. 5·7 (a). There is no reflected wave, consequently the reflexion coefficient ρ_l is everywhere zero and the impedance $z = 1/y$ is everywhere equal to unity.

Wave guide terminated in a metal plate. This is shown in fig. $5\cdot 7\,(b)$. The field in the wave guide is

$$E = A \sin\left(\frac{\pi x}{a}\right) e^{j\omega t}[e^{j\gamma l} + \rho_l e^{-j\gamma l}],$$

and at the surface of the metal plate $(l = 0)$ the field E is zero for all waves of x and y. This requires

$$\rho_l = -1 = 1e^{j\pi}.$$

Thus
$$|\rho_l| = 1; \quad \phi = \pi.$$

This gives
$$z_l = \frac{1}{y_l} = 0.$$

From (15)
$$z = \frac{1}{y} = j\tan\gamma l = j\tan\frac{2\pi l}{\lambda_g}.$$

The impedance at distance l is therefore a pure reactance.

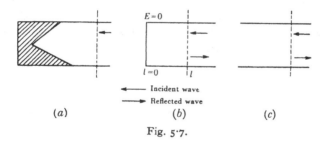

Incident wave
Reflected wave

(a) (b) (c)

Fig. 5·7.

Wave guide open at the end (fig. $5\cdot 7\,(c)$). It might be thought, by analogy with the usual assumption about an open-circuited termination to a transmission line, that the reflexion coefficient of the open end of the wave guide would be $\rho = +1$ and its impedance $z = \infty$. This surmise would be erroneous for the following reasons: first, the dimensions of the open end are not negligible in comparison with the wave-length λ, with the result that power is radiated from the end into space; secondly, since the electromagnetic field at the end of the wave guide is seriously distorted from the simple pattern of the H_{10}-wave, it comprises within the guide, in addition to the incident and reflected H_{10}-waves, a series of evanescent modes. At a sufficient distance l from the end the field becomes simply the sum of the fields of the H_{10}-waves.

However, whereas the power lost into space puts a resistive term in the impedance z_t, the storage field of the evanescent modes contributes a reactive term to it. The terminating impedance z_t would be expected to be of the form $z_t = r_t + jx_t$.

Some experimental results with a $2\frac{1}{2} \times 1$ in. wave guide are given in illustration.

Oscillator TEM wave-length $\lambda = 9$ cm.

Wave-length $\lambda_g = 13\cdot8$ cm. (twice the distance between adjacent minima).

Standing wave ratio $S = 2\cdot83$.

Distance l_1 of nearest maximum $E_{\text{max.}}$ from the open end $= 6\cdot6$ cm. $= 0\cdot478\lambda_g$.

From the circle diagram we find

$$z_t = 2\cdot46 - 0\cdot89j \quad \text{or} \quad y_t = 0\cdot36 + 0\cdot13j.$$

This impedance is by no means infinite, and the result illustrates the importance of not making hasty assumptions on the basis of analogies with the behaviour of transmission lines at low frequencies.

The equivalent circuit representation of the wave guide and its open end is a transmission line with any characteristic impedance Z_0 terminated by a load comprising a resistance $2\cdot46Z_0$ and a capacitance $j \times 0\cdot89Z_0$ in series; or, alternatively, a conductance $0\cdot36Y_0$ and a susceptance $0\cdot13Y_0j$ in parallel. This termination possesses the same reflexion coefficient on the transmission line for principal waves, as the open end possesses for H_{10}-waves in the wave guide.

5·4. The addition of equivalent lumped circuit elements in parallel or in series with a wave guide

The development of the theory of transmission in wave guides based on the concept of reflexion coefficients is still incomplete, because we have not yet discussed what elements in wave guides play the part of the lumped circuit elements that are commonly placed in shunt across a transmission line.

It will appear that an obstacle placed in a wave guide which is able to scatter an H_{10}-wave, when the incident wave falls on it without absorbing power into itself, is completely equivalent to a lumped reactive element placed in shunt across a transmission line. In fig. 5·8 (a) the obstacle is supposed, for simplicity, to be a thin metal cylinder with its axis parallel to the electric vector of the H_{10}-wave, although the treatment is applicable to any object whose thickness in the direction of the axis is small in comparison with the wave-length λ_g. The wave guide is terminated in a reflexionless load as shown.

When the object is introduced, the original field in its vicinity is distorted from that of the incident H_{10}-wave and the new field resolves into the following constituents:

(*a*) The original H_{10} incident wave (indicated by the long arrow in fig. 5·8 (*a*) and (*b*)).

(*b*) A pair of H_{10} *scattered waves*, the one propagated into the reflexionless load and the other towards the generator. These waves are indicated in the figure by the short arrows.

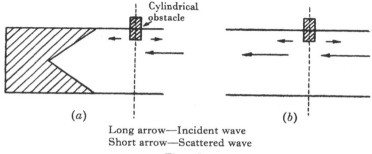

Long arrow—Incident wave
Short arrow—Scattered wave

Fig. 5·8.

(*c*) A series of evanescent modes which are prominent near the obstacle but whose fields are negligible at a sufficient distance from it. These arise because the wave guide is constructed so that the H_{10} is the only possible progressive mode. The storage field in these evanescent modes give the obstacle its reactive character. When the excited modes are of the E (TM) type the stored energy is predominantly electric and the obstacle behaves as a capacitance; but when of the H (TE) type, the stored energy is mainly magnetic and the obstacle behaves as an inductance. If the object also absorbs power then a resistive term is also required to describe its behaviour.

At an adequate distance l away from the obstacle in the direction of the generator the evanescent modes may be neglected and the resultant field is that of the incident and scattered waves superimposed. Take the cross-section through the axis of the obstacle to be at the position $l = 0$ and represent the incident and scattered waves respectively by

$$E_1 = A \sin\left(\frac{\pi x}{b}\right) e^{j(\omega t + \gamma l)}$$

and

$$E_2 = |h| A \sin\left(\frac{\pi x}{b}\right) e^{j(\omega t - \gamma l + \theta)} \qquad (1)$$

$$= hA \sin\left(\frac{\pi x}{b}\right) e^{j(\omega t - \gamma l)},$$

in which we define the complex number

$$h = |h| e^{j\theta} \qquad (2)$$

to be the *scattering coefficient* of the obstacle with respect to an H_{10}-wave.

It is possible, where the obstacle possesses a simple geometrical form, to calculate, in many cases, its scattering coefficient h, but in general h would be determined experimentally.

In the arrangement shown in fig. 5·8 the section $l = 0$, before the introduction of the obstacle, possesses a reflexion coefficient $\rho_l = 0$ and an impedance $z = 1/y = 1$. After the introduction of the obstacle, however, it acquires a reflexion coefficient $\rho_l = h$ and an admittance $y_l = (1 - h)/(1 + h) = g_l + jb_l$.*

It is tempting to assume that the obstacle itself possesses an admittance y_1 such that when it is introduced into the section whose admittance is unity the resultant admittance, corresponding to the reflexion coefficient $\rho_l = h$, becomes

$$y_l = (1 + y_1).$$

We shall show that not only is this the case, but when the same obstacle is introduced in exactly the same way into a section whose admittance alone is y the resulting admittance, as judged from the new reflexion coefficient of the section, becomes

$$y_l = y + y_1.$$

It is found that at the same frequency and in the same wave guide the obstacle preserves its individual admittance y_1 whatever the

---

* To avoid frequent repetition of the word 'normalized' it will be omitted in most of what follows, and it should be assumed that small letters refer to normalized admittances and impedances and capitals to true admittances and impedances.

value of y. It is therefore possible to describe the result of introducing an obstacle into a wave guide in the language of circuit theory and to represent the obstacle by a lumped admittance $y_1 Y_0$ placed in shunt across a transmission line whose characteristic impedance is $Z_0 = 1/Y_0$.

We proceed to justify these statements. It is convenient to classify the possible forms of scattering that can arise under the following headings:

(a) *Symmetrical scattering* in which the electric fields E in the scattered H_{10}-waves are similarly directed and of equal amplitude at equal distances $\pm l$ on opposite sides of the scattering source. (The transverse magnetic fields are oppositely directed.)

(b) *Anti-symmetrical scattering.* Here the amplitudes of the electric fields in the scattered waves at a pair of distances $\pm l$ are of equal amplitude but the oscillations are in antiphase. (The magnetic fields oscillate in phase.)

(c) *Unsymmetrical scattering.* The amplitudes and/or the phases in the waves at any pair of distances $\pm l$ from the scattering centre are unequal.

We shall discuss cases (a) and (b) in turn.

Symmetrical scattering. Let a partial standing wave already exist within the wave guide, and suppose the obstacle to be introduced at a section where the reflexion coefficient is ρ and the admittance is

$$y = (1-\rho)/(1+\rho) = 1/z.$$

Let the scattering coefficient of the obstacle be h. The problem is to find the effective reflexion coefficient of the section with the obstacle present and from it to derive the new admittance of the section.

Let the incident wave be

$$E_1 = A \sin\left(\frac{\pi x}{b}\right) e^{j(\omega t + \gamma l)} \quad (\gamma = 2\pi/\lambda_g),$$

and suppose the obstacle to lie at the section $l = 0$ whose reflexion coefficient alone is ρ. We shall suppose that the coefficient $A = 1$, so that the electric field possesses unit amplitude where $x = \frac{1}{2}b$.

When this incident wave of unit amplitude first impinges on the obstacle it excites a scattered wave whose complex amplitude at $l = 0$ is h. Thus a wave of amplitude h returns to the generator and a wave with amplitude $(1+h)$, composed of the incident and scattered waves superimposed, proceeds to the left (fig. 5·9).

Since the reflexion coefficient of the section is ρ, a wave $(1+h)\rho$ returns to the obstacle from the left and excites a further scattered wave $h(1+h)\rho$. The additional wave returning to the generator is $h(1+h)\rho+(1+h)\rho=\rho(1+h)^2$ and that to the left $h(1+h)\rho$.

A succession of scattered wave trains is thus excited, and it can be seen that because $|h|<1$, the successive scattered wave trains become progressively weaker and the final combined wave train that returns to the generator may be represented by the following

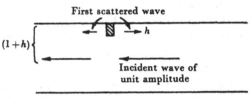

Fig. 5·9.

convergent series whose sum is the new reflexion coefficient of the section:

$$\rho_t = h+\rho(1+h)^2+h\rho^2(1+h)^2+h^2\rho^3(1+h)^2+\ldots,$$

whence
$$\rho_t = h+\frac{\rho(1+h)^2}{(1-\rho h)} = \frac{h+\rho+2\rho h}{(1-\rho h)}. \qquad (3)$$

The corresponding admittance of the section containing the obstacle is

$$y_t = \left(\frac{1-\rho_t}{1+\rho_t}\right)$$
$$= \left(\frac{1-h-\rho-3\rho h}{(1+h)(1+\rho)}\right). \qquad (4)$$

If the obstacle does in fact possess a self-admittance y_1, as we are attempting to show, then when it is introduced at a section whose admittance is zero ($\rho = +1$) we should expect, if the hypothesis is correct, that y_1 is the value of y_t in (4) obtained by putting $\rho = +1$.

This gives
$$y_1 = \frac{-2h}{(1+h)}. \qquad (5)$$

We have, of course, here supposed the wave guide to be closed by a short-circuiting plate and the obstacle to be placed at a section an odd number of quarter wave-lengths—$\frac{1}{4}(2n+1)\lambda_g$—away from it.

If the supposition is correct, that the self-admittance of the obstacle merely adds to that of the section, then y_1 in (5) is the self-admittance expressed in terms of the scattering coefficient.

It remains to show that when y_1 is introduced at a section whose admittance has an arbitrary value y, then the resultant admittance y_t as obtained from (4) is actually $y_t = (y_1 + y)$.

The admittance of the section without the obstacle is

$$y = (1 - \rho)/(1 + \rho).$$

We assume the self-admittance of the obstacle to be

$$y_1 = -2h/(1 + h),$$

whence

$$y_1 + y = \frac{1 + h - \rho - \rho h - 2h - 2\rho h}{(1 + \rho)(1 + h)}$$

$$= \frac{(1 - h - \rho - 3\rho h)}{(1 + \rho)(1 + h)} = y_t \quad \text{(from (4)).} \tag{6}$$

This proves that y_1 as given by (5) is the true self-admittance of the obstacle, and that an obstacle does in fact possess a self-admittance with respect to a particular wave guide. In another wave guide y_1 would in general be different.

Special cases. The validity of equations (4), (5) and (6) may be readily tested by means of simple examples.

For instance, suppose the wave guide to be closed at a distance $\frac{1}{2}\lambda_g$ from the section containing the obstacle. Then $\rho = -1$; from both (4) and (6)

$$y_t = \infty = y.$$

When the wave guide is terminated in a reflexionless load, then $\rho = 0$, $y = 1$,

$$y_t = (1 + y_1).$$

Stub matching. To remove the reflected wave between $l = 0$ and the generator we first find by means of the standing wave indicator a position where the admittance is

$$y = 1 \pm jb_1.$$

(How to do this was explained in § 5·2·8.)

When an obstacle whose admittance $y_1 = \mp jb_1$ is introduced at this place the resulting admittance of the section becomes

$$y_t = 1 \pm jb_1 \mp jb_1 = 1.$$

Its reflexion coefficient ρ_t is zero and no reflected wave returns from it to the generator.

The self-admittance y_1 of an obstacle is simply measured at a position of $E_{\text{max.}}$ in a *complete* standing wave pattern where $y = 0$. The new impedance $y_t = y_1$ may be found by means of a standing wave indicator. Alternatively, the wave guide can be terminated in a reflexionless load and $y_t = 1 + y_1$ measured in the same way.

It is found that any thin flat metallic object scatters symmetrically if its plane lies in the cross-section of the guide. Also, a thin object with an axis of symmetry such as a cylinder, scatters symmetrically if its axis lies in the cross-section. Such obstacles therefore are equivalent to lumped circuit elements connected in shunt across a transmission line.

Anti-symmetrical scattering. Here the wave is scattered on the first scattering with a coefficient $+h$ back to the generator but with $-h$ in the opposite direction.

We find, by the method used above, that the reflexion coefficient of a section whose own reflexion coefficient is ρ becomes ρ_t, where

$$\rho_t = h + \rho(1-h)^2 + h\rho^2(1-h)^2 + \ldots$$
$$= \frac{h+\rho-2\rho h}{(1-\rho h)}. \tag{7}$$

Here, however, the circuit equivalents are expressed more simply as combinations of impedances rather than of admittances.

The impedance of the section without the scattering source is

$$z = \left(\frac{1+\rho}{1-\rho}\right), \tag{8}$$

and with the scattering source

$$z_t = \left(\frac{1+\rho_t}{1-\rho_t}\right)$$
$$= \frac{1+h+\rho-3\rho h}{(1-h)(1-\rho)}. \tag{9}$$

To find the self-impedance z_1 of the scattering source we suppose the wave guide to be terminated, at a distance $\frac{1}{2}\lambda_g$ on the side away from the generator, in a short-circuiting metal plate.

The reflexion coefficient ρ of the section then becomes $\rho = -1$. Put $\rho = -1$ in (9) and write $z_t = z_1$. We find

$$z_1 = 2h/(1-h). \tag{10}$$

From equations (8) and (10) it follows that, in general,

$$z + z_1 = \left(\frac{1+\rho}{1-\rho}\right) + \frac{2h}{(1-h)} = \frac{1+h+\rho-3\rho h}{(1-\rho)(1-h)}$$

$$= z_l. \tag{11}$$

Thus the self-impedance z_1 of an anti-symmetrical scattering centre combines with the normalized impedance of the section to which it is added according to the law of combination of series impedances. The equivalent circuit representation is a transmission line of characteristic impedance Z_0 with an impedance $Z_0 z_1$ in series with it at a section where the line impedance is $Z_0 z$. To avoid unbalance of the line the correct representation is twin lines with $\frac{1}{2} Z_0 z_1$ in series with each conductor. For simplicity the impedance is usually shown in one conductor only.

The criterion of symmetrical or anti-symmetrical scattering of the electric field therefore determines whether the scattering source is to be considered as being added in shunt or in series. This criterion corresponds exactly to that for waves on a transmission line where a shunt-circuit element produces no discontinuity in the voltage but does so in the current, but a series-circuit element produces a discontinuity in the line voltage but not in the current. In the wave guide we merely replace voltage by the electric field of the H_{10}-mode and current by the magnetic field of this mode to complete the correspondence.

Unsymmetrical scattering. Here the equivalent circuit representation requires both series and shunt elements, and, if so desired, could be reduced to an equivalent filter section of T, π or lattice type. We shall not, however, discuss these possibilities (see § 6·12).

Illustrative experiments on wave-guide matching. To illustrate what has been said about the self-admittance of an obstacle and the elimination of the reflected wave by the introduction of an obstacle into a wave guide, we give some experimental results obtained with a $2\frac{1}{2} \times 1$ in. wave guide and a wave-length of 9 cm. We return to the example at the end of § 5·3, where measurements of the impedance of the open end of the wave guide were discussed. The standing wave ratio was $S = 1/s = 2\cdot83 = 1/0\cdot354$. The admittance at a position of $E_{\text{max.}}$ is a conductance $0\cdot354$ (points A in fig. 10(a) and (b)). The point B in the circle diagram of admittances (fig. 5·10(a) and (b)) where the admittance in the wave guide is of the form $(1 + jb_1)$ lies on the n-arc whose n value is $n_1 = 0.163$; further, the length of CB is 1.07. (These values were obtained from a circle diagram and not from fig. $5.10\,(a)$.) [See fig. 5.59.]

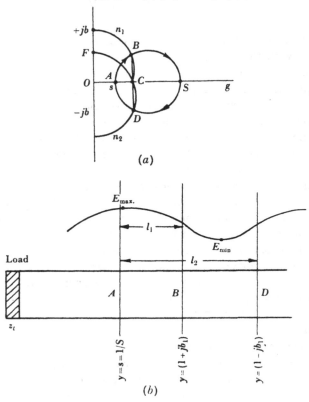

(a)

(b)

Fig. 5·10.

We conclude that at a section a distance

$$l_1 = n_1 \lambda_g = 0.163 \times 13.8 = 2.25 \, \text{cm}.$$

from any position of $E_{\text{max.}}$ towards the generator the admittance of the section is there equal to $y = 1 + j \times 1.07$.

At the section D (fig. $5.10\,(b)$) the corresponding point on the circle diagram is D whose admittance is $(1 - jb_1) = 1 - j \times 1.07$. The section lies at distance $l_2 = n_2 \lambda_g$ from $E_{\text{max.}}$ (represented by A in fig. $5.10\,(a)$), where $n_2 = 0.337$. Thus

$$l_2 = 0.337 \times 13.8 = 4.65 \, \text{cm}.$$

A metal screw was introduced into the wave guide through a narrow slot on the centre line of the broad face, at the position l_2, and its projection into the guide parallel to the electric field was gradually increased. It was found that with a particular length of screw inside the guide the reflected wave vanished and the standing wave ratio in the guide between the screw and the generator was closely $S = 1$. It is therefore possible to use a metal object to eliminate a reflected wave by using a procedure entirely analogous to stub matching in transmission lines.

In the example under discussion, the self-admittance y_1 of the screw which scatters symmetrically is such that

$$1 - j \times 1 \cdot 07 + y_1 = 1,$$

that is $\qquad y_1 = +j \times 1 \cdot 07.$

The screw, in the matching position, therefore possessed a capacitive susceptance $b_1 = 1 \cdot 07$. As a check, the end of the wave guide was closed by a metal plate in order to produce a complete standing wave in the wave guide and the screw was introduced, to exactly the same extent, at a position of E_{max} where the guide admittance $y = 0$. As a result, between the screw and the generator the standing wave was displaced, as a whole, through a distance $1 \cdot 8$ cm. away from the generator; that is, the position of each E_{max} was moved $1 \cdot 8$ cm. towards the end of the wave guide.

Form $n' = 1 \cdot 8 / \lambda_g = 0 \cdot 13$ and locate the point F (fig. $5 \cdot 10\,(a)$) on the reactive axis at the end of the n-arc whose n value is equal to $n' = 0 \cdot 13$. The susceptance of the screw is $OF = j \times 1 \cdot 07$ (as read from the chart). The experiment confirms the theory in showing that the obstacle possesses a self-admittance $y_1 = j \times 1 \cdot 07$.

Thus, to summarize, although the quantities y and z that we call the admittances and impedances of obstacles and wave guides are in fact no more than transformed reflexion and scattering coefficients, they are of the greatest practical convenience; first, their laws of combination are much simpler than the laws of combination of the reflexion and the scattering coefficients; secondly, these laws are identical with the simple laws for the addition of admittances and impedances in shunt and series respectively, so that it is often possible, as we have seen, to find exact circuit equivalents to describe transmission phenomena in wave guides.

5·5. Examples of obstacles whose self-admittances can be calculated

When the obstacle possesses a simple geometrical form it is some-times possible to calculate its self-admittance by applying the principles of electromagnetism to the problem. In such applications the investigator, operating untrammelled on the intellectual plane, is at liberty to introduce any concept such as voltage, current, electric field, magnetic field, true impedance, wave impedance, or power, that will lead to a solution of the problem. Whatever the

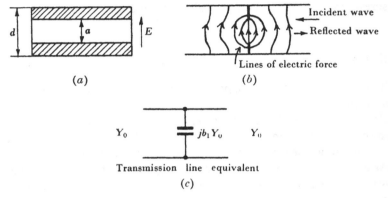

(a)

(b)

Lines of electric force

Incident wave

Reflected wave

Y_0 jb_1Y_0 Y_0

Transmission line equivalent

(c)

Fig. 5·11.

method, the aim is, in effect, to find a scattering coefficient or a reflexion coefficient that can be converted into the required self-impedance.

Unfortunately, even the simplest of these calculations is too long for inclusion in the middle of this account, and we shall do no more than quote results. The principal methods of attack are those of Schwinger (U.S.A.), who employs the calculus of variations, and of Macfarlane (T.R.E.), who has been able to extend the powerful method of conformal transformation (of two-dimensional electro-statics) to find the composition of the electromagnetic field near simple obstacles in wave guides. Such calculations are valuable because, from their results, it is possible to design structures which possess specified self-susceptances.

Irises. A common form of obstacle is the so-called iris (an in-appropriate name), which is merely a metal strip or pair of strips of negligible thickness lying in the cross-section of the wave guide. Common types are:

(*a*) *The capacitive iris*, an example of which is shown in fig. 5·11, comprises a thin metal strip or pair of strips that stand perpendicular to the broad face of the wave guide and lie in the cross-section.

In the example of fig. 5·11 (*a*), *d* is the narrow dimension of the wave guide and *a* is the separation of the edges of the strips. The direction of the electric field is indicated at *E* and a longitudinal central section of the field is shown in fig. 5·11 (*b*).

The susceptance of this iris is b_1, where

$$y_1 = jb_1 = j\frac{4d}{\lambda_g}\log_e \operatorname{cosec}\left(\frac{\pi a}{2d}\right). \tag{1}$$

The equivalent transmission-line circuit is shown in fig. 5·11 (*c*).

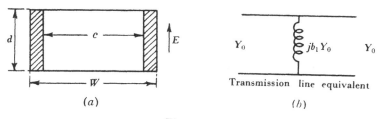

Transmission line equivalent

(*a*)   (*b*)

Fig. 5·12.

The capacitive iris is not employed in wave guides in which high powers are transmitted because the intense electric field at the edge of the strips causes electrical breakdown of the air.

(*b*) *The inductive iris.* An example is shown in fig. 5·12. Here the edges of the strips run parallel to the electric field. If *W* and *d* are the long and short dimensions respectively of the wave-guide cross-section and *c* is the distance between the edges of the strips, the admittance is

$$y_1 = jb_1 = -j\cot^2\left(\frac{\pi c}{2W}\right). \tag{2}$$

Since this iris is commonly used for matching, formula (2) is exhibited graphically in fig. 5·13.

Other irises and reactive obstacles. Figs. 5·14–5·17 summarize in diagrammatic form a variety of information about irises and reactive obstacles and their equivalent circuit representations. These

figures are taken from an unpublished article by Dr G. G. Macfarlane. The diagrams are numbered consecutively throughout the
series. Attention is drawn to no. 10, which represents a simple
inductive obstacle formed by stretching a wire across the guide
with its axis parallel to the electric field. If, however, the wire does

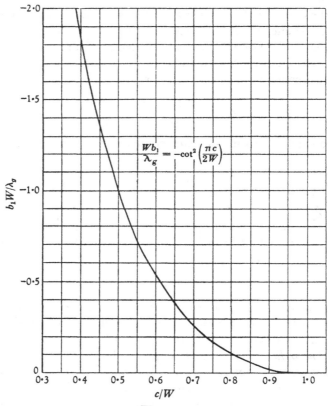

$$\frac{W b_1}{\lambda_g} = -\cot^2\left(\frac{\pi c}{2W}\right)$$

Fig. 5·13.

not reach the whole way across, as shown in no. 8, the susceptance
is part capacitive and part inductive, becoming predominantly
capacitive when the post is short. It should be noted that the formulae refer to thin irises and are only approximately correct for
thick irises. The effect of thickness is touched upon in § 6·12.

   Irises in circular wave guides. In circular wave guides designed to
carry only the dominant H_{11}-mode as a progressive wave, irises take

the forms shown in nos. 6, 14, 15 and 16, to which further reference is made below. Formulae do not appear to be available for the susceptances of these structures.

List of irises in wave guides with equivalent circuits

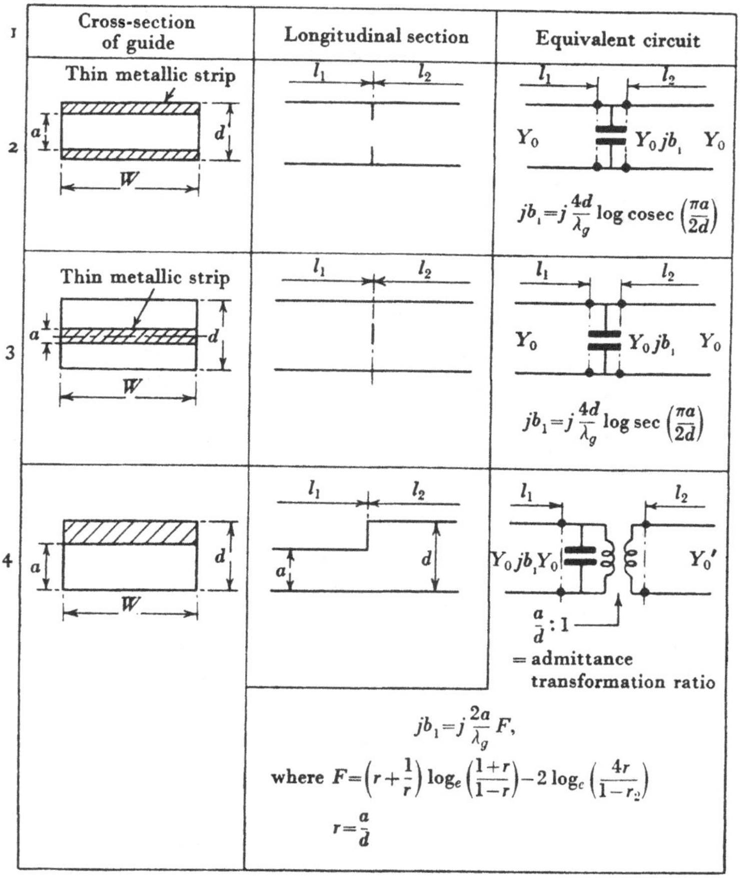
1
Cross-section of guide
Longitudinal section
Equivalent circuit
2
Thin metallic strip
a
d
W
l1
l2
Y0
Y0 jb1
Y0
jb1 = j 4d/λg log cosec (πa/2d)
3
Thin metallic strip
a
d
W
l1
l2
Y0
Y0 jb1
Y0
jb1 = j 4d/λg log sec (πa/2d)
4
a
d
W
l1
l2
Y0 jb1 Y0
Y0'
a/d : 1
= admittance transformation ratio
jb1 = j 2a/λg F,
where F = (r + 1/r) loge ((1+r)/(1−r)) − 2 loge (4r/(1−r2))
r = a/d

Fig. 5·14.

5·6. Resistive impedances

Metallic obstacles behave as shunt susceptances because they distort the field of the H_{10}-wave and thus excite a series of evanescent modes whose storage fields resemble those of condensers and inductances. There is no resistive term because the obstacle itself absorbs no power. A piece of loss-free dielectric which distorts the field also behaves as a shunt susceptance. Conversely, a structure which is to behave as a pure shunt conductance must absorb power but produce no distortion in the field configuration. The simplest

List of irises in wave guides with equivalent circuits

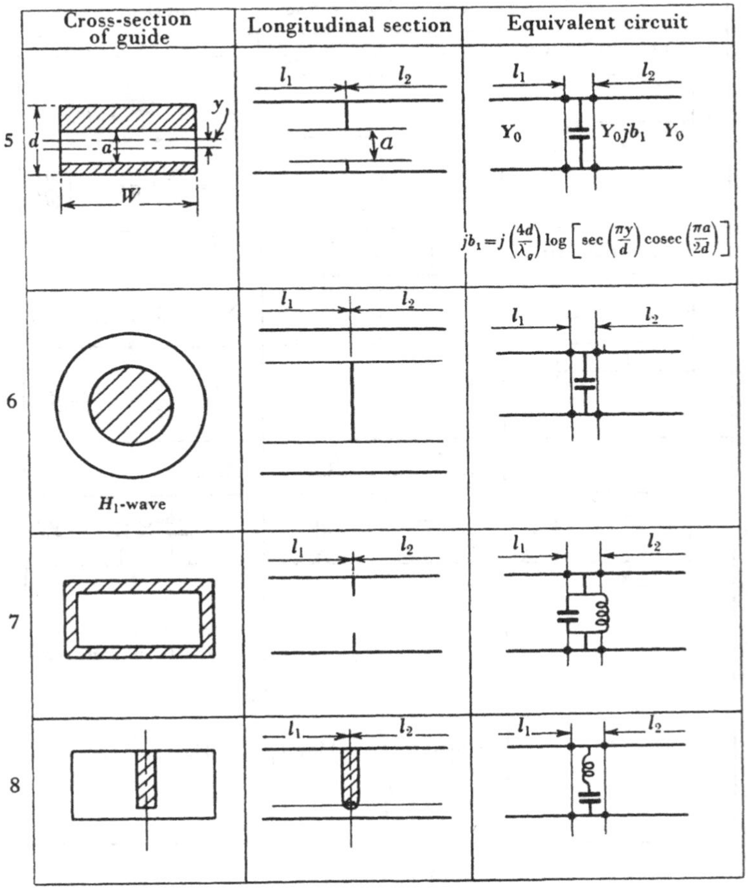

Cross-section of guide	Longitudinal section	Equivalent circuit

Fig. 5·15.

resistive device is a resistive film with uniform conductance G mhos per square at all points on its surface. Suppose this film to occupy a cross-section of a wave guide that is terminated in a reflexionless load beyond the film, and let an H_{10}-wave whose transverse electric and magnetic fields at any chosen point on the film are E_0 and H_0, be incident on it.

List of irises in wave guides with equivalent circuits

	Cross-section of guide	Longitudinal section	Equivalent circuit
9	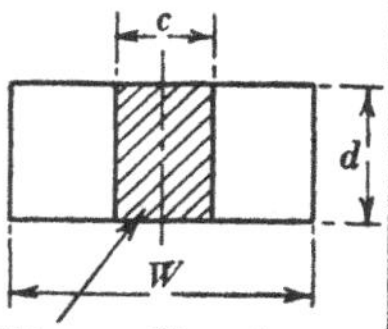 Thin metallic strip	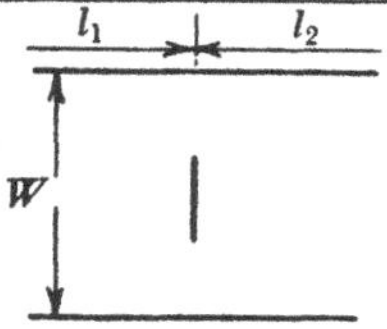	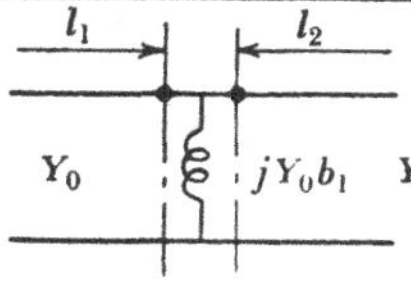 $\frac{1}{jb_1}=j\frac{W}{2\lambda_g}\left[\cosh^{-1}\left(\csc\frac{\pi c}{4W}\right)\right]-2$
10	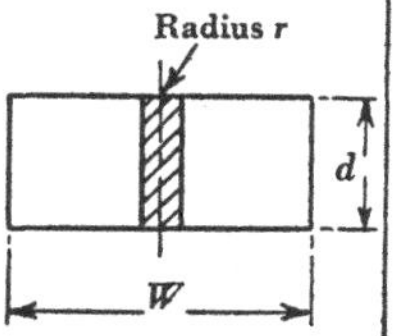	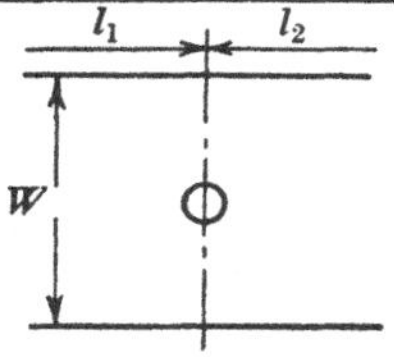	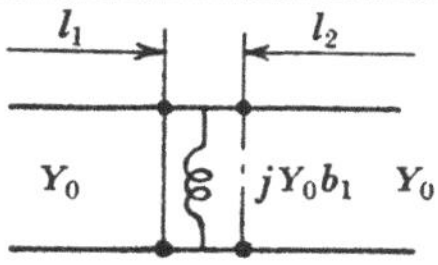 $\frac{1}{jb_1}=j\frac{W}{2\lambda_g}\left[\log\left(\frac{2W}{\pi R}\right)-2+2\sum_{2}^{\infty}\left(\frac{1}{\sqrt{\left\{1-\left(\frac{2d}{2n-1\lambda}\right)^2\right\}}}-1\right)\frac{1}{2n-1}\right]$
11	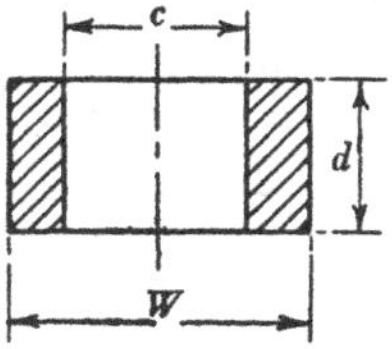	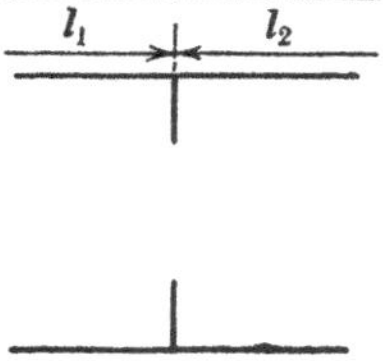	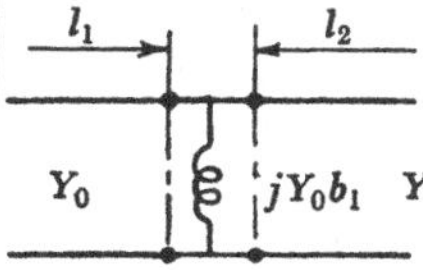 $\frac{1}{jb_1}=j\frac{W}{\lambda_g}\tan^2\left(\frac{\pi c}{2W}\right)$
12	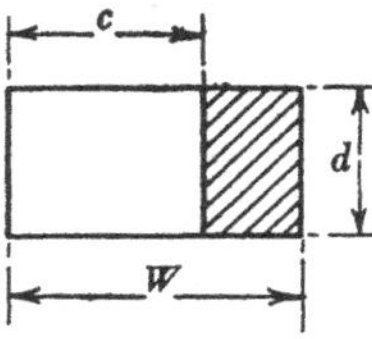	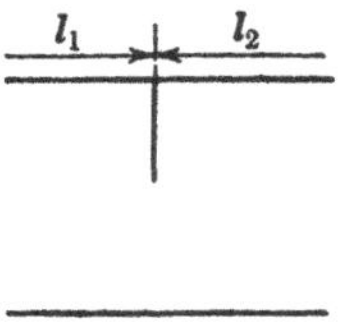	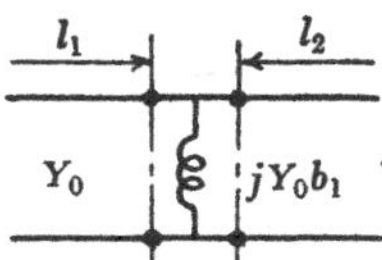 $\frac{1}{jb_1}=j\frac{W}{\lambda_g}\frac{\tan^2\left(\frac{\pi c}{2W}\right)}{1+\csc^2\left(\frac{\pi c}{2W}\right)}$

Fig. 5·16.

Let E_1 and H_1 be the corresponding fields in the scattered wave. The total electric field tangential to the film is (E_0+E_1) V./m. at this point, and the local current density in the film is $G(E_0+E_1)$

amp./m. The magnetic field on the side of the film facing the generator is $(H_0 - H_1)$ and on that facing the reflexionless load $(H_0 + H_1)$. According to equation 1·4 (2)

$$G(E_0 + E_1) = (H_0 - H_1) - (H_0 + H_1),$$

List of irises in wave guides with equivalent circuits

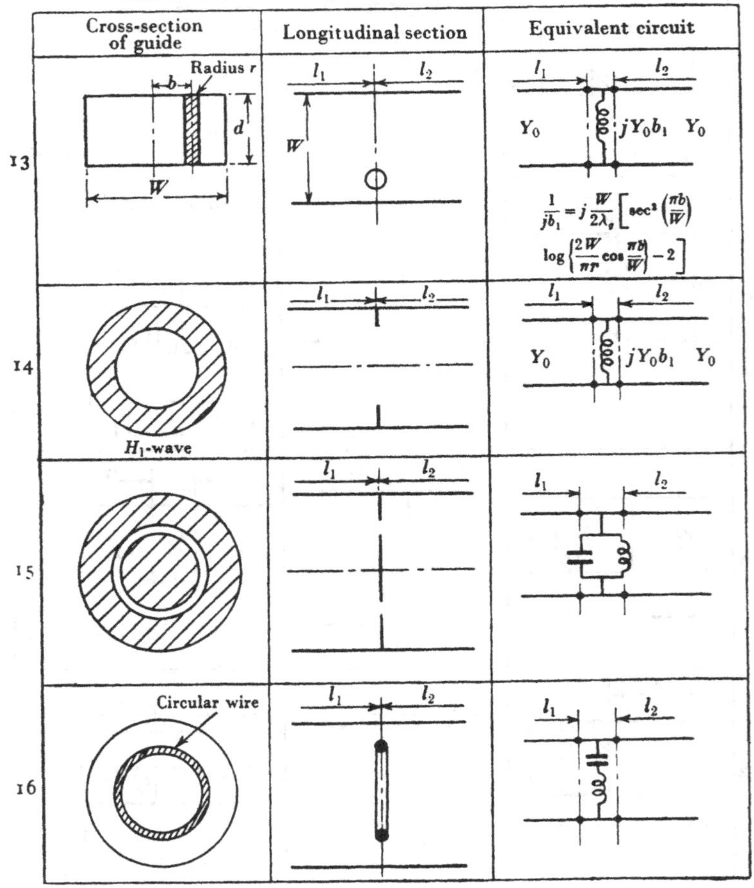

Fig. 5·17.

that is
$$G(E_0 + E_1) = -2H_1. \qquad (1)$$

But, from 3·4·1 (3) and (4),
$$E_1 = Z_H H_1 \quad \text{or} \quad H_1 = Y_H E_1, \qquad (2)$$

where $$Z_H = \sqrt{\left(\frac{\mu}{\epsilon}\right)}\frac{\lambda_g}{\lambda} = \frac{1}{Y_H}$$

is the intrinsic impedance of the wave guide, and Y_H is the intrinsic admittance.

Equation (1) may therefore be written

$$\left(\frac{E_0}{E_1} + 1\right) = -\frac{2Y_H}{G}. \tag{3}$$

But $E_1/E_0 = h$ the scattering coefficient, consequently

$$\left(\frac{h+1}{h}\right) = -\frac{2Y_H}{G},$$

or, from 5·4 (5), the self-admittance of the resistive obstacle is

$$y_1 = \frac{-2h}{(1+h)} = \frac{G}{Y_H} = g. \tag{4}$$

Thus, the self-admittance of a uniform resistive film is its surface conductance G per square divided by the intrinsic admittance of the wave guide. Alternatively, its self-impedance $z_1 = 1/y_1$ is its surface resistance R per square divided by the intrinsic impedance Z_H. The equivalent circuit representation is clearly a transmission line with characteristic impedance Z_H shunted by a resistance R.

In the example under discussion the total admittance of the section containing the film becomes

$$y_t = (1 + y_1) = (1 + g).$$

Replace the reflexionless load by a short-circuiting plate and introduce the film at a position of $E_{\max.}$ where the admittance of the section is zero. The resulting admittance is

$$y_t = 0 + g = g.$$

If the resistance R per square of the film is made equal to the intrinsic impedance Z_H of the wave guide, then $g = 1 = y_t$ and the wave guide is matched. Thus, the theoretically ideal reflexionless termination is a resistive film whose surface resistance is equal to the intrinsic impedance, backed by a $(2n+1)\frac{1}{4}\lambda_g$ short-circuited extension of the wave guide. In the special case of a transmission line the intrinsic impedance is the same as the wave impedance of a

TEM-wave, namely $\sqrt{\dfrac{\mu}{\epsilon}} = 120\pi \sqrt{\dfrac{K_m}{K_e}}$ ohms, and the resistive film would be given a specific surface resistance of this value. When backed by a $\frac{1}{4}\lambda$ short-circuited extension such a film forms a reflexionless termination.

It is interesting to note that the total resistance of the film measured between the conductors of the transmission line is equal to the characteristic impedance Z_0 of the line whatever may be the geometry of the system. Reflexionless terminations of this type are not used in practice.

It is useful to review at this point what is implied by the terms capacitive, inductive or resistive, as applied to an obstacle in a wave guide. We have seen that we are able to distinguish the electrical behaviour of the obstacles only by the differences in the waves scattered or reflected from them. By means of a standing wave indicator we observe the nature of the standing wave pattern produced, for instance, when the obstacle is introduced into the previously matched wave guide. If the partial standing wave of the electric field is the same, as regards standing wave ratio and the position of the maxima and minima, as the standing wave of voltage produced at the same frequency on a transmission line by connecting a reactance in shunt across it, then the obstacle in the wave guide is called reactive, and its normalized reactance or susceptance (whether capacitive or inductive) is calculated, as explained above, from the standing wave ratio and the location of the maxima or minima. Similar remarks apply to the resistive film. When the obstacle possesses a simple geometrical form it may be possible to apply the principles of electromagnetism to calculate its scattering coefficient h with respect to an H_{01}-wave and to transform this scattering coefficient into a normalized susceptance, as explained in § 5·4. An example of such a calculation is given in § 7·11.

It is found that the total electromagnetic field in the immediate vicinity of the obstacle is quasi-static, that is, it resembles a certain electrostatic or magneto-static field. For instance, a thin wire stretched across a wave guide parallel to the electric field of the incident wave carries an oscillating current which produces a preponderance of magnetic storage field near the wire.

The wire, as would be anticipated, behaves as an inductance which stores magnetic energy. Similarly, the capacitive iris of fig. 5·11 accumulates a concentration of charge on its edges which excites a local storage field of an electrostatic character. Such an iris possesses a capacitive reactance. In this way it is possible to judge the nature of a reactive obstacle from the character of the storage field; however, the numerical value of its normalized reactance must be obtained as already explained from its scattering coefficient h, whether calculated or measured indirectly by means of a standing wave indicator.

5·7. Reflexion from a plane interface

Another type of problem for whose investigation wave (intrinsic) impedances prove useful, is that of finding the reflexion from a plane

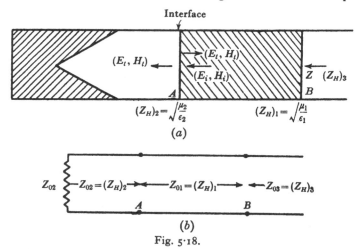

Fig. 5·18.

interface between two dielectric media. Consider, as a specific example, the situation shown in fig. 5·18 (a) which represents a wave guide partially filled with a dielectric and terminated in a reflexionless load. We require to find the reflexion and transmission coefficients of the interface A.

Let $(Z_H)_1 = \sqrt{\left(\dfrac{\mu_1}{\epsilon_1}\right)} \dfrac{\lambda_{g1}}{\lambda}$ and $(Z_H)_2 = \sqrt{\left(\dfrac{\mu_2}{\epsilon_2}\right)} \dfrac{\lambda_{g2}}{\lambda}$ be the intrinsic impedances of the two portions of the wave guide (§ 3·4·1).

Let the transverse field components in the incident and reflected waves be respectively E_i, H_i and E_r, H_r, and those in the transmitted wave E_t and H_t.

HWG

The field components are related as follows:

$$\frac{E_i}{H_i} = -\frac{E_r}{H_r} = (Z_H)_1{}^*$$

$$\frac{E_t}{H_t} = (Z_H)_2.$$

The boundary conditions at the interface require

$$E_t = E_i + E_r, \quad H_t = H_i - H_r$$

or
$$\frac{E_t}{(Z_H)_2} = \frac{(E_i - E_r)}{(Z_H)_1}.$$

Whence
$$\frac{(Z_H)_2}{(Z_H)_1} = \left(\frac{E_i + E_r}{E_i - E_r}\right) = \left(\frac{1+\rho}{1-\rho}\right),$$

or
$$\rho = \frac{[(Z_H)_2/(Z_H)_1 - 1]}{[(Z_H)_2/(Z_H)_1 + 1]},$$

(1)

where $\rho = E_r/E_i$. The transmission coefficient is

$$\frac{E_t}{E_i} = (1+\rho) = \frac{2(Z_H)_2/(Z_H)_1}{[(Z_H)_2/(Z_H)_1 + 1]}.$$

When we compare equation (1) with equations 5·2·2 (6) and (7) we note that the process of reflexion at the interface is exactly analogous to that on a transmission line with characteristic impedance $(Z_H)_1$ terminated in an impedance (Z_H).

The whole wave guide may be represented by the equivalent triple transmission-line system shown in fig. 5·18 (b) and the reflexion coefficient at the terminals calculated as an exercise in the application of the standard transmission-line formulae. For instance, if the length AB is equal to $\frac{1}{2}\lambda_{g1}$ the section AB behaves as a half-wave transformer. If, therefore, $(Z_H)_3 = (Z_H)_1$, there is no reflexion from the interface B. The idea of using wave impedance to reduce a problem of wave reflexion at an interface to an equivalent problem in the theory of transmission lines has been extended to the case of plane waves in space incident normally or obliquely on an interface and to other types of wave propagation in one dimension. The reader who is interested in these matters should consult Schelkunoff's treatise, *Electromagnetic Waves*.

* Negative sign expresses the fact that the incident and reflected waves travel in opposite senses.

5·8. Resonant obstacles

Fig. 5·15, panel 8, shows the equivalent circuit of a metal post that projects into the wave guide with its axis parallel to the electric field. The equivalent circuit comprises a series combination of an inductance and a capacity placed in shunt across a transmission line. This suggests that if it were possible to alter the values of the inductive and capacitive reactances it might be possible to attain a

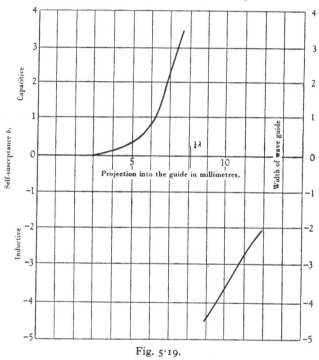

Fig. 5·19.

condition of resonance, at the frequency of the H_{10}-wave, when the total impedance shunting the transmission line would vanish. There would then be complete reflexion at the shunt circuit and no transmission beyond it. The values of the reactances in the equivalent circuit can be altered as required, by inserting the post progressively farther into the guide. With a definite length of the post inside the wave guide complete reflexion occurs and the self-susceptance of the rod becomes infinite. Fig. 5·19 exhibits graphically the results of measurements of the self-susceptance of a rod as a function of length within the wave guide. The method employed was that described at the end of §5·4, and a straight wire of diameter $\frac{1}{2}$ mm. was inserted into an X-band $1 \times \frac{1}{2}$ in. wave guide operated at a wave-length $\lambda = 3·25$ cm. The wave-length of the H_{10}-wave was $\lambda_g = 4·2$ cm.

The curve shows clearly that the susceptance becomes very large when the length of the rod within the guide is approximately equal to $\frac{1}{4}\lambda$, and that in passing through this value it changes from a capacitive to an inductive susceptance.

The results obtained with a wire twice as thick were very similar. We conclude therefore that a *thin* metal post becomes resonant within a wave guide when its length is $\frac{1}{4}\lambda$ and that it then throws a short circuit across the wave guide.

A different type of resonance is that in which the equivalent inductance and capacity are thrown in parallel across the transmission line as shown in fig. 5·15, panel 7.

On resonance the tuned circuit is a rejector circuit with zero admittance. Consequently the principal wave passes over the resonant obstacle without reflexion. An obstacle that behaves in this way is shown in fig. 5·15, panel 7. It is called a resonant iris and is clearly a combination of the capacitive iris of fig. 5·16, panel 2, and the inductive iris of fig. 5·16, panel 11.

The storage fields of the inductive (H-modes) and capacitive (E-modes) portions of the iris can, by choosing the dimensions correctly, be made to store equal energies in the magnetic and electric forms so that the stored energy is merely exchanged between these forms during oscillation without drawing energy from and restoring it to the incident H_{10}-wave.

A rough method of finding the resonant dimensions of this iris, based on an approximate theory, is given in Slater's book, *Microwave Transmission*, p. 186. It is shown in fig. 5·20. *ABCD* (fig. 5·20 (b)) represents the cross-section of the wave guide. The curves *ALD* and *BMC* are the two branches of a hyperbola that pass through the corners of *ABCD* and such that their poles L and M are $\frac{1}{2}\lambda$ apart ($\lambda =$ TEM wave-length). If the corners of the composite iris are made to fall on this hyperbola, then the iris is approximately resonant. The dotted rectangle is an example. Evidently there is an infinity of such resonant structures for any given wave guide. If W and b, and W' and b' are the dimensions respectively of the cross-sections of the wave guide and of the window in the iris, then the geometrical construction given above is equivalent to the following relation between these dimensions:

$$\frac{W}{b}\sqrt{\left[1-\left(\frac{\lambda}{2W}\right)^2\right]}=\frac{W'}{b'}\sqrt{\left[1-\left(\frac{\lambda}{2W'}\right)^2\right]}.$$

It is clear from fig. 5·20(b) that when b'/W' is made small, then W' is approximately equal to $\frac{1}{2}\lambda$; that is, the resonant length of any

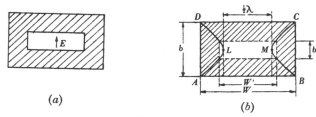

(a)   (b)

Fig. 5·20.

narrow slot centrally placed in a diaphragm with its length perpendicular to the electric field is very nearly half a wave-length. The remarkable feature of these slots is that, although the section of the wave guide is almost entirely occupied by metal, yet the power is transmitted through the slot without reflexion.

The formation of a resonant iris in a circular wave guide carrying the H_{11}-mode as a progressive wave is shown in fig. 5·21.

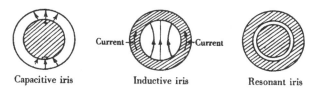

Capacitive iris   Inductive iris   Resonant iris

Fig. 5·21.

For this iris, when the circular gap is thin its resonant length, measured on the inner circumference, is almost equal to λ. Fig. 5·22(a) shows a set of resonant structures that will transmit the dominant mode without reflexion. Structures III and IV have already been discussed. Iris I is a resonant slot for transmitting an H_{11}-wave in a circular guide. Its length at resonance is about 4 % shorter than $\frac{1}{2}\lambda$ when $\lambda = 9\cdot1$ cm. The slot with the broadened ends shown at II is less frequency sensitive than the simple slots.

The Q-factor of a resonant iris. The voltage $|V|$ across a circuit whose admittance is $Y = G + jB$, when fed from a constant current generator with current of amplitude $|i|$, is,

$$|V| = |i|/\sqrt{(G^2 + B^2)}$$
$$= V_{\mathrm{max.}}/\sqrt{(1 + B^2/G^2)}.$$

At the resonant frequency f_0 the susceptance B is zero, but at some pair of frequencies $(f_0 \pm \Delta f)$ the susceptance B has become equal to

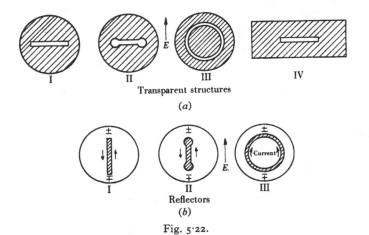

Transparent structures

(a)

Reflectors

(b)

Fig. 5·22.

the conductance G. It follows that Δf is the mistuning required to reduce the amplitude V to $V_{\mathrm{max.}}/\sqrt{2}$. The Q-factor of the circuit is defined to be

$$Q = f_0/2\Delta f = -\lambda_0/2\Delta\lambda. \tag{1}$$

Thus when Q is large the circuit possesses a sharp response curve around the resonant frequency. This definition also serves for the Q of a resonant iris if f_0 is taken to be the resonant frequency at which the transmitted wave has maximum amplitude and Δf the mistuning required to reduce the amplitude of the transmitted wave by a factor of $1/\sqrt{2}$. When Q is small the iris remains transparent over a reasonably large range of frequencies.

Consider, for example, the irises shown in fig. 5·22 (a). The experimental values of the Q-factors at $\lambda = 9·1$ cm. were $Q = 25$ for slot I with a width of 0·5 mm. but $Q = 50$ with a width of 0·1 mm. Thus the narrower the slot the larger the value of Q. The dumb-bell slot of figure II has a much smaller Q value. In a slot whose straight portion was 4·2 mm. long with 1·4 mm. diameter circles at the ends Q was equal to 10. Such an iris is transparent over a broad band of frequencies. The Q values of the ring iris of figure III were as follows:

Ring width in mm.	0·1	0·5	0·8
Q	40	20	16

Too much stress should not be placed on the precise numerical values, since the thickness of the foil forming the iris also controls the Q values.

Reflecting irises. The metal post, discussed at the beginning of this section, provided an example of a resonant obstacle that was completely reflecting, unlike the resonant slots which are completely transparent. There are, however, irises that are completely reflecting at resonance, examples of which are shown in fig. 5·22 (*b*). If each is compared with the resonant iris above it in fig. 5·22 (*a*), it will be seen that the lower iris is obtained from the upper by interchange of the metal and open portions, followed by a rotation through a right angle. A pair of diaphragms related in this way are termed *complementary*. Not only are their geometrical properties complementary, but their electromagnetic properties also. There is an underlying theoretical explanation for this statement based on an extension of Babinet's principle, as used in optical theory, to embrace electromagnetic waves. Although it is not rigorously applicable to iris theory it serves as an approximate method for studying the properties of irises.

The effect of the interchange and rotation is to give an equivalent circuit representation in which the rejector circuit of inductance and capacity in shunt across a transmission line becomes one in which they are in series. Compare, for instance, in fig. 5·17, panel 15 with panel 16. We have seen that the circular slot is resonant and transparent when its inner circumference is almost equal to λ; conversely, the ring of fig. 5·22 (*b*) III is resonant when its circumference exceeds λ slightly (1·1λ). It then becomes a complete reflector. Similarly, the half-wave strips or rods of fig. 5·22 (*b*) I and II are reflectors. The ring may, as an alternative, be regarded as a pair of half-wave reflectors bent to form a circle whose circumference is therefore equal to λ (approximately). We note that the half-wave reflectors lie parallel to the electric field. The elucidation of the properties of resonant irises was due, in the main, to the pioneer experimental investigations of W. D. Allen at T.R.E.

5·9. Applications of resonant obstacles

5·9·1. *Resonant slots as switches and protective devices*

The electric field at the centre of a resonant slot is many times more intense than in the H_{10}-wave at a distance from the slot. Consequently, when high powers are transmitted, an electrical discharge may occur across the gap. This property is used in the gas-filled resonant cell. Here the slot is enclosed in a glass capsule which is filled with argon or some other monatomic gas at a pressure of 70 mm. of mercury.

Fig. 5·23 (*a*) is a section through a resonant gas cell containing a resonant ring, but in fig. 5·23 (*b*) the iris is a resonant slot. The metal

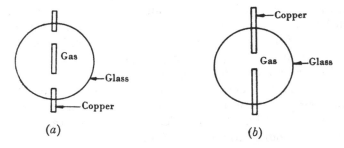

(*a*)   (*b*)

Fig. 5·23.

of the irises projects beyond the glass walls so that they may be fitted tightly into a circular wave guide. Such cells are transparent at low powers, as, for instance, during the reception of a radar echo, but at high power they spark over and become completely reflecting. Instead of placing zero admittance across the guide they now throw a short circuit across it, and thus prevent all but a small fraction of the power from passing beyond them. The equivalent circuit is a transmission line (fig. 5·24) with a rejector circuit in shunt across it. The rejector circuit carries a spark gap G in shunt, and when the gap strikes the shunt admittances rises immediately from zero to a large value. In the CV115 cell, used at X-band wave-lengths, the resonant slot is of the low-Q form shown in fig. 5·22 (*a*) II with larger end circles and a much shorter intervening straight portion.

These cells are used both as automatic switching and protective devices in common T.R. systems (see later).

A modification of these cells which provides better protection is illustrated in fig. 5·25. It is the American 1 B 24 cell for use at X-band frequencies and it breaks down at lower powers than the CV 115, acting more quickly and providing better protection to the

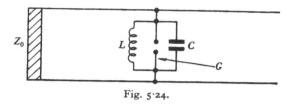

Fig. 5·24.

receiver against the transmitter power. The cell is placed in series with the wave guide and is held in position by bolts through flanges on the wave guide. The cell cavity is separated from the wave-guide interior by glass windows, and the cell itself contains water vapour

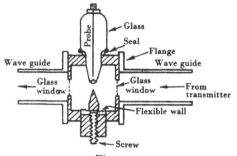

Fig. 5·25.

and hydrogen at a pressure of a few millimetres of mercury. It is tuned to resonance by adjusting the separation of the tips of the spikes that project into it, and the adjustment is made by means of a screw that pushes against a flexible area of the wall carrying one of the spikes. To avoid lag in the striking of the discharge when a powerful wave reaches the cell, a glow discharge is kept running between the probe and the inner surface of the upper spike which is made hollow. Electrons diffuse through a hole in the end of the spike into the cavity and thus provide initial ionization from which the discharge can build up without delay. A high resistance is included in series with the probe in order to limit the glow discharge, but its value must be chosen to avoid an intermittent discharge which impairs the protective action of the cell. The cell thus transmits weak, but blocks powerful signals. This cell replaces the earlier British types CV 221 and CV 114 which were fitted to circular wave guides and contained a mixture of argon and water vapour at a pressure of about 6 mm. of mercury for each gas.

5·9·2. *Ring reflectors as switches*

Reflecting rings form very convenient mechanical switches for diverting power from one branch in a wave guide to another. The small inertia of the switches permits rapid change-over, and therefore reduction in the period during which the wave guide is mismatched. This is important in high-power systems where a serious mismatch may damage subsidiary components such as magnetrons and T.R. cells.

Numerous wave-guide switches have been developed, and fig. 5·26 shows typical examples. In rectangular wave guides the switching loop may be circular or rectangular.

When a loop is rotated through 90° from the position at which it produces maximum reflexion about an axis in the H-plane it permits the power to pass it without reflexion. (The E-plane is a central section of the wave-guide system parallel to the electric field of the H_{10}-wave, and an H-plane is a section parallel to the magnetic field. The nature of the section is indicated in each part of fig. 5·26.) The figures (*a*) to (*c*) are almost self-explanatory. The switch of figure (*a*), although simple in principle, requires subsidiary matching devices that make it frequency-sensitive. The switch of figure (*b*) has the advantage that the rings are carried on a common axis and can be continuously rotated. The discrimination between the arms is good and the change-over period is short. The single loop Y-junction switch of figure (*c*) is the most satisfactory of the switches designed for X-band applications. The loop should not be rotated continuously. Its performance figures over $\pm 3 \%$ frequency band about resonance are:

Standing wave ratio in main wave guide: $s = 1/S = 0.94$ to 0.97.
Power ratios in the transmitting and non-transmitting branches (discrimination): 25–35 db.
Power handling capacity: 50 kW. peak at least.

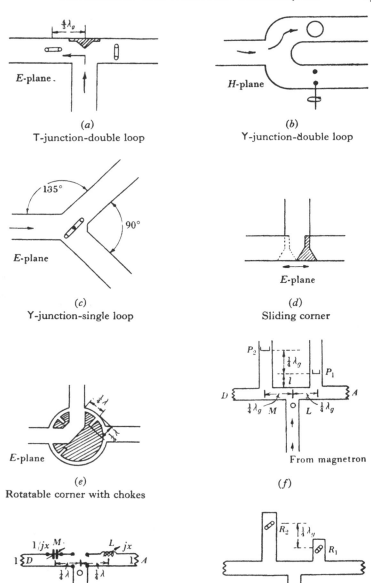

(a)
T-junction-double loop

(b)
Y-junction-double loop

(c)
Y-junction-single loop

(d)
Sliding corner

(e)
Rotatable corner with chokes

(f)

(g)

(h)

Fig. 5·26.

The switches shown in figures (d) and (e) are not ring switches and are unsuitable for use at high powers because their inertia makes the change-over time too great for safety, but they are used in low-power test equipment. Reflecting $\frac{1}{4}\lambda$ probes have also been used in switches requiring a rapid rate of switching of the order of 1000 per min.

An important switching system is that shown in fig. 5·26 (f). Its purpose is to obtain a division of power between the matched loads D and A, such that the proportion of the total power reaching one of the loads may be varied continuously between zero and unity, while at the same time no mismatch is introduced at the junction with the main wave guide run from the magnetron. Although it employs wave-guide series stubs, which we have not yet discussed, it is convenient to consider the device in this section. In fig. 5·26 (f), D and A are loads that match their respective branches—for instance, D may be a dummy load and A the output to a scanner (aerial). The pistons P_1 and P_2, although movable in the side arms, are linked, so that when P_1 is at a distance l from the opening into the wave guide, P_2 is at $(l + \frac{1}{4}\lambda)$ whatever value l may have. Since these side arms are series stubs they throw the following normalized reactances in series with the wave guide at the centres of the openings:

Series reactance introduced by stub 1 is

$$jx = j \tan 2\pi l/\lambda_g.$$

Series reactance introduced by stub 2 is

$$jx' = j \tan 2\pi(l + \frac{1}{4}\lambda_g)/\lambda_g$$
$$= -j \cot 2\pi l/\lambda_g = 1/jx.$$

The equivalent transmission line representation is given in figure (g). The resultant impedances of the sections at L and M are respectively $(1 + jx)$ and $(1 + 1/jx)$ which become transformed, through the $\frac{1}{4}\lambda_g$ section, to impedances $1/(1 + jx)$ and $jx/(1 + jx)$ at O. The total impedance terminating the main wave guide run at O is, therefore,

$$\frac{1}{(1 + jx)} + \frac{jx}{(1 + jx)} = 1.$$

Thus, no reflexion arises in the main wave guide at the junction O

whatever the position of the pistons. The fraction of the total power reaching load A is $1/(1+x^2)$, and load D is $x^2/(1+x^2)$, and the first of these fractions changes from unity to zero as the distance l is increased from zero to $\frac{1}{4}\lambda_g$.

As movable pistons are inconvenient they are replaced, in practice, by resonant rings R_1 and R_2, whose distances from the openings of their respective side arms again differ by $\frac{1}{4}\lambda_g$. We have seen that when the ring is in the resonant position it throws a short-circuit across the wave guide, but when turned through $90°$ its shunting effect vanishes. The variation of resultant impedance at the section of the wave guide containing the loop is similar to that produced by moving a piston from zero distance back to a distance $\frac{1}{4}\lambda_g$ from the section. Thus a synchronous rotation of the loops R_1 and R_2 simulates the original motion of the pistons P_1 and P_2 (fig. 5·26 (h)).

5·10. T-junctions

T-junctions, which are of great practical importance, are of two kinds—the shunt or H-plane junction and the series or E-plane junction. These are shown respectively in fig. 5·27 (a) and (b). The shunt T-junction comprises a wave guide with a side arm leading out from a narrow face. The break in the wall affects the flow of the transverse wall currents but does not interrupt the longitudinal currents in the broad face. As it does not introduce a discontinuity in the electric field of the H_{10}-modes in the main wave guide the type of scattering produced at the junction is symmetrical (§ 5·4). These properties of the junction justify the nomenclature 'shunt' for it. The transmission-line analogue is a straight portion of line with a branch line tapped to it in parallel as shown in fig. 5·27 (a).

The series T-junction and its transmission-line analogue are shown in fig. 5·27 (b).

In this junction, the longitudinal current in the broad face is interrupted and a discontinuity is thus introduced in the electric field of an H_{10}-wave incident on the junction; the scattering process is therefore unsymmetrical. The junction consequently behaves like a series element in a transmission line and is therefore called a series junction.

It might be thought that the simple transmission-line representations shown in fig. 5·27 would suffice to describe the behaviour of each of these junctions, but unfortunately these representations are inadequate and the actual behaviour is more complicated. This complexity is to be attributed to the fact that the junctions are major discontinuities in the wave guide at which, in addition to the scattered and reflected components of the progressive H_{10}-wave,

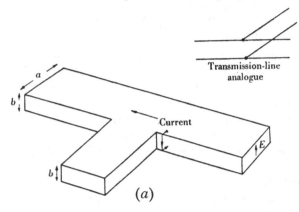

Shunt or H-plane T-junction

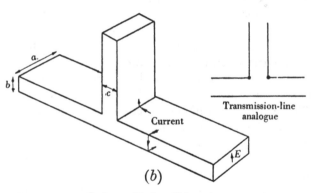

Series or E-plane T-junction

Fig. 5·27.

evanescent modes are also strongly excited. The storage field of these modes behave as equivalent series and parallel reactive elements which must be taken into account in an equivalent circuit that represents the junction.

The electromagnetic theory of the series junction was first given by Frank and Chu (M.I.T.). Because of the somewhat more extensive use that has been made of the series compared with the shunt junction and of the fact that the results of the theory can be presented in the form of a convenient equivalent circuit, we shall, in what follows, consider the series T-junction first. A more exact theory of both junctions has also been given by Allanson, Cooper and Cowling, but we shall defer discussion of their treatment.

The case studied by Frank and Chu is that of the general series T-junction in which the broad dimensions a of the main guide and branch are the same, but the narrow dimensions b and c may be different as indicated in fig. 5·27 (b). We suppose an H_{10}-wave to be excited in any one of the limbs at a point where the storage field of the junction is negligible, and that the other two limbs are terminated in arbitrary loads. In general, therefore, partial, or complete, standing waves exist in each arm.

Suppose the distribution of the electric field of the H_{10}-mode in each limb to be found over regions free from the effects of the storage fields of the junction or of the terminations by means of a standing wave indicator. From these measurements the distribution of normalized admittance y along each limb can be determined and extrapolated into the regions where the storage field is important. The theory of Frank and Chu predicts the distribution of these admittances y in each limb whatever the nature of the terminations. Fig. 5·28 (a) represents a central section of the T-junction. Let the admittance y in the respective limbs be extrapolated to the sections OC and AOB in the limbs of the main guide and of the side limb respectively. Replace the T-junction by an equivalent transmission-line system (fig. 5·28 (b)) whose branches are connected at the terminals 1, 2, 3 and 4 and whose characteristic admittances are each equal to unity, and suppose the distribution of normalized admittance y along each of the three branches to be the same as that in the corresponding limbs of the wave-guide system. Thus, the extrapolated admittances at the terminals are the same as those at the sections OC and OAB (fig. 5·28 (a)).

We call these admittances, presented at the respective pairs of terminals, y_{12}, y_{23} and y_{14}. The algebraic sign of the admittances y_{23} and y_{14} are determined by the convention (indicated in fig. 5·28 (b)) that in the formulae given below, y_{23} and y_{14} carry a positive sign when the terminals are viewed from the left, but a negative sign when viewed from the right.

According to Frank and Chu, the three admittances at the terminals are related by the following formulae:

$$y_{12} = \frac{1}{f_3}\left[-jf_1 - \frac{(y_{23}-y_{14})-2jf_2 y_{23} y_{14}}{2+jf_2(y_{23}-y_{14})}\right], \tag{1}$$

$$y_{14}^{23} = \frac{[1+f_2(-f_1+jf_3 y_{12})]y_{23}^{14} \mp 2(jf_1+f_3 y_{12})}{1+f_2(-f_1 \mp jf_3 y_{12}+j2y_{23}^{14})}. \tag{2}$$

These formulae are equivalent, and permit any one of the quantities to be expressed in terms of the other two. The quantities

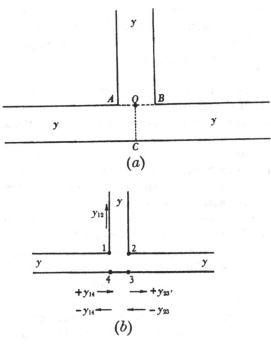

Fig. 5·28.

f_1, f_2 and f_3 are always positive and are functions of b/λ_g and c/b only, where λ_g is the smaller of the H_{10} wave-lengths in either limb.

In formula (2) y_{14}^{23} means either, y_{23} or y_{14} and y_{23}^{14} indicates the reverse order of choice. If y_{23} appears on the left then the negative sign is used of the alternatives $\mp$ on the right, but with y_{14} on the left then the positive sign in the formula is on the right.

In most T-junctions the two limbs are pieces of identical wave guide and therefore $b = c$. Frank and Chu have given curves expressing f_1, f_2 and f_3 as functions of b/λ_g for *the case $b = c$*, and these

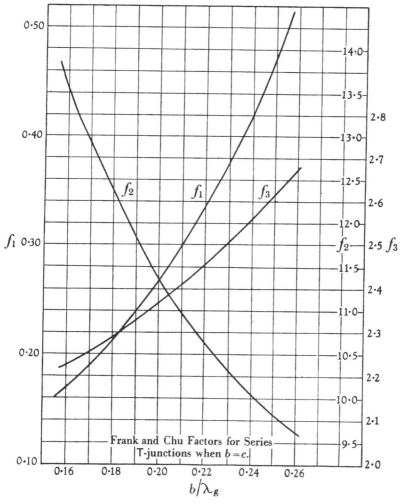

Fig. 5·29.

curves are reproduced in fig. 5·29. The limiting values attained by f_1, f_2 and f_3 as $b/\lambda_g \to 0$ are

$$f_1 \to 0, \quad f_2 \to \infty, \quad f_3 \to 2.$$

To illustrate the use of the formulae we consider some specific examples:

Power is fed in along the left-hand limb and the right-hand limb is perfectly matched, but the side limb is closed by a short-circuiting piston, that is,

$$y_{23} = +1, \quad y_{12} = -j \cot 2\pi l/\lambda_g,$$

where l is the distance of the piston from AOB (fig. 5·28 (a)) and λ_g refers to that limb. This gives from (2)

$$y_{14} = 1 + 2j \frac{f_1 - f_2 - f_3 \cot 2\pi l/\lambda_g}{1 + f_2(-f_1 + f_3 \cot 2\pi l/\lambda_g) + 2jf_2}. \tag{3}$$

In the limit $b/\lambda_g \to 0, f_1 \to 0, f_2 \to \infty, f_3 \to 2$, formula (3) gives

$$y_{14} = \frac{1}{(1 + j \tan 2\pi l/\lambda_g)} = \frac{1}{z_{14}},$$

which is what would be expected from the simple transmission-line analogue of fig. 5·27 (b) where the storage field becomes relatively unimportant.

The left-hand limb is matched at the junction when $y_{14} = +1$. From (3) this requires

$$\cot 2\pi l/\lambda_g = (f_1 - f_2)/f_3, \tag{4}$$

whereas the condition $\cot 2\pi l/\lambda_g = \infty$ $(l = \frac{1}{2}\lambda_g$ or $0)$, for the idealized case, neglects the storage field at the junction. Thus, the position of the piston for transmission past the junction without reflexion from the left-hand limb to a matched load at the end of the right-hand limb is no longer at a distance $\frac{1}{2}\lambda_g$ or zero but at

$$l = \frac{\lambda_g}{2\pi} \cot^{-1}\left(\frac{f_1 - f_2}{f_3}\right).$$

Suppose $b/\lambda_g = 0·19$, which is its value for a 9·1 cm. wave in a $2\frac{1}{2} \times 1$ in. wave guide. From the curves of fig. 5·29 we find

$$f_1 = 0·24, \quad f_2 = 11·9, \quad f_3 = 2·325,$$

whence $\cot 2\pi l/\lambda_g = -4·96$

or $\dfrac{2\pi l}{\lambda_g} \times \dfrac{360}{2\pi} = 168·6°, \quad \dfrac{l}{\lambda_g} = 0·47,$

where the simple theory requires $l/\lambda_g = 0·5$.

To obtain the position of the plunger in order that no power shall reach the load in the right-hand limb we require complete reflexion at terminals 1—4. The condition for complete reflexion is that y_{14} shall be a pure susceptance. It can be shown from equation (3) that the position of the plunger must satisfy

$$\cot \frac{2\pi l}{\lambda_g} = \frac{(1 + f_1 f_3)}{f_2 f_3} \tag{5}$$

and

$$y_{14} = \frac{1}{j f_2}. \tag{6}$$

With the same values of f_1, f_2 and f_3 we find

$$\cot \frac{2\pi l}{\lambda_g} = 0 \cdot 138,$$

$$\frac{2\pi l}{\lambda_g} \times \frac{360}{2\pi} = 82 \cdot 2° \quad \text{or} \quad (82 \cdot 2 + 180),$$

$$\frac{l}{\lambda_g} = 0 \cdot 228 \quad \text{or} \quad 0 \cdot 728$$

and from (6) $y_{14} = -\dfrac{j}{11 \cdot 9} = -0 \cdot 084 j.$

Thus, the position of the piston up the side limb is not $l = \frac{1}{4}\lambda_g$ or $\frac{3}{4}\lambda_g$ as the simple theory suggests but at somewhat smaller distance.

This example serves to illustrate the application of the general formula to a specific problem.

We proceed to consider a very convenient equivalent circuit for the T-junction proposed by N. Elson and based on formula (1) of Frank and Chu.

This formula may be rewritten

$$f_3 y_{12} + j f_1 = \frac{2 y_{23} y_{14} - (y_{23} - y_{14})/j f_2}{2/j f_2 + (y_{23} - y_{14})}.$$

Add $1/f_2$ to each side and divide throughout by 2. Then, after reduction,

$$\frac{f_3 y_{12}}{2} + j \left(\frac{1 + f_1 f_2}{2 f_2} \right) = \frac{-1}{\dfrac{1}{(1/j f_2 + y_{23})} + \dfrac{1}{(1/j f_2 - y_{14})}},$$

whence

$$\left[\frac{f_3 y_{12}}{2} + j \left(\frac{1 + f_1 f_2}{2 f_2} \right) \right]^{-1} + \left[y_{23} + \frac{1}{j f_2} \right]^{-1} + \left[-y_{14} + \frac{1}{j f_2} \right]^{-1} = 0. \tag{7}$$

Equation (7) is Elson's transformation and is represented by the equivalent circuits shown in fig. 5·30 (a) and (c).

We note that the admittance y_{12} now couples in series with the transmission line through a transformer which increases its value

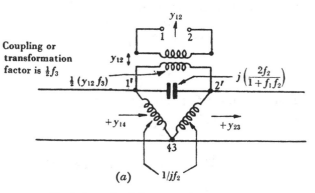

Coupling or transformation factor is $\frac{1}{2}f_3$

Equivalent circuit in terms of admittances

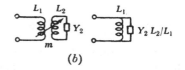

(b)

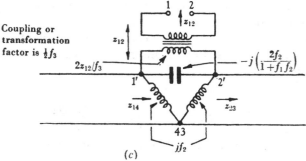

Coupling or transformation factor is $\frac{1}{2}f_3$

Equivalent circuit in terms of impedances

Fig. 5·30.

by the factor $\frac{1}{2}f_3$. It is thrown in shunt with a capacitive susceptance $j\left(\dfrac{1+f_1f_2}{f_3}\right)$. In addition, the transmission line is shunted by a pair of inductive susceptances $1/jf_2$ as shown.

Fig. 5·30(b) shows how an admittance Y_2 is transformed by a transformer to produce an equivalent circuit comprising the admittance $Y_2 L_2/L_1$ in shunt with the primary inductance.

Fig. 5·30(c) is the same as fig. 5·30(a) but expressed in terms of admittances.

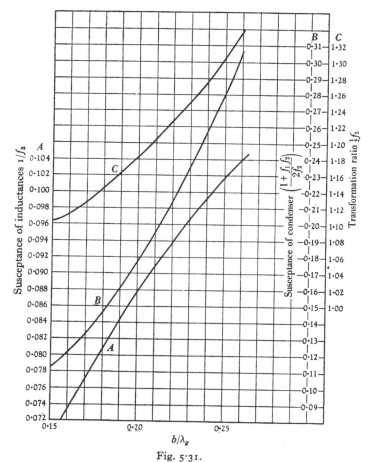

Fig. 5·31.

The behaviour of the T-junction is readily understood by reference to the equivalent circuit. Fig. 5·31 gives curves for the quantities $1/f_2$, $\frac{1}{2}f_3$, $\left(\dfrac{1+f_1 f_2}{f_3}\right)$ that appear in the circuit as functions of b/λ_g in a T-junction for which $b = c$.

We consider some applications of the equivalent circuit:

(a) Suppose the side limb of the T-junction to be closed by a sliding piston. If the piston is adjusted so that $y_{12} = j\left(\dfrac{f_2-f_1}{f_3}\right)$ then $y_{14} = 1$ when $y_{23} = 1$ and no reflexion occurs in the main wave guide. This result agrees with (4). If the piston is adjusted so that $y_{12} = -j\left(\dfrac{1+f_1f_2}{f_2f_3}\right)$, then the capacitive susceptance is neutralized and an open circuit is placed across terminals $1'$—$2'$. It follows that $y_{14} = 1/jf_2$ whatever may be the value of y_{23}. These conclusions agree with (5) and (6).

(b) If $y_{23} = -(1/jf_2)$, which requires a reflecting piston in the right-hand limb, then $y_{14} = 1/jf_2$ whatever the value of y_{12}. Similarly the admittance $(-y_{12})$ (looking into the system) at terminals 1—2 is $j\left(\dfrac{1+f_1f_2}{f_2f_3}\right)$ for all values of $-y_{14}$.

(c) It is impossible for y_{14} to equal unity if y_{23} is a pure reactance.

The T-junction thus affords a good example of how it is possible to discuss the properties of a complex electromagnetic field in terms of an equivalent circuit.

The values of the circuit elements in the equivalent circuits of fig. 5·30 (a) may be obtained from the curves in fig. 5·31 when b/λ_g is known for the case $b = c$. When, however, the side limb and the main wave guide have different dimensions b and c, it would be necessary to return to the basic theory of Frank and Chu and to calculate f_1, f_2 and f_3 in terms of b/λ_g and b/c.

Alternatively, the circuit elements can be found experimentally. The quantities required are:

The capacitive series susceptance

$$jb_s = j\left(\frac{1+f_1f_2}{2f_2}\right),$$

the inductive parallel susceptances

$$jb_p = 1/jf_2,$$

and the transformation or coupling factor of the transformer

$$n = \tfrac{1}{2}f_3.$$

Two *experimental methods* have been used by E. B. Callick to find these quantities.

In the one, the right-hand limb of the main wave guide is terminated in a reflexionless load and the side limb in a movable piston. Measurements of y_{14} are made for a range of positions of the piston. In the second method, the reflexionless load remains in

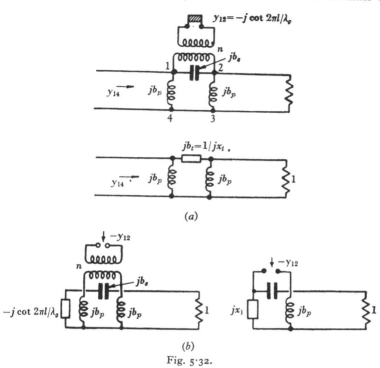

Fig. 5·32.

position but the piston is transferred to the right-hand limb of the main wave guide and the input admittance is measured in the side limb. The equivalent circuits for the two methods are shown in fig. 5·32 (a) and (b) respectively.

First method. From the circuit it may be deduced that

$$y_{14} = jb_p + \cfrac{1}{\cfrac{1}{1+b_p^2}+j\left(x_t-\cfrac{b_p}{1+b_p^2}\right)} = g_{14}+jb_{14}, \qquad (8)$$

in which the total series impedance x_t and susceptance b_t are

$$\frac{1}{jx_t} = jb_t = j(ny_{12}+b_s) = j\left(b_s-n\cot\frac{2\pi l}{\lambda_g}\right). \qquad (9)$$

From (8),

$$g_{14} = \left(\frac{1}{1+b_p^2}\right)\bigg/\left[\left(\frac{1}{1+b_p^2}\right)^2 + \left(x_t - \frac{b_p}{(1+b_p)^2}\right)^2\right], \quad \left.\begin{array}{c} \\ \\ \\ \\ \end{array}\right\}$$

$$b_{14} - b_p = -\left(x_t - \frac{b_p}{(1+b_p^2)}\right)\bigg/\left[\left(\frac{1}{1+b_p^2}\right)^2 + \left(x_t - \frac{b_p}{1+b_p^2}\right)^2\right]. \quad (10)$$

On eliminating x_t from (10) the following equation is obtained:

$$(b_{14} - b_p)^2 + \left(g_{14} - \frac{1+b_p^2}{2}\right)^2 = \left(\frac{1+b_p}{2}\right)^2. \qquad (11)$$

If, therefore, the representative points of y_{14} are plotted in the complex plane (admittance diagram) for a range of positions of the stub, these points, according to (11), all lie on a circle whose centre lies at $\{\frac{1}{2}(1+b_p^2), b_p\}$ and whose diameter is $(1+b_p)^2$. The circle therefore passes through the origin.

The parallel susceptance b_p can therefore be obtained by finding the centre and diameter of the circle through the experimental points.

To find n and b_s, we note from the curves in fig. 5·31 that b_p is small compared with b_s and unity, and that in equations (10) it is justifiable to neglect b_p^2 in comparison with unity. When this is done equations (10) become

$$g_{14} = \frac{1}{1+(x_t-b_p)^2}, \quad \left.\begin{array}{c} \\ \\ \\ \end{array}\right\}$$

$$(b_{14} - b_p) = -\frac{(x_t-b_p)}{1+(x_t-b_p)^2}. \qquad (12)$$

From the first of equations (12) and from equation (9) we obtain

$$\left(n\cot\frac{2\pi l}{\lambda_g} - b_s\right) = 1\bigg/\left\{b_p \pm \sqrt{\left(\frac{1}{g_{14}}-1\right)}\right\}. \qquad (13)$$

When g_{14} is measured for two values of l two equations are obtained from which n and b_s may be obtained, since b_p has already been determined. It is convenient to choose for the two values of l those that make $\cot 2\pi l/\lambda_g$ equal to $+1$ and -1 respectively, that is, $l = \frac{1}{8}\lambda_g$ and $\frac{3}{8}\lambda_g$. The second of equations (12) may similarly be used to provide n and b_s.

Second method. From fig. 5·32 (b) we deduce

$$ny_{12} = jb_s + \frac{1}{\dfrac{1}{(1+b_p^2)} + j\left(x_1 - \dfrac{b_p}{1+b_p^2}\right)} = g_{12}' + jb_{12}' = n(g_{12}+jb_{12}), \quad (14)$$

where $\qquad jx_1 = 1/j(b_p - \cot 2\pi l/\lambda_g)$

whence
$$g'_{12} = \frac{1}{(1+b_p^2)}\bigg/\left[\left(\frac{1}{1+b_p^2}\right)^2 + \left(x_1 - \frac{b_p}{1+b_p^2}\right)^2\right],$$

$$b'_{12} - b_s = -\left(x_1 - \frac{b_p}{1+b_p^2}\right)\bigg/\left[\left(\frac{1}{1+b_p^2}\right)^2 + \left(x_1 - \frac{b_p}{1+b_p^2}\right)^2\right]. \tag{15}$$

These again, by elimination of x_1, provide a circular locus on the admittance diagram for the representative points of y_{12}.

The equation of the circle is

$$\left(b_{12} - \frac{b_p}{n}\right)^2 + \left(g_{12} - \frac{1+b_p^2}{2n}\right)^2 = \left(\frac{1+b_p^2}{2n}\right)^2. \tag{16}$$

Thus, if n has been determined by the first method, b_p may be found from the coordinates of the centre and from the diameter of this circle when the experimental points are plotted.

If the reflexionless termination is replaced by a piston at $\frac{1}{2}\lambda_g$ from the axis of symmetry, it throws a short circuit across the right-hand inductive susceptance $1/jf_2$.

It follows that

$$y_{14} = ny_{12} + jb_s + jb_p = -jn\cot 2\pi l/\lambda_g + j(b_s + b_p).$$

If l is chosen to make $\cot 2\pi l/\lambda_g$ in turn equal to 0, $+1$, -1, then n and $(b_s + b_p)$ can be determined. If b_s is assumed from method 1 then b_p is known.

T-*junction with unit coupling factor* $n = 1$. In order to couple the side limb strongly into the main wave guide the T-junction should be designed to place the maximum series impedance z_{12}/n across the terminals $1'$—$2'$ (fig. 5·30 (c)), where $n = \frac{1}{2}f_3$.

Since the smallest value of $\frac{1}{2}f_3$ is unity, the greatest coupling is obtained when $n = 1$. This cannot be achieved in a T-junction in which the narrow dimensions b and c (fig. 5·27 (b)) are equal. However, if b and c are unequal it is possible to design a series T-junction for which $n = 1$. Fig. 5·33 is a curve due to Macfarlane, based on the theory of Frank and Chu, from which the correct values of b/λ_g and c/λ_g for unit coupling factor can be read. In Callick's experiments a T-junction for use at $\lambda = 9\cdot1$ cm. was constructed with the main wave guide of standard $1 \times 2\frac{1}{2}$ in. cross-section ($b = 2\cdot54$) and with the dimension c of the side limb $1\frac{5}{16}$ in. as obtained from fig. 5·33, with $\lambda_g = 13\cdot2$ cm. (wave-length of

H_{10}-wave in the $2\frac{1}{2} \times 1$ in. wave guide). Measurements on this T-junction gave:

$$b_p \qquad -0\cdot11$$
$$b_s \qquad +0\cdot18$$
$$n \qquad 1\pm0\cdot02$$

Compensated T-*junction*. The equivalent circuit of fig. 5·30 is valuable in that it suggests how, by the use of compensating reactive

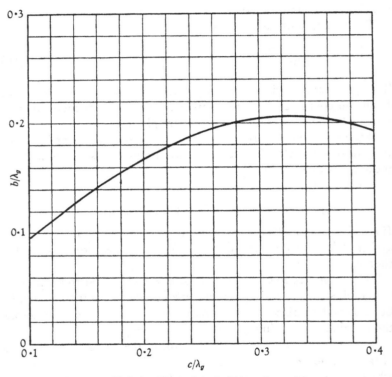

Fig. 5·33. Relative dimensions in T-junctions with unit transformation ratio.

irises, it is possible to neutralize the storage field at the junction, that is, to compensate for the reactances b_s and b_p. Fig. 5·34 shows how this may be done.

It is not possible, however, to predict the dimensions of the irises exactly from the formulae given in §5·5, because these formulae refer to the behaviour of the iris in the field of an incident H_{10}-wave.

If an iris is placed in a region where a storage field already exists, its apparent susceptance will not be the same. The capacitive irises in the main wave guide in fig. $5\cdot34\,(b)$ are not used in practice

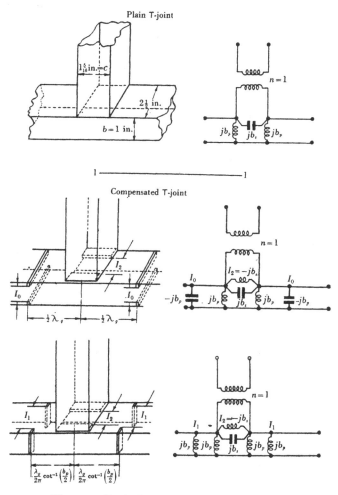

Fig. $5\cdot34$. Equivalent circuits of T-junctions in rectangular wave guide.

because of electrical breakdown at high powers, and the inductive irises of fig. $4\cdot34\,(c)$ are preferred. These are placed at a distance l from the central section such that $\cot 2\pi l/\lambda_g = \tfrac{1}{2}b_p$. It can be seen from the transmission-line equation that at this distance an inductive iris whose susceptance is $\frac{1}{2}jb_p$ will cancel the junction susceptance $\frac{1}{2}jb_p$.

A compensated junction behaves like the ideal series junction in a transmission line, shown in fig. 5·27 (b). For instance, when $y_{23} = \infty$ and $y_{12} = 1$ then $y_{14} = 1$ and there is no reflexion at the junction, the whole of the power passing to the load in the side arm. Fully compensated junctions are used in the switching device shown in fig. 5·26 (h).

5·11. Shunt or H-plane junctions

In this section we shall consider briefly the treatment of T-junctions given by J. T. Allanson, R. Cooper, and T. G. Cowling,* in which both series and shunt T-junctions and Y-junctions are investigated.

As already mentioned, the purpose of a theory of a three-limb junction is to permit the standing wave pattern and the distribution of normalized admittance y with respect to the dominant mode to be calculated when the distributions of the admittances y in the other two limbs are known. For this purpose it suffices to derive a formula relating the admittance at a fixed chosen section in the one limb with the admittances at similar fixed positions in the other two limbs. In the investigations of Frank and Chu the reference positions coincide at the junction (fig. 5·28) and the admittances y_{12}, y_{23} and y_{14} (with respect to H_{10}-waves) are extrapolated back to these sections.

In the treatment due to Allanson, Cooper and Cowling the admittances are taken at positions which they call characteristic points, which are not, in general, located at the junction of the limbs. To appreciate the significance of the characteristic points we refer to fig. 5·35 which shows a series and a shunt T-junction with the characteristic points (more accurately, characteristic sections) as dotted planes P_1, P_2 and P_3, whose positions relative to the junction depend on the dimensions of the wave guides and the type of junction.

We consider first the properties of these planes P_1, P_2 and P_3. Let limb 2, in either type of junction, be terminated in a perfectly reflecting piston, and suppose an H_{10}-wave to be propagated towards

---

* 'The Theory and Experimental Behaviour of Right-Angled Junctions in Rectangular-Section Wave Guides', *J. Inst. Elect. Engrs*, vol. 93, part III, no. 23, May 1946.

the junction along limb 1. When the piston is set with its reflecting face in the plane P_2 it is found that the wave in limb 1 is completely reflected at the junction and no power proceeds into limb 3.

Further, a complete standing wave is produced in limb 1 with nodal planes at spacings of $\frac{1}{2}\lambda_g$, over which the electric field of the H_{10}-mode vanishes. Let P_1 be one of these nodal planes; then P_1

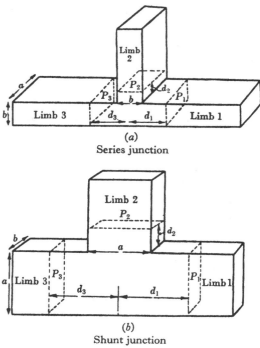

(a)

Series junction

(b)

Shunt junction

Fig. 5·35. Characteristic points.

is also a characteristic plane of limb 1, for according to theory (and confirmed in practice) if the piston is transferred from limb 2 to limb 1, when power is fed into limb 2, complete reflexion occurs at the junction provided the face of the piston lies at the plane P_1 (or any other of the nodal planes of limb 1). Moreover, the plane P_2 is one of the nodal planes of limb 2.

Next, suppose the piston to be placed at P_2 and that power is fed into limb 3. Complete reflexion again occurs and the nodal planes in limb 3 are also its characteristic planes. Thus, to generalize, if power is fed into one limb and a reflecting piston is placed at a characteristic plane of a second limb, no power proceeds down the third limb.

In formulating the results of the theory it is convenient to measure distance along the respective limbs, from a characteristic section as origin (usually the section in each limb nearest to the junction). When the limbs are given arbitrary terminations partial H_{10} standing waves and a distribution of admittance y will exist, in general, in each limb. The values of the admittances in each limb extrapolated back to the characteristic sections, according to Allanson, Cooper and Cowling, are always related as follows:

$$a_1^2 y_1 + a_2^2 y_2 + a_3^2 y_3 = -jK, \qquad (1)$$

where y_1, y_2 and y_3 are the admittances at the characteristic sections in the respective limbs, and a_1, a_2, a_3 and K are real constants. The coefficients of the admittances are therefore always positive.

The convention concerning the signs of the admittances is that in formula (1) the admittances are to be prefixed with a positive sign when the observer is looking towards the junction, but with a negative sign if away from the junction.

The theory provides the values of the constants in equation (1). We consider some applications.

Let limb 2 contain a reflecting piston, and suppose limb 3 to be terminated in a reflexionless load, and denote distances measured along the limbs from the characteristic sections by l_1, l_2 and l_3 respectively. When, therefore, the piston is at distance l_2 from P_2, the admittance y_2, looking away from the junction, is

$$(-y_2) = -j \cot 2\pi l_2 / \lambda_g,$$

and, since limb 3 is matched,

$$y_3 = -1.$$

The admittance Y_1 (looking towards the junction) is, from (1),

$$a_1^2 y_1 + a_2^2 j \cot 2\pi l_2 / \lambda_g - a_3^2 = -jK, \qquad (2)$$

or
$$y_1 = \frac{a_3^2}{a_1^2} - j\left(\frac{a_2^2 \cot 2\pi l_2 / \lambda_g + K}{a_1^2}\right).$$

On introducing new constants g_1, A and B this expression for y_1 becomes
$$y_1 = g_1 - j[A \cot 2\pi l_2 / \lambda_g + B], \qquad (3)$$

where, $g_1 = a_3^2/a_1^2$ (and is essentially positive), $A = a_2^2/a_1^2$ and $B = K/a_1^2$. Thus y_1 is of the general form $y_1 = g_1 + jb_1$.

In the special junctions shown in fig. 5·35, where limbs 1 and 3 are symmetrical with respect to limb 2, the constants a_1 and a_3 in (1) are equal and g_1 in (3) is equal to unity.

In this case the stub behaves as a variable susceptance

$$- [A \cot 2\pi l_2/\lambda_g + b]$$

placed in shunt across the characteristic section of limb 1. We have therefore

$$y_1 = 1 - j[A \cot 2\pi l_2/\lambda_g + B]. \qquad (4)$$

The condition for matching at the junction is that $y_1 = 1$, and this is achieved by adjusting the piston to the distance l_2 from P_2 such that

$$\cot 2\pi l_2/\lambda_g = -B/A. \qquad (5)$$

The condition for complete reflexion at the junction is $y_1 = \infty$, and from (4) the piston must be placed at one of the distances l_2 from P_2, $l_2 = 0$, $\frac{1}{2}\lambda_g$, λ_g, etc.

The values of B and A as functions of b/λ_g or a/λ_g have been calculated by Allanson, Cooper and Cowling for both forms of T-junction, and their results are exhibited in figs. 5·36 (a) and (b).

Let $d_1 = d_3$ and d_2 be the distance of the nearest characteristic points in limbs 1 and 2 from the central sections of the junction and the plane of entry respectively (fig. 5·35). The distance of the face of the reflecting piston in limb 2 from the plane of entry when $y_1 = 1$ (no reflexion at junction) we call L_2, that is, from (5),

$$\cot [2\pi(L_2 - d_2)/\lambda_g] = -B/A.$$

The quantities d_1/λ_g, d_2/λ_g and L_2/λ_g are exhibited as functions of b/λ_g and a/λ_g for the series and shunt junctions respectively by the curves in fig. 5·37 (a) and (b).

For example, in a series junction made from $2\frac{1}{2} \times 1$ in. wave guide with $\lambda_g = 13\cdot2$ cm., $b/\lambda_g = 1\cdot9$. From fig. 5·37 (a) we find $L_2 = 0\cdot47$, which agrees with the result derived from 5·10 (4).

When the piston is placed in limb 3 and the matched load in limb 2, the equation (1) for y_1 becomes

$$a_1^2 y_1 + a_2^2(-1) + a_3^2[-(-j \cot 2\pi l_3/\lambda_g)] = -jK,$$

or

$$y_1 = \frac{a_2^2}{a_1^2} - j\left[\frac{a_3^2 \cot 2\pi l_3/\lambda_g + K}{a_1^2}\right]$$

$$= A - j[\cot 2\pi l_3/\lambda_g + B], \qquad (6)$$

where A and B have their previous meanings and $a_3^2/a_1^2 = 1$ (symmetrical junction).

From fig. 5·36 (a) and (b) it is evident that A cannot be equal to unity in practice, and that therefore it is impossible to transfer power to the load in limb 2 from a source feeding into limb 1 without reflexion. On the other hand, complete reflexion occurs at the

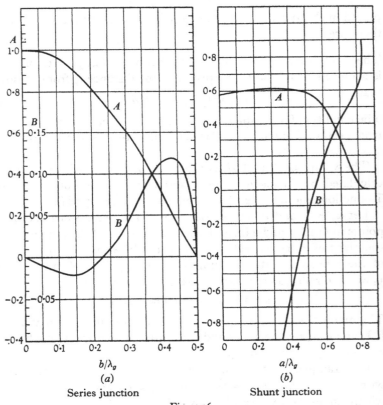

b/λ_g

(a)

Series junction

(b)

Shunt junction

Fig. 5·36.

junction when $l_3 = 0$, $\frac{1}{2}\lambda_g$, λ_g, etc., that is, when the piston lies at the characteristic sections of limb 3.

The theory has been tested experimentally by its authors over a wave-length (TEM) range $\lambda = 7·05$–$10·97$ cm., and good agreement was obtained between theory and experiment for the series junction, but less good for the shunt junction.

It has been very carefully tested at a wave-length of 3·2 cm. by W. J. Whitehouse, who obtained good experimental confirmation of the theory both for the series and the shunt junction and showed that it may be safely used in the design of compensated junctions at X-band wave-lengths.

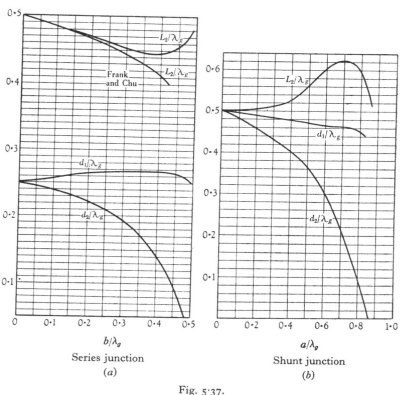

Series junction
(a)

Shunt junction
(b)

Fig. 5·37.

5·12. Junctions of circular with rectangular wave guides

It is necessary on occasions to employ junctions comprising a limb of circular wave guide associated with rectangular limbs. Fig. 5·38 illustrates types of combination that are met in practice.

Fig. 5·38 (a) is a series T-junction of the type used in British X-band T.R. systems. The purpose of the plate is both to increase the coupling of the side limb, and to match the junction. Fig. 5·38 (b) is a shunt connexion of a circular guide to a rectangular guide which is essentially a form of cut-off corner. The design of these devices is carried out empirically.

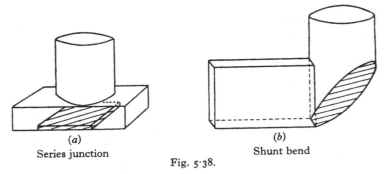

(a)
Series junction

(b)
Shunt bend

Fig. 5·38.

5·13. Applications of T-junctions

5·13·1. *Common aerial working—T.R. systems*

In most radar systems advantage is taken of the fact that the polar diagram of an aerial is the same in transmission as it is in reception, to employ a single aerial for both purposes. To do so, however, demands the use of special ancillary automatic switching and protective devices to protect the receiver during transmission and to isolate the transmitter and to reconnect the receiver during reception. Such systems are known as T.R. (transmit-receive) systems and they were first developed by British scientists in 1940 for use with the 1½ m. C.H.L. and (later) G.C.I. equipments.

T.R. systems were subsequently developed for use at microwavelengths (also first in Britain) both with coaxial feeders and with wave guides. The advantages of employing T.R. systems in radar equipments are outstanding; for instance, in ground equipments operating at a wave-length of 1½ m. or less the use of a single aerial makes possible the use of a P.P.I. display in cases where the aerials are too cumbersome for two of them to be mounted on a single axis. In airborne microwave equipments T.R. arrangements are a necessity; first, the effective range is doubled at least, because it is possible to fit into the available space a scanner of at least twice the aperture of two separate scanners; and secondly, the mechanical difficulties of synchronizing the scanning of two separate aerials would be considerable, if not insuperable. They also permit P.P.I. and similar forms of display to be employed.

The T.R. devices must be so devised that in transmission the aerial feeder system presents a matched load to the transmitter while at the same time the receiver is protected, from paralysis or damage by power from the transmitter. Further, during the interval between emission of the transmitted pulse and the return of the echo (in some instances only a microsecond), the system must switch so as to isolate the transmitter from the aerial and to provide a matched path from the aerial to the receiver.

It is not possible to describe in detail the great number of individual T.R. systems, and it will suffice to give an account of the basic principles common to them all. Since these principles are most easily appreciated in T.R. systems employing transmission lines we shall first consider briefly the T.R. system employed in C.H.L. equipment at a wave-length of $1\frac{1}{2}$ m. as it is probably the simplest of them. It is illustrated in fig. 5·39 (a). (It is assumed that the reader is familiar with the elements of the theory of transmission lines.) *ABCD*, the main feeder line from the transmitter, is an unscreened twin feeder line of 200 lb. copper wire, with a characteristic impedance of 350 ohms. The whole feeder-line system is strained in order to preserve the spacing between the conductors without the need of using a large number of spacers. At *B* a composite transformer system is placed in shunt across the main feeder. It comprises a half-wave transformer *DE* ($Z_0 = 350$ ohms), followed by a second half-wave transformer *EF* with a larger characteristic impedance (600 ohms). *EF* is bridged at its middle by a spark gap G_T. This double transformer and spark gap is called the transmitter gap unit (T.G.U.) or anti-T.R. unit, and a similar unit is connected to the main feeder at *C*, the receiver gap unit (R.G.U.) or T.R. unit.

The distance *BC* is an odd number of quarter wave-lengths and the R.G.U., unlike the T.G.U., is not short-circuited at its end, but feeds into a balance-to-balance transformer (trombone) which allows the 80-ohm coaxial cable to the receiver to be connected without mismatch to the twin feeder of the R.G.U. The spark gap is shown separately in fig. 5·39 (b).

The cycle of operations is the following. The transmitter emits a pulse and power is carried by a TEM-wave along *ABCD*. At *B* power flows into *BEF* and the spark gap strikes, placing a low resistance of a few ohms across the line. This low impedance is transformed to a very high impedance at E which is transferred unchanged to B as a large shunt impedance across the line. Thus, little power, just sufficient to maintain the discharge, is diverted into BEF. In precisely the same manner, the spark gap G_R strikes very rapidly before sufficient power can pass to the receiver to damage or paralyse it, and the shunt impedance at C also becomes large. Consequently, all but an inappreciable proportion of the power proceeds to the

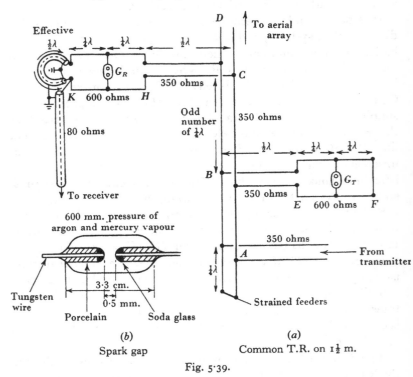

(b)

Spark gap

(a)

Common T.R. on $1\frac{1}{2}$ m.

Fig. 5·39.

aerial, which is matched to the main feeder. Further protection is given to the receiver by a relay which disconnects it in the event of failure of G_R. When the transmitter pulse has ceased, the spark gaps extinguish and the system rapidly reverts to the receiving condition. When, upon extinction, G_T again becomes a high impedance the short circuit at F is transferred to B as a short circuit and the input impedance of the line CB, in the direction CB, becomes very high. On the other hand, the extinction of G_R leaves a matched path from

C to the receiver, and the power received by the aerial when the echo returns traverses the path $DCHR$ and on to the receiver without reflexion at any of the junctions. The whole process is repeated some 500–600 times per second according to the pulse recurrence frequency in use. The same essential ingredients—anti-T.R. unit, T.R. unit and spark gaps—are to be found in almost all T.R. systems. We proceed to consider the principles of T.R. systems for microwave equipments, omitting, however, T.R. systems for coaxial S-band feeders and proceeding direct to the discussion of T.R. with wave guides.

Wave-guide T.R. systems. The demands on a T.R. system at microwave-lengths are far more severe than at $1\frac{1}{2}$ m. wave-lengths for the following reasons:

(*a*) The first stage in a microwave receiver is a crystal mixer (§ 4·5) whose crystal must be protected against 'burn out' by power from the transmitter pulse. Thus, although the power in the pulse proceeding down the main wave guide to the aerial may be of the order of 100 kW. or much greater, yet the power permitted to enter the crystal mixer, lodged in a side arm of the main guide, must not exceed $\frac{1}{10}$ W.

(*b*) In centimetre-wave airborne radar search equipments, called A.I. (aircraft interception) fitted to night fighters it is important to achieve the smallest possible minimum range in order that the night fighter may, by means of the A.I. equipments, approach within visual range of the bomber. This minimum range should be of the order of 500 ft. or less. To achieve a minimum range of this order it is necessary for the T.R. system to revert from the transmitting to the receiving condition within 1 μsec. after the start of the transmitter pulse.

One of the most straightforward T.R. systems for wave guides is that shown in fig. 5·40. It is used in the American X-band A.S.V. (aircraft to surface vessel) Equipment AN/APS 15 (or A.S.V. Mark 10—British nomenclature) which is also used as an H_2S equipment in bombers.

The unit comprises a straight run of American size (0.9×0.4 in.) X-band ($\lambda = 3.1$–3.3 cm.) wave guide (fig. 5·40 (*a*)) into which the magnetron feeds on the left, and to which side arms are attached by series T-junctions. The one side arm leads through a 1 B 24 protective cell (see § 5·9·1) to the crystal mixer and forms the T.R.

unit. The other side arm leads to a resonant cell, the 724A, which is essentially a resonant cylinder with coaxial tips projecting from its flat ends. These tips, and the central portion of the cavity, are enclosed in a glass envelope as indicated in fig. 5·40 (*b*) (not to scale), which also shows how the cell is coupled to the end of

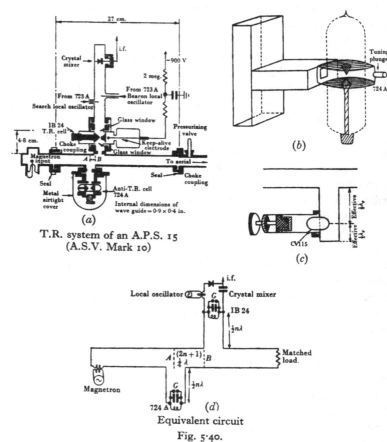

T.R. system of an A.P.S. 15
(A.S.V. Mark 10)

Equivalent circuit

Fig. 5·40.

the wave-guide side arm through a square window. The cavity is tuned to resonance by the tuning plunger shown in fig. 5·40 (*b*).

In practice the glass tube shown in fig. 5·40 (*b*) has an airtight metal block and covering built round it as shown in fig. 5·40 (*a*).

The resonant cavity is a rhumbatron of the type discussed in §6·8 (iii) and shown in fig. 6·10 (*a*) and (*b*). The equivalent representation by a transmission line and circuits, of the wave-guide system, is shown in fig. 5·40 (*d*). The resonant cells are represented by resonant *L-C* circuits, and their positions from the openings of their T-junctions are such that the effective lengths of the transmission-line series stubs are $\frac{1}{2}n\lambda$, where *n* is an integer, which the present example is unity. The pair of conical projections in each resonant cell are represented by spark gaps *G* in parallel with the tuned circuits.

We shall use the equivalent circuit to discuss the action of the T.R. system. Suppose the magnetron to emit its high-frequency pulse. The power which enters the side arms causes a high-frequency discharge in the gas between the conical projections, and the effect of this discharge is to reduce the *Q* of each cavity to a small value, that is, in the equivalent circuits the gaps *G* become bridged by a low resistance. Thus a low impedance is transferred to the junctions of the stubs with the transmission line. In consequence, the power proceeding down the stubs is limited to the small amount required to keep the discharge running, and practically the whole of the transmitter power proceeds to the aerial. By choosing the correct pressure for the gas in the 1 B 24 cell the power reaching the crystal mixer can be limited to be less than $\frac{1}{10}$ W. which causes 'burn out'. The gas employed is a mixture of hydrogen and water vapour at a pressure of 6 mm. of mercury. At the end of the transmitter pulse the discharges stop and the cells revert to high *Q* resonant circuits and are ready to receive the returning echo. The anti-T.R. cell presents a high series impedance at the end of its stub, and this is transferred by the $\frac{1}{2}\lambda$ length of line to *A* where it puts an open circuit in the main transmission line. This in turn is transformed by the $\frac{1}{4}\lambda$ length of line into a short circuit at *B*. On the other hand, the 1 B 24 is now transparent to the weak echo signal. The power received by the aerial therefore flows direct into the crystal mixer where it is combined with the local oscillator e.m.f. to produce the i.f. output e.m.f. which is abstracted and taken to the pre-amplifier.

The local oscillators are 723 A klystrons which can be tuned both mechanically and by variation of the voltage on the reflector electrode (§ 6·8 (iii)). The one oscillator is for normal use when the equipment functions as a search device, but the second klystron is employed instead when the equipment is interrogating a centimetre-wave beacon which replies on a different frequency from that of the pulse transmitted by the airborne radar set.

In early researches on T.R. systems it was found that the damage done to crystals was due principally to the power which leaked through the T.R. cell before the discharge had struck. To ensure early and certain striking of the discharge one of the conical projections is made hollow and with a hole in its tip. A wire, called the 'keep-alive' electrode, runs from through the end of the glass envelope almost to the tip of the hollow electrode. By applying a d.c. voltage of about 300 negative (dropped through current limiting resistors) between the end of the 'keep alive' and the interior of the hollow cone a d.c. discharge can be maintained, from which electrons diffuse into the space between the tips and so provide an initial ionization from which the h.f. discharge can rapidly build up. In this way the damaging 'spike' leakage power is kept within safe limits.

Another early difficulty was the persistence of electrons in the gap after the cessation of the discharge. By using water vapour, either alone or as an admixture, free electrons are rapidly converted to heavy ions by attachment to water molecules. In this way the absorption of the echo power in the T.R. box, near minimum range, is prevented and good minimum range is restored. The choice of gas pressure compromises between low striking voltage and rapid recovery time, the one requiring gas at low pressure and the other at high pressure. These various components of the system are named on fig. 5·40 (a). Fig. 5·40 (c) shows how the British CV 115 cell (§ 5·9·1) is used in an anti-T.R. unit. In this case a series junction is shown, although shunt arrangements are also used. Equivalent spacings are indicated. The piston in the circular side tube is adjusted to produce an antinode of electric field across the gap of the CV 115, which therefore breaks down under the transmitter power but behaves as an open circuit in the wave-guide wall during reception.

The action of the unit as a whole should now be self-evident, and the explanation is left to the reader. Anti-T.R. cells do not usually contain water vapour since rapid recovery is not of prime importance. The British equivalents of the 1 B 24 are the CV 221 and CV 114 (obsolete), with which they are identical in principle but differ in constructional details, mainly in being designed to fit into a side limb made of circular wave guide.

We conclude this sketch of T.R. systems with a brief account of systems employing shunt instead of series T-junctions. These in general are used at S-band wave-lengths, but recent X-band systems have been designed which employ one series and one shunt junction in order to economize in space.

At S-band wave-lengths ($\lambda = 9$–11 cm.) the T.R. cell is essentially a Sutton tube (reflector klystron) from which the electron gun assembly has been omitted and the reflector replaced by a 'keep-

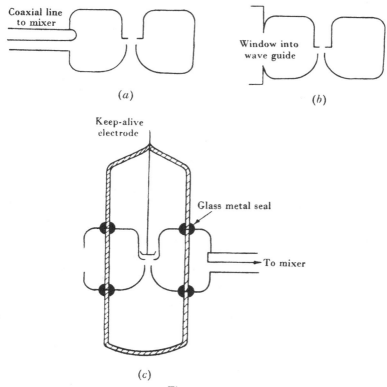

Coaxial line to mixer

Window into wave guide

(a)

(b)

Keep-alive electrode

Glass metal seal

To mixer

(c)

Fig. 5·41.

alive' electrode. The glass envelope contains water vapour at a pressure of 6 mm. of mercury. A typical example is the British CV 43, which operates at a wave-length of 9 cm. but can be tuned externally by means of screw plungers. The resonant cavity (rhumbatron) is of the type shown in fig. 6·10 (a) and (b), and the general construction of the cell is indicated in fig. 5·41. Since these cells contain gas they are also known as 'soft rhumbatrons'.

Fig. 5·41 (a) represents the rhumbatron together with the coupling loop and coaxial output of the crystal mixer (fig. 4·8), whereas fig. 5·41 (b) indicates the position of the rectangular window through which the field within the rhumbatron is coupled to that in the wave guide. The whole assembly with glass envelope and 'keep-alive' electrode is given in fig. 5·41 (c). This form of T.R. cell is usually directly coupled to the narrow face of the wave guide without the intervention of a shunt side arm, and the size of window to give best protection and a large coupling factor is previously determined by experiment. A pair of such cells used in a T.R. system is shown in fig. 5·42 (a). The T.R. and anti-T.R. cells are identical except for the absence of an output loop and coaxial in the latter, the output orifice of the rhumbatron being closed with a plunger.

The equivalent circuit is shown in fig. 5·42 (b). It should be noted that the rhumbatron is represented by a series L-C circuit which offers no impedance to the passage of a current along a transmission line at resonance. When, however, the spark gap G breaks down, the impedance at the capacity becomes small and resonance is destroyed. The full impedance of the inductance remains uncompensated and the power that proceeds down the branch is small. By placing the windows at a separation of an odd number of quarter wave-lengths $[(2n+1)\frac{1}{4}\lambda_g]$ between their centres as measured along the centre line of the broad face, a satisfactory T.R. system is achieved. The reader may easily follow the action of the system from what has been described previously and from fig. 5·42 (b).

Fig. 5·42 (c) is an alternative circuit in which the resonant systems are represented as tuned rejector circuits in shunt with the side arms, but connected at $\frac{1}{4}\lambda$ from the main transmission line. As already mentioned, composite T.R. systems have been designed for X-band operation, in which one of the side arms makes a series, and the other a shunt connexion with the main wave guide. X-band T.R. systems incorporate a system of automatic frequency control (A.F.C.) by using as local oscillators the American SCR 723 A klystrons which can be tuned, over a limited range, by changing the d.c. potential of the reflector (§ 6·8 (iii), last paragraph).

The incorporation of A.F.C. is required to keep the receiver tuned to the magnetron frequency although the latter should wander. The A.F.C. is usually achieved by leading a fraction of the i.f.

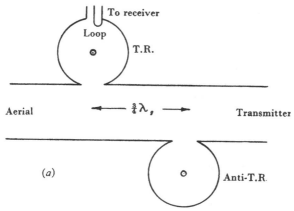

Wave-guide T.R. with shunt coupling

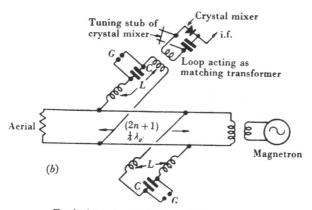

Equivalent circuit of shunt T.R. system

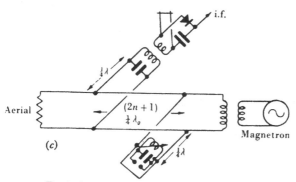

Equivalent circuit of shunt T.R. system

Fig. 5·42.

output of the pre-amplifier through a frequency discriminator which provides a d.c. potential of positive or negative sign according as the local oscillator frequency differs from the magnetron frequency by more or by less than the correct i.f. frequency. This d.c. voltage can be applied to the reflector of the klystron to tune it to the correct frequency.

The sketch that has been given of T.R. systems has stressed principles rather than details of design, for to describe the latter would require more space than would be justified in an introductory survey of this character.

5·13·2. *The magic tee*

This is the ambitious title of the double T-junction illustrated in fig. 5·43. It comprises a section of wave guide AB which carries both a series side limb C and a shunt limb D (§ 5·10), which form with it a double T-junction. Its alleged magical properties are these: when limbs A, B, and C are *terminated in reflexionless loads* and a wave is launched into D, then power is propagated down limbs A and B, but none along limb C. Similarly, when power is introduced into C with A, B and D terminated in reflexionless loads, then power proceeds along A and B but none along D. If the power is

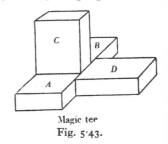

Magic tee
Fig. 5·43.

introduced at A or B an H_{10}-wave proceeds down each of the three remaining limbs when they are matched. This behaviour of the junction is readily understood by considering the relative directions of the fields and currents associated with H_{10}-waves in the respective branches. For instance, when an H_{10}-wave proceeds down C towards the junction, the longitudinal currents in the broad face of C flow across the junction as longitudinal currents in the upper broad faces of A and B and thus excite H_{10}-waves in both these branches. At the centre of the broad face of D at the junction, the current in C is that in the narrow face, from which no longitudinal current flows along the centre of the broad face of D so that no H_{10}-wave is excited in D. Similarly, an H_{10}-wave approaching the junction along D does not excite an H_{10}-wave in C, since the currents in its broad face fail to provide the necessary transverse currents in the narrow face of C. It is easy to see that an H_{10}-wave in A or B excited waves in both C and D.

Consider, however, the case in which B and D are matched, A is unmatched and a wave is introduced at C. The wave from C excites waves that travel towards the terminations of A and B, of which that in A is partially or completely reflected at its end.

The reflected component on returning to the junction excites waves in all three limbs B, C and D. We conclude, therefore, that only when both A and B are matched does no power proceed along D when it is introduced at C, and that a mismatch in A and/or B gives rise to a wave in D. Similarly, a wave introduced at D excites a wave in C provided A and/or B are mismatched.

Irises may be incorporated in limbs A and B to produce a 'magic-tee' into which power may be introduced along any one limb with no standing wave in that limb when all three remaining limbs are matched.

Applications of the magic-tee. (i) *Balanced mixer.* Let crystal mixers (§4·5) be fitted to the ends of branches A and B which they are tuned to match (give no reflexion), and suppose that the mixers possess equal sensitivities. Let the local oscillator feed into limb D and suppose the signal (radar echo) to enter along C, so that each crystal mixer receives the same signal power and the same power from the local oscillator.

The power supplied from the local oscillator is modulated by 'noise' and this noise modulation appears in the output of each crystal mixer. Fortunately, when the outputs of the mixers are so combined that the i.f. signals add, it is found that the noise components are in antiphase and cancel. In this way local oscillator noise can be reduced and a considerable improvement in the radar receiver performance achieved.

It is difficult to obtain crystal mixers with identical sensitivities, so that complete cancellation of local oscillator noise is not achieved in practice. Nevertheless, a worth-while reduction in noise is achieved by the balanced mixer.

(ii) *Magic-tee impedance bridge.* Suppose branch A to be matched and that power is introduced into C. The signal in D is determined by the load at the end of B and vanishes only when B is matched. Thus the output from a mixer attached to D can be calibrated to measure directly the normalized impedance of the termination of B.

(iii) *Magic-tee T.R. system.* A balanced T.R. system for providing greater protection of the crystal mixer against break-through of the transmitter pulse is shown in fig. 5·44. The magnetron pulse proceeds down C and divides equally into the branches A and B of the magic-tee and breaks down both T.R. cells. Since C is a series branch, the two H_{10}-waves in A and B leave the junction with their electric fields in antiphase.

Since the T.R. cell in B is $\frac{1}{4}\lambda_g$ farther from the junction than that in A, the two waves reflected from the cell recombine in phase at the junction and send the power through D to the aerial and none back along C.

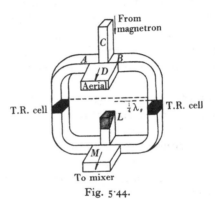

Fig. 5·44.

A very small fraction of the power leaks through each cell and proceeds to the lower magic-tee, where the waves combine to send the residual power into the absorbing load L and none into the mixer M. In reception, however, the signal enters along D, divides into two waves which start from the junction in phase and reach the lower magic-tee in phase, unlike the leakage waves which arrive at ML in antiphase. Consequently, the signal from D proceeds along M and no power runs into L. None of the signal from the aerial proceeds along the branch C to the magnetron. This system affords good protection for the mixer against the transmitter pulse.

5·14. Measurement of power

Power measurements at microwave-lengths are effected by converting the whole, or a known fraction, of the energy carried by the wave into heat. The power is then obtained from the rate of generation of heat. A straightforward method of finding the power output of a magnetron transmitter is illustrated in fig. 5·45. Power from the magnetron enters at the left and proceeds to the flow calorimeter at the wave-guide termination. A rectangle of trolitul is employed as a matching reactance whose value can be controlled by altering the angle between the long axis of the rectangle and the axis of the wave guide. The position of the trolitul block can also be altered by moving its support along the longitudinal slot.

It is thus possible to match the calorimeter to the wave guide and to ensure that no reflected wave returns to the magnetron. The measured power is then the same as that produced by the magnetron

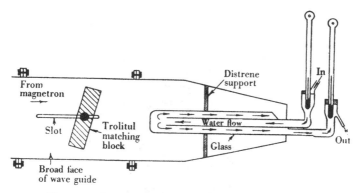

Fig. 5·45. Measurement of power from a magnetron.

under normal working conditions. The rate of conversion of energy into heat is obtained from the rate of flow of the stream of water and the rise in temperature indicated by the thermometers. The mean power is thus obtained directly. To obtain the peak pulse power it is necessary to measure the pulse duration and pulse envelope on a monitor, and the pulse recurrence frequency on a frequency meter.

The power level is too high to permit the direct use of the standing wave indicators described in §4·4 to show when matching is achieved, and the usual method is to adjust the trolitul block until the maximum generation of heat in the calorimeter is obtained. Recently, however, a convenient high-power standing wave indicator has been developed which allows the matching adjustments to be readily made and also indicates the mean power. It is shown in fig. 5·46. It comprises a set of six evenly spaced glass tubes filled with neon at reduced pressure, each of which protrudes into the wave guide through one of six holes on the centre line of the broad face as shown. Each tube is sheathed by a pair of metal semi-cylinders which are spaced to leave a longitudinal slit in front and behind, through which the tubes can be viewed. The tubes and sheaths are marked *A*, *B*, *C*, *D*, *E*, *F* in the figure, and the sheaths are shown shaded. A perspex cover-plate carrying numbered horizontal lines

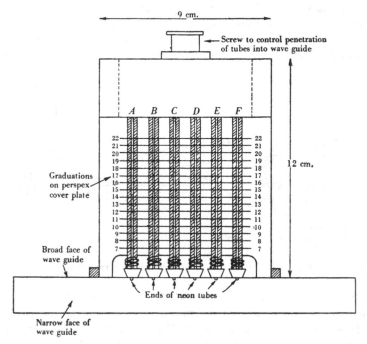

Fig. 5·46. *S*-band output tester.

is fixed in front of the set of neon tubes. The section of wave guide with the neon tubes is included in the wave-guide run leading to the load.

When power passes down the wave guide high-frequency discharges can be seen to extend different distances up the tubes when a partial or complete standing wave exists in the wave guide. When the calorimeter or load is matched then all the discharges terminate at the same level, whose value, read from the numbered horizontal graduation lines, gives a measure of the power passing to the load, if the scale of levels has previously been calibrated against a flow calorimeter. In the case of a mismatch the standing wave ratio S can be found by comparing the scale readings of the longest and shortest of the discharges in the system of tubes. This indicator can conveniently be included in the wave-guide run of an S-band radar installation to indicate the degree of mismatch presented by the aerial and the mean power radiated.

The following results obtained with a flow calorimeter serve to illustrate this method of measuring the output of a magnetron:

Pulse recurrence frequency = 840 pulses per sec.
Pulse duration (width) = $1\cdot4\,\mu$sec.
Temperature difference between inflow and outflow of calorimeter = $14\cdot8°$.
Rate of flow = 100 c.c. in 274 sec.
1 joule = 0·239 cal.
In 1 sec. power is passing into the water for $840 \times 1\cdot4 = 1176\,\mu$sec.
Therefore mean power converted to heat is

$$\frac{100}{274} \times \frac{14\cdot8}{0\cdot239}\ \text{W.}$$

Peak power (power during pulse) is

$$\frac{100}{274} \times \frac{14\cdot8}{0\cdot239} \times \frac{10^{6}}{1176} = 19\cdot2\,\text{kW.}$$

The frequency, found by a wave meter, was

$$3297\cdot7\ \text{Mc./sec.}$$

The flow calorimeter is unsuitable for use at low powers and it is replaced by a thermistor, a device which has been developed principally in the United States.

Thermistors are of the two types shown in fig. 5·47 (a)—the bead thermistor and the disk thermistor. The former, which is used to measure power, comprises a bead of a substance (an oxide of titanium) whose specific resistance is highly temperature-dependent. The bead is held by fine supporting wires within a glass envelope, as shown, and placed within a wave guide and matched (that is, it replaces the crystal in the equivalent of a crystal mixer). The electric field of the wave drives current in the supporting wires and through the bead; consequently, when proper matching is achieved, the whole of the power entering the wave-guide branch containing the thermistor is dissipated as heat in the bead.

The bead thermistor is made one arm of a Wheatstone's network as shown in fig. 5·47 (*b*). With no power entering the wave guide the bridge is balanced by adjustment of the zero setting resistor which controls the total current flowing into the network and in particular that through the thermistor. Since the resistance of the thermistor

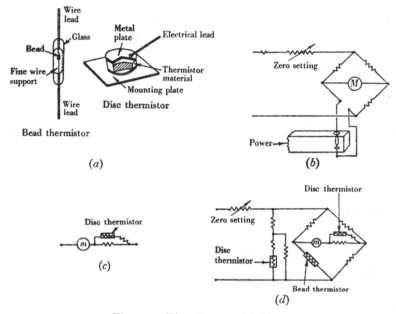

(*a*)

(*b*)

(*c*)

(*d*)

Fig. 5·47. Thermistors and bridges.

depends on its temperature, which is itself changed when the current through it is changed, the zero setting resistance can be used to balance the bridge. When the power is absorbed by the thermistor bead from the wave in the guide its temperature changes and the bridge is thrown out of balance. The deflexion of the meter M is proportional to the power in the wave and may be calibrated to give this power direct.

Variations in ambient temperature affect the simple bridge of fig. 5·47 (*b*) in two ways. First, the balance of the bridge is upset even though there is no power entering the wave guide, and it is necessary to readjust the zero setting resistance. Secondly, the sensitivity of the thermistor (change in resistance in ohms per watt absorbed) diminishes as its temperature rises; it is therefore impossible, with the simple bridge, to calibrate the meter M to read power direct, in a manner independent of the ambient temperature. The simplest method of improving the stability of the bridge is to lag the wave guide with thermally insulating material, and this is generally a satisfactory procedure. A more ambitious method is to employ the disc thermistor, shown in fig. 5·47 (a), as a compensating device.

The disc thermistor has a large thermal capacity, and its resistance is therefore largely independent of the current through it, but is determined principally by the ambient temperature. By associating the disc thermistor with a series resistance and a shunt resistance, with suitable values, a resultant resistance can be obtained which is temperature-dependent, in such a manner that when placed in series with the bridge meter M, as shown in fig. 5·47 (c), the meter sensitivity becomes temperature-dependent in the opposite sense to that of the bead thermistor. The calibration of the bridge in terms of power absorbed from the wave then becomes largely independent of changes in the ambient temperature.

It is also necessary to introduce a similar temperature-compensating device in shunt across the input terminals to the bridge in order to obtain a zero setting resistance which is virtually self-adjusting with alteration in the ambient temperature. The final bridge is shown in fig. 5·47 (d). The meter M is a microammeter, and the order of magnitude of the maximum power that the bridge can handle is about 2 mW., which corresponds to a meter current of the order of 100 μA. Thermistor bridges are used extensively for measurements at X-band. Larger powers may be measured by diverting a known small fraction of the power from the main wave guide, by means of a directive feed (fig. 4·3) or a radiating hole, into the matched thermistor mount, where it is measured as described above.

5·15. A wave-guide quarter-wave transformer

The difference in dimensions of standard British and American X-band wave guides demands a suitable coupling section for joining the one type to the other. A tapered section (§4·11) can be used for this purpose but in order to avoid reflexions the taper should be gradual, a requirement that makes the tapered section too long for some applications.

A convenient alternative is a coupling section which is analogous to the quarter-wave transformer of transmission-line practice, where two transmission lines with characteristic impedances Z_{01} and Z_{02} can be matched by means of a $\frac{1}{4}\lambda$ section of transmission line with characteristic impedance $Z_{03} = \sqrt{(Z_{01}Z_{02})}$. It would be anticipated, by analogy, that two wave guides with intrinsic impedances (wave impedances, §3·4·1) Z_{H1} and Z_{H2} could be matched through an intermediate section with length $\frac{1}{4}\lambda_{g3}$ and intrinsic impedance $Z_{H3} = \sqrt{(Z_{H1}Z_{H2})}$. This surmise is true in so far as the effects of geometrical discontinuities at the junctions are disregarded, but these, as we have seen, excite storage fields, which throw effective susceptances in shunt across the junctions.

The practical problem is so to design the quarter-wave transformer that the effects of the storage fields at the junctions are unimportant.

The intrinsic impedances of the three types of wave guide are

$$Z_{H1} = \frac{120\pi b_1}{a_1\sqrt{\{1-(\lambda/2a_1)^2\}}}, \quad Z_{H2} = \frac{120\pi b_2}{a_2\sqrt{\{1-(\lambda/2a_2)^2\}}},$$

and

$$Z_{H3} = \frac{120\pi b_3}{a_3\sqrt{\{1-(\lambda/2a_3)^2\}}}, \tag{1}$$

where a_1, a_2 and a_3 are the respective long dimensions and b_1, b_2 and b_3 the short dimensions of the wave-guide cross-sections.

The problem is, given a_1, b_1 and a_2, b_2, to choose a_3 and b_3 so that

$$Z_{H3}^2 = Z_{H1}Z_{H2}. \tag{2}$$

There is an infinity of pairs of values of a_3 and b_3 that satisfy this requirement, so that it is possible to choose a pair for which the influence of the storage field at the junctions is unimportant. It proves better to have a small step at both walls of a junction rather than a larger step on one wall and none on the other. First, the steps on the two walls contribute moderate reactances of opposite sign with resulting partial compensation, rather than a single large uncompensated reactance. Secondly, the reflected wave from the reactances at the other junction tend to cancel those excited at the first junction.

The design of wave-guide quarter-wave transformers for connecting British and American X-band wave guides has been studied by B. G. Loach, who has taken advantage of the specific dimensions of these wave guides to construct a very satisfactory wave-guide quarter-wave transformer.

From equations (1) and (2) it follows that

$$\frac{b_3}{a_3\sqrt{\{1-(\lambda/2a_3)^2\}}} = \frac{\sqrt{(b_1 b_2)}}{\sqrt{\{a_1 a_2\sqrt{[1-(\lambda/2a_1)^2]}\sqrt{[1-(\lambda/2a_2)^2]}\}}}. \quad (3)$$

A particular pair of values (a_3, b_3) that satisfies (3) is the following:

$$\left.\begin{aligned} b_3 &= \sqrt{(b_1 b_3)}, \\ a_3^2 &= (\tfrac{1}{2}\lambda)^2 + a_1 a_2 \sqrt{\left[1 - \frac{\lambda}{2a_1}\right]}\sqrt{\left[1 - \frac{\lambda}{2a_2}\right]}. \end{aligned}\right\} \quad (4)$$

It happens that the internal dimensions $1 \times \tfrac{1}{2}$ in. of the British wave guide are the same as the external dimensions of the American wave guide whose internal dimensions are 0.9×0.4 in. (wall thickness 0.05 in.). Thus the American wave guide can be fitted into the British wave guide. To obtain a_3 and b_3 for the quarter-wave transformer we write

$$a_1 = 2.54 \text{ cm.}, \qquad b_1 = 1.27 \text{ cm.} \quad \text{(British)},$$
$$a_2 = 2.285 \text{ cm.}, \qquad b_2 = 1.015 \text{ cm.} \quad \text{(American)}.$$

From (3)

$$b_3 = 1.138 \text{ cm.} = 0.447 \text{ in.}, \quad a_3 = 2.400 \text{ cm.} = 0.946 \text{ in.}$$

With $\lambda = 3.2$ cm.

$$\lambda_{g3} = \lambda/\sqrt{\{1 - (\lambda/2a_3)^2\}} = 4.30 \text{ cm.}$$

Thus the length of the transformer $= \tfrac{1}{4}\lambda_{g3} = 1.075$ cm. $= 0.423$ in. Since

$$a_1 \text{ (American wall thickness)} = 1 \text{ in.} - 0.05 \text{ in.} = 0.95 \text{ in.} \doteqdot a_3,$$

and

$$b_1 \text{ (American wall thickness)} = 0.5 \text{ in.} - 0.05 \text{ in.} = 0.45 \text{ in.} \doteqdot b_3,$$

the transformer is very conveniently constructed by removing 0.423 in. of one narrow face and one broad face from the end of a piece of American wave guide and inserting it into a length of British wave guide as shown in fig. 5·48.

This transformer gave a standing wave ratio $s = 0.98$ over a wavelength band $\lambda = 3.20 \pm 2\,\%$ and 0.99 at $\lambda = 3.20$ cm. in one wave guide when the other was matched. The power-handling capacity is high and of the order of 400 kW. peak power.

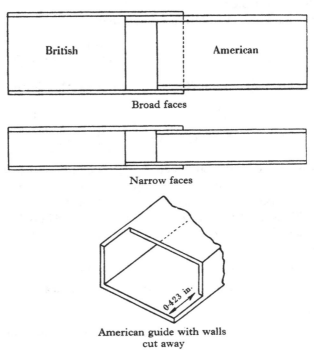

Fig. 5·48. Wave-guide quarter-wave transformer.

5·16. Corrugated wave guides

This section comprises a very brief account of a novel type of wave guide—the corrugated wave guide—which promises to be of considerable practical importance. For certain applications, the fact that the phase or pattern velocity in a wave guide exceeds the free space or TEM wave velocity is an inconvenience, and methods for reducing the wave velocity in a wave guide have received considerable attention. It is known that the phase velocity of a wave along an actual or an artificial transmission line is $v = \sqrt{(L/C)}$, where L and C are the series inductance and shunt capacitance per unit length or section.

Since in this instance v is reduced by increasing L, that is, by increasing the magnetic storage field, it is suggested that the physical equivalent of inductive loading would also diminish the

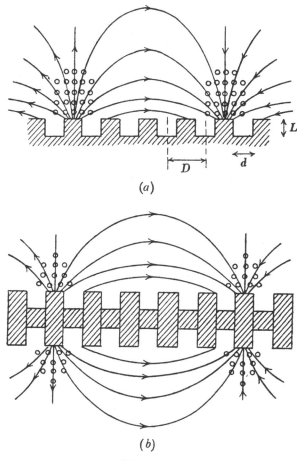

(a)

(b)

Fig. 5·49.

phase velocity of a wave along a wave guide or parallel strip transmission line.

The simplest example of an electromagnetic wave guided by an inductively loaded system is that shown in fig. 5·49(a).

This represents the field pattern of an electromagnetic wave propagated over a corrugated metal sheet of infinite extent. The corrugations are rectangular slots which run at right angles to the direction of propagation. When the depths L of the slots are less than $\frac{1}{4}\lambda$, where λ is the free-space wave-length, the input impedance across each slot, regarded as a parallel strip transmission line, is an inductive and the storage field is predominantly magnetic. It is found in these circumstances that the electromagnetic wave travels over the surface with a phase velocity less than the free space velocity $1/\sqrt{(\mu\epsilon)}$. Further features of the wave are the following:

There is a longitudinal component E_z of the electric field in the direction of propagation. After a certain distance all components of the electromagnetic field diminish rapidly with distance from the corrugated surface so that the flow of energy is effectively confined to the immediate vicinity of the guiding surface. It is found that the waves still adhere to the surface even when it is curved. The phase velocity and wave-length are less than those of a wave of the same frequency guided by a plane conducting surface.

Such a surface, for which $0 < L < \frac{1}{4}\lambda$, or $\frac{1}{2}\lambda < L < \frac{3}{4}\lambda$, etc., where L is the slot depth, is termed an inductive surface.

If, however, $\frac{1}{4}\lambda < L < \frac{1}{2}\lambda$, $\frac{3}{4}\lambda < L < \lambda$, etc., the surface is termed capacitive. Such a surface is unable to guide a wave, and the flow of power is away from the surface into the space above and is no longer confined to the vicinity of the surface.

When $L = \frac{1}{4}\lambda$ the oscillations in adjacent slots are in antiphase and a complete standing wave exists on the surface. The wave-length of this standing wave is twice the distance between the centres of adjacent slots.

If the corrugated surface is gradually transformed to a flat surface, across an intermediate region where the slot depth is steadily diminished from its full value on one side to nothing on the other, then the guided wave expands from the adhesive wave on the corrugated surface to an ordinary TEM wave on the flat surface.

Any irregularity on the corrugated surface, such as a metal projection, radiates strongly, and a row of evenly spaced projections or recesses can be used to produce a beamed radiation.

An alternative method of guiding power near a surface is that shown in fig. 5·49 (b) which represents a metal rod with ring slots. The simplest form of wave pattern along such a rod is that shown in fig. 5·49 (b).

If the surface is alternately raised and depressed at spacings of $\frac{1}{2}\lambda_r$, (λ_r is the wave-length on the rod) then a conical beam of radiation is produced. When the corrugated rod is transformed gradually into a smooth rod the system behaves as an end-fire array.

We next consider a rectangular wave guide in which one of the broad faces is corrugated as indicated in fig. 5·50.

It would be anticipated from what has been said above that the energy flow along the axis of the wave guide is concentrated near the corrugated surface, and that the phase velocity and wave-length are smaller than those in a smooth wave guide of the same dimensions. This wave guide possesses two cut-off wave-lengths—the one

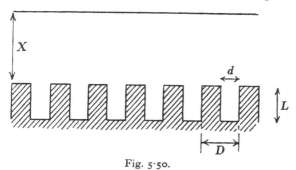

Fig. 5·50.

$\lambda_{c1} = 2a$ corresponds to the normal cut-off in a rectangular wave guide with broad dimension a, and the other λ_{c2} corresponds to the transition of the corrugated surface from an inductive to a capacitive surface. That is, $\lambda_{c2} \doteqdot 4L$, where L is the slot depth. The wavelength λ_g of the wave is then equal to twice the distance between the centres of adjacent slots, since the fields in adjacent slots then oscillate in antiphase.

The corrugated wave guide behaves, therefore, as a band pass filter. The dependence of λ_g upon λ (TEM wave-length) is indicated by the curve in fig. 5·51.

There is a longitudinal component of E in the wave, and this fact gives the corrugated wave guide its practical importance.

The principal uses for corrugated wave guides are as flexible wave guides (bellows type), delay lines, transformers for changing the mode of transmission of a wave (for instance E_{01} to E_{02} in a circular guide), as phase shifters and as filters. A further application, of outstanding importance, is the acceleration of electrons or ions.

Since the wave in a corrugated wave guide possesses an E_z-component in the direction of propagation and also a phase velocity which can be made less than $c = 3 \times 10^8$ m./sec., it is possible to cause an electron or ion to travel down the axis of corrugated wave guide at the same speed as the E_z-component of the field, by which it is continuously accelerated. It is necessary to accelerate the phase velocity of the wave so that the particle always remains in the same longitudinal field, and this acceleration is achieved by progressively grading the depths of the corrugations.

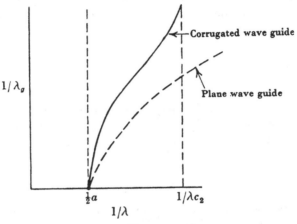

Fig. 5·51.

Corrugated circular wave guides have been used for this purpose, and in the extreme case where the partitions between the corrugations are narrow compared with the width of the corrugations and the depth of the corrugations an appreciable fraction of the radius of the wave guide, the system may be regarded as a series of cylindrical resonant cavities coupled through holes in their flat ends and oscillating in an E_{010} resonant mode (fig. 6·11).

Although it would be out of place to enter deeply into the theory of corrugated wave guides, it is permissible, briefly, to indicate the theoretical procedure adopted by Slater, Goldstein, Cuttler and others.

It is assumed that the row of slots (rectangular wave guide) is of infinite extent and that the electric field across the mouth of a slot is of simple character. For instance, the field component E_z may be assumed to be of constant amplitude (independent of z) across the slot and zero on the walls between the slots. The phase of the oscillation at the centre of each slot is supposed to advance by β from one slot to the next. The field in the wave guide is then resolved into an infinite set of modes whose relative amplitudes are determined from the fact that the field in the wave guide reduces to the Fourier expansion, mentioned above, at the slotted boundary. A transcendental equation is also obtained for the phase constant β which gives the phase velocity of the composite wave, since all the modes are here interlocked so that they are propagated at the same speed.

The complicated transcendental equation which results from the assumption that E_z is constant (independent of z) across the mouth of the slot is

$$\frac{d}{D} \sum_{m=-\infty}^{m=\infty} \frac{k' \tan k'L}{k_{xm} \tanh k_{xm} X} \left(\frac{\sin k_{gm} \frac{1}{2} D}{k_{gm} \frac{1}{2} D} \right)^2 = 1,$$

where D is the distance between the centres of adjacent slots,

d is the width of the mouth of a slot,

$$k' = 2\pi \bigg/ \frac{1}{\lambda'} \left(\frac{1}{\lambda'} = \sqrt{\left(\frac{1}{\lambda^2} - \frac{1}{(2a)^2} \right)} \right); \ a \text{ is broad dimension of guide} \bigg),$$

X is the distance from the mouth of a slot to the uncorrugated opposite wall of the same guide,

$k_g = 2\pi/\lambda_g$ (λ_g is wave-length of wave in corrugated guide),

$k_{gm} = k_g + 2\pi m/D$ ($m = 0, \pm 1, \pm 2, \ldots$),

$k_{xm} = \sqrt{(k_{gm}^2 - k'^2)}$.

The integers m correspond to the different modes in the wave pattern.

When $\lambda' \gg 4L$ the transcendental equation may be approximately represented by ($m = 1$)

$$\frac{d}{D} \frac{k' \tan k'L}{k_x \tanh k_x X} = 1.$$

This equation gives k_x from the dimensions of the wave guide and those of the corrugations. λ_g is then deduced from k_x.

5·17. Resonant slots in the walls of wave guides

5·17·1. *Introduction*

The elucidation of the properties of slots in the walls of wave guides is due principally to the work of Watson and Guptill and their collaborators. This section briefly reviews the results obtained by these Canadian workers.

5·17·2. *Slots in wave guides*

Examples of slots in wave guides are given in fig. 5·52. When these slots are given a total length of approximately $\frac{1}{2}\lambda$ they become resonant and may be excited by the surface currents associated with the H_{10}-wave in the wave guide, provided the slots interrupt the flow of current. For instance, the two slots shown in fig. 5·52 (a) are not excited by an H_{10}-wave in the guide and do not radiate, since the currents in the walls flow parallel to the edges of the slots as indicated by the arrows (vide fig. 2·9), whereas the slots of fig. 5·52 (b) and (c) are excited and radiate, those of fig. 5·52 (b) being more strongly coupled to the H_{10}-wave than those of fig. 5·52 (c). The electric field of the hemispherical wave radiated from the slot is polarized at right angles to the long axis of the slot. In addition to radiating externally a slot scatters a wave internally within the wave guide and also excites a storage field in its vicinity. It therefore throws an impedance either in series or in parallel with the wave guide. The series slots are marked A in fig. 5·52 (b) and (c). It will be noted that at their centres the longitudinal current in the wave guide is intercepted. It will also be found that the internal scattering is antisymmetrical with respect to the electric field in the scattered wave (§ 5·4). The equivalent circuit is a transmission line with a series impedance $Z = R + jX$ in series (with one or both lines) as shown in fig. 5·52 (d). The resistive term R accounts for the power lost in external radiation.

In contrast, the slots marked B interrupt lateral current and scatter symmetrically and are represented by an admittance $Y = G + jB'$ in shunt across a transmission line as shown in fig. 5·52 (d). These slots are therefore called shunt slots.

A slot, such as slot C in fig. 5·52 (c), which is both twisted and displaced from the centre line, intercepts at its centre, both lateral and longitudinal, components of the current and also scatters unsymmetrically.

Both series and shunt components appear in the equivalent circuit, and it is probably most simply represented by a lattice section inserted in a transmission line.

We have noted two principal methods for controlling the coupling of the simple series and shunt slots, namely, rotation of the axis

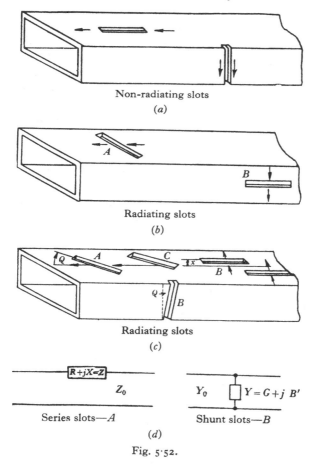

Non-radiating slots
(a)

Radiating slots
(b)

Radiating slots
(c)

Series slots—A Shunt slots—B
(d)

Fig. 5·52.

and displacement of the centre. Suppose the slots in fig. 5·52 to be cut to resonance so that the reactive component of the impedance or admittance is small or zero, and the normalized resistance or admittance may be found from measurements with a standing wave indicator in the usual manner.

Consider the shunt slots B in fig. 5·52 (c). Watson showed that the normalized conductance g of the slot in the broad face was related to its displacement x from the centre line as follows:

$$g = K_1 \sin^2\left(\frac{\pi x}{a}\right), \tag{1}$$

where a is the broad dimension of the wave guide.

This is what would be expected if the amplitude of the electric field across the centre of the slot is proportional to the component of the wall current perpendicular to the slot, since the power radiated is then proportional to the square of the electric field strength at the centre.

Watson and Guptill found that the shunt conductance of a resonant slot $\frac{3}{16} \times 2$ in. cut in the broad face of a standard American S-band wave guide ($3 \times 1\frac{1}{2}$ in. O.D.) was represented accurately by formula (1) with $K_1 = 0·475$. They also found that the resonant length was 6 % greater at the edge than near the centre ($x \to 0$).

The series normalized resistance r of slot with its centre on the axis (slot A, fig. 5·52 (c)), but with an angle ϕ between its axis and the centre line of the broad face, is

$$r = K_2 \sin^2 \phi$$

for small values of ϕ.

The shunt conductance of the inclined resonant slot in the narrow face (slot B, fig. 5·52 (c)) is

$$g = K_3 \sin^2 \phi.$$

The experimental conductances of an oblique resonant slot $\frac{1}{16}$ in. wide in the wall of an American X-band wave guide were represented accurately by this formula with $K_3 = 0·715$.

The series resistance r of the transverse slot (slot A, fig. 5·52 (b)) whose centre lies at distance x from the *nearest edge* of the wave guide is

$$r = K_4 \sin^2\left(\frac{\pi x}{a}\right).$$

The constants K depend on the geometry of the wave guide, the wave-length and the slot width.

The shunt conductances of narrow shunt slots are proportional to their widths.

One of the most important applications of slots is as radiating elements in wave-guide linear arrays, the two principal types of which are shown in fig. 5·53 (*a*) and (*b*).

It can be seen that shunt slots are used in each case, but in the array of fig. 5·53 (*a*) the electric field in the radiation vibrates parallel to the long axis of the wave guide and in the second array, perpendicular to the long axis. Let us suppose the centres of

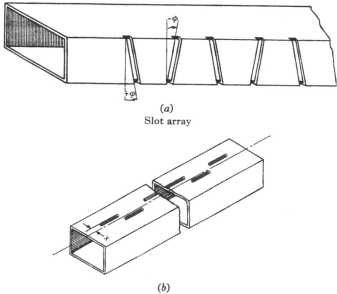

(*a*)

Slot array

(*b*)

Slot array with transverse polarization

Fig. 5·53.

adjacent slots to be $\frac{1}{2}\lambda_g$ apart, then by alternating the inclinations ϕ and the lateral displacements x respectively, between positive and negative values from slot to slot, the slots in each array radiate in phase, and the beam proceeds at right angles to the axis of the wave guide. Linear arrays may be resonant, or non-resonant. In the former the end of the wave guide is short circuited at a distance $\frac{1}{4}\lambda_g$ from the last slot, but as these arrays are not extensively employed because they are frequency-sensitive, we shall not discuss them. In non-resonant arrays the end of the wave guide distant from the transmitter is terminated in a reflexionless load, and a travelling H_{10}-wave proceeds to it down the wave guide, becoming progressively weaker as it is robbed of power by the radiating slots. It is evident that if the slots were all coupled equally strongly into this wave, more power would be radiated from the slots near the input, where the amplitude of the wave is great, than from the slots near the termination. To ensure that equal power is radiated from each element the coupling of the elements is increased progressively down the wave guide by increasing ϕ or x as the termination is approached. The residual power wasted in the load is about 5 % of the total power radiated.

It has been assumed so far that the centres of adjacent slots are separated by $\frac{1}{2}\lambda_g$. This spacing is, however, not adopted in practice because the scattered waves from the individual slots reinforce to give a powerful reflected wave at the input which has a deleterious effect on the performance of the magnetron. The slots are therefore separated by $0.55\lambda_g$, and as a result the direction of maximum radiation is no longer normal to the wave-guide axis but is shifted by about 5° from the normal. When this shift has been determined it is easy to allow for it in any direction-finding equipment in which the array is incorporated.

Since our subject is wave guides and not microwave aerials, it is necessary to resist the temptation to expatiate further on linear arrays whose study is now a highly specialized pursuit.

5·17·3. *Coupling of wave guides through resonant slots*

The commonest types of simple coupling are shown in fig. 5·54. It will be noted that they are of the following types:

Series-series in which the slot is a series slot in both wave guides (fig. 5·54 (a) and (e)).

Shunt-shunt in which the slot is a shunt slot in both wave guides (fig. 5·54 (b) and (c)).

Shunt-series (or series-shunt) are mixed couplings in which the slot is a series slot in the one wave guide but a shunt slot in the other (fig. 5·54 (d) and (f)).

The coupling properties of the slots in these arrangements were unravelled in an extensive series of researches by Watson, Guptill and their associates at McGill University, whose results are schematically summarized in fig. 5·55.

A marked difference exists between the unmixed couplings (series-series and shunt-shunt) and the mixed couplings (shunt-series and series-shunt). Consider, for instance, the series-series coupling represented in fig. 5·55 (*a*). When the lower wave guide is closed by a shorting piston at *B*, at a distance $\frac{1}{2}\lambda_g$ from the centre of the slot, and the upper wave guide is terminated at *C* and *D* so that the impedances in the two branches have arbitrary values z_2

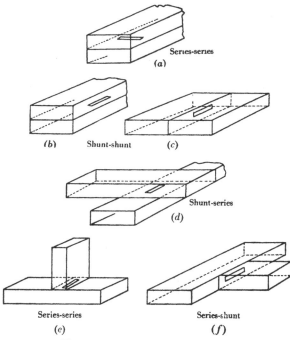

Fig. 5·54. Resonant slot couplings.

and z_2' when extrapolated back to the centre of the slot, then the input impedance in the arm *A* of the lower guide, at the slot, is

$$z_m = z_2 + z_2',$$

in complete analogy with circuit concepts.

Similarly, the shunt-shunt coupling of fig. 5·55 (*b*), when the shunt impedance presented by the arm *B* at the centre of the slot is infinite, and arbitrary shunt impedances z_2 and z_2' are presented at the slot by *C* and *D*, then the input impedance z_m, in arm *A*, is

$$\frac{1}{z_m} = \frac{1}{z_2} + \frac{1}{z_2'}.$$

With the mixed couplings, however, shown in fig. 5·55 (c) and (d), it is necessary to introduce transformation factors K_1 and K_2 in order to obtain the correct value of z_m. Thus, as indicated in the figures, for the shunt-series connexion

$$z_m = K_2(z_1 + z_1'),$$

and for the series-shunt connexion

$$\frac{1}{z_m} = K_1\left(\frac{1}{z_1} + \frac{1}{z_1'}\right).$$

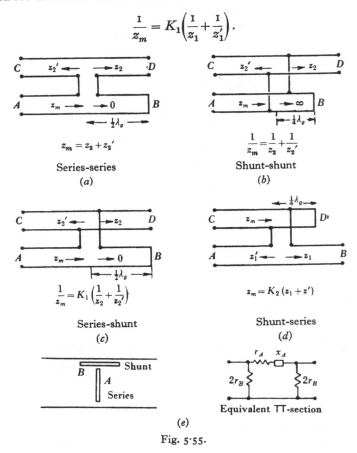

$$z_m = z_2 + z_2'$$

Series-series
(a)

$$\frac{1}{z_m} = \frac{1}{z_2} + \frac{1}{z_2'}$$

Shunt-shunt
(b)

$$\frac{1}{z_m} = K_1\left(\frac{1}{z_2} + \frac{1}{z_2'}\right)$$

Series-shunt
(c)

$$z_m = K_2(z_1 + z')$$

Shunt-series
(d)

(e)

Equivalent Π-section

Fig. 5·55.

For instance, in the case of the slot-coupled T-junction of fig. 5·54 (e) and (f), when the series-series side limb is shorted by a plunger at distance l from the slot, the series impedance presented to the main wave guide by the slot is $j \tan 2\pi l/\lambda_g$. When, however, the side limb of the series-shunt T-junction is closed by a plunger, the shunt admittance placed across the main wave guide is

$$-jK_1 \cot 2\pi l/\lambda_g.$$

To obtain a pure shunt susceptance free from the transformation factors K, it is necessary to employ the shunt-shunt coupling of fig. 5·54 (b) and (c). When z_2 (fig. 5·55 (b)) is made infinite by means of a plunger in D at $\frac{1}{4}\lambda_g$ from the slot, and another plunger is placed at distance l from the slot in C, then $z_2' = j \tan 2\pi l/\lambda_g$.

If the piston at B is replaced by a reflexionless load, then

$$\frac{1}{z_m} = \left(1 + \frac{1}{j \tan 2\pi l/\lambda_g}\right).$$

A simple combination of a pair of slots studied by Watson and Guptill is shown in fig. 5·55 (e).

Slot A is a series slot and slot B a shunt slot, and it was shown that the combination behaved as a π-section. The slots were covered in turn and their separate impedances were measured and were

$$z_A = r_A + jx_A = 0\cdot45 + j0\cdot45,$$
$$z_B = r_B + jx_B = 0\cdot89 + j\times0.$$

The calculated impedance of the π-section followed by a matched wave guide is $0\cdot705 + j0\cdot17$.

The measured impedance of the combination was found to be $0\cdot67 + j0\cdot14$.

The radiation from this combination of slots is elliptically polarized.

5·18. Methods for feeding microwave aerials from wave guides

This section is a brief description of some typical methods for feeding aerials from wave guides. These are illustrated in fig. 5·56.

Fig. 5·56 (a) shows a simple *front feed* from the open end of the wave guide which is matched by an inductive iris. When the long edge of the end of the wave guide is vertical, then the radiation in the beam from the aerial is horizontally polarized.

In fig. 5·56 (b) and (c) are shown examples of rear feeds. In the one the power is fed into a box and escapes through a pair of resonant slots, and in the other it is scattered by a dipole placed in front of a reflecting plate and supported on a metal strip. These three methods are used at *X*-band wave-lengths.

The second three of fig. 5·56 show methods of feeding *S*-band aerials.

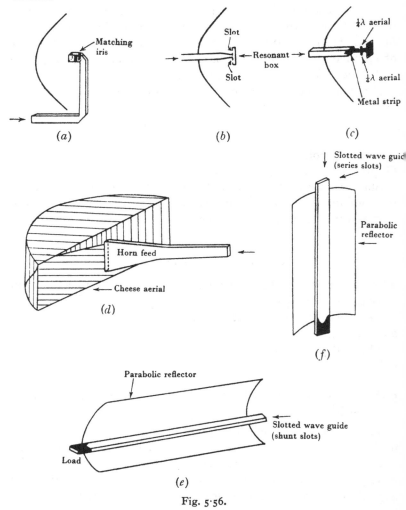

Fig. 5·56.

A cheese aerial is usually fed from a horn, and parabolic cylinders from wave-guide slot arrays which act as line sources at the focus. The aerial of fig. 5·56 (*e*) is of the type used in early warning and

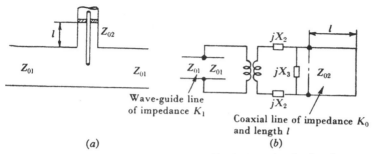

(a)

Variable reactive probe in wave
guide. Probe is inserted P_{10}
transverse electric field

(b)

Equivalent circuit of probe
system of fig. (a)

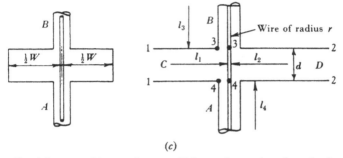

(c)

Coaxial-wave guide transformer which may be used to shunt load
a rectangular H_{10}-wave guide or a circular H_1 wave guide

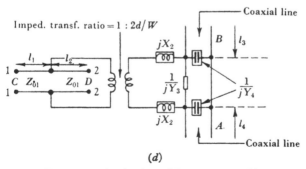

(d)

Equivalent circuit of coaxial-rect. wave guide.
Transformer of fig. (c).

Fig. 5·57.

coverage ground equipments, and in order to obtain a beam with
the electric field horizontally polarized an array of shunt slots cut
in the narrow face of the wave guide is employed as a line source
(fig. 5·53 (e)).

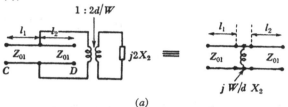

(a)

Equivalent circuit of fig. 5·57 (c) when terminals A and B
are shorted

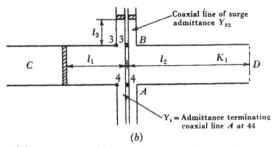

(b)

Coaxial-rect. wave guide transformer used as feeding system

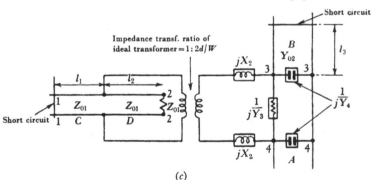

(c)

Equivalent circuit of fig. (b)

Fig. 5·58.

Conversely, the height-finding aerial of fig. 5·56(f) is fed from
an array of series slots of the type shown in fig. 5·53 (b) in order to
preserve the horizontal polarization of the electric field.

5·19. Miscellaneous equivalent circuits

This section merely refers to figs.* 5·57 and 5·58, which represent some useful equivalent circuits.

Even when all the constituent elements cannot be calculated they may often be measured by giving special values to those components that are known and are variable. Fig. 5·59 is a circle diagram and is included as being generally useful.

* Taken from an unpublished article by Dr G. G. Macfarlane.

REFERENCES

§ 5·5.   Macfarlane, G. G. *J. Instn Elect. Engrs*, 1946, vol. 93, part III A, no. 4, p. 703. See also § 7·11.
           Watson, W. H. *The Physical Principles of Wave Guide Transmission and Antenna Systems*, chap. X. Oxford: Clarendon Press.

§ 5·7.   Booker, H. G. The Elements of Wave Propagation using the Impedance Concept. Paper read on 4th Dec. 1946 at a meeting of the Institution of Electrical Engineers.

§ 5·8.   Booker, H. G. Slot aerials and their relation to complementary wire aerials (Babinet's principle). *J. Instn Elect. Engrs*, 1946, vol. 93, part III A, no. 4, p. 620.
           Bailey, C. E. Ibid., p. 615.
           Watson, W. H. Loc. cit., chap. V. See also § 7·13.

§ 5·13·1. Banwell, C. J. The use of a common aerial for radar transmission and reception on 200 Mc./sec. *J. Instn Elect. Engrs*, 1946, vol. 93, part III A, no. 3, p. 545.
           Maclese, A. and Ashmead, J. *J. Instn Elect. Engrs*, 1946, vol. 93, part III A, no. 4, p. 700.

§ 5·17.   Watson, W. H. Loc. cit., chaps. VI, VII and VIII.
           Watson, W. H. Resonant slots. *J. Instn Elect. Engrs*, 1946, vol. 93, part III A, no. 4, p. 747.

GENERAL

J. Instn Elect. Engrs, vol. 93, part III A, nos. 1 and 4, 1946.
Schelkunoff, S. A. *Electromagnetic waves.* Van Nostrand Co. Inc.

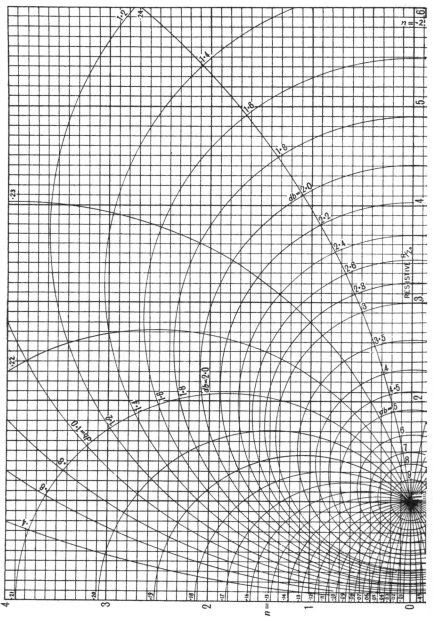

X/Z_0 capacitive

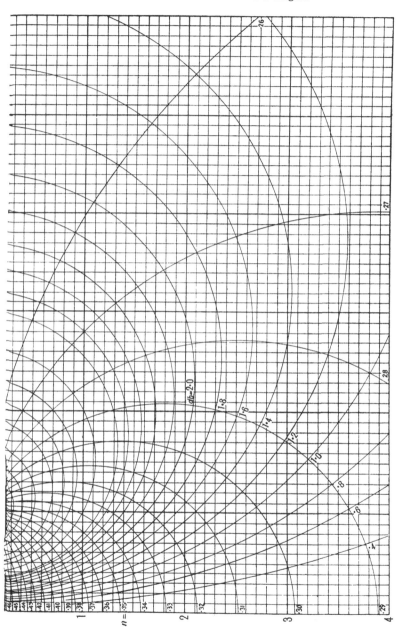

Fig. 5·59. Circle diagram

Chapter 6*

CAVITY RESONATORS

6·1. Importance of cavity resonators

It is found that electromagnetic oscillations can be excited within an empty cavity bounded by conducting walls in much the same way as hollow gas-filled vessels can be excited into acoustical resonance. A given cavity resonator will resonate at a number of discrete frequencies each corresponding to a particular 'mode' of oscillation with its own characteristic electromagnetic field pattern. These field patterns, like those of progressive waves in wave guides, can be classified into E and H types.

Since the free-space wave-length that corresponds to the mode with the lowest frequency is of the order of magnitude of the greatest linear dimension of the cavity, it follows that at centimetre wave-lengths a cavity resonator has a small physical size which renders it convenient for laboratory use and for incorporation in equipments. The principal uses for cavity resonators are as:

(i) tuned elements in oscillators in place of the conventional L-C-R circuits which are physically unrealizable at centimetre wave-lengths,

(ii) accurate wave-meters,

(iii) echo boxes, which are equivalent to ringing circuits.

The electromagnetic oscillations in cavity resonators in the form of hollow rectangular boxes or cylinders have field patterns, with one exception, that are the same as those belonging to standing waves in rectangular and circular wave guides. It is therefore possible to employ elementary methods to deduce many of the important features of oscillations in cavity resonators.

We begin with the simple case of a rectangular resonator.

6·2. Stationary waves in a wave guide

Consider a rectangular wave guide in which a complete standing wave is produced by allowing two trains of H_{01}-waves with the same wave-length λ_g and field strengths to be propagated in it in

---

* This chapter is included in order to illustrate in a simple manner the relation of the theory of resonators to that of wave guides and to mention some applications of resonators. It is not intended to be a complete account of the subject.

opposite directions. The progress of the individual waves is indicated successively in the first two of each of fig. 6·1 (a), (b) and (c). The third diagram in each case gives the resultant field derived by superimposing the field patterns of the individual waves. We take the time $t = 0$ to correspond to the instant at which the magnetic loops in the two oppositely travelling patterns superimpose exactly, so that at $t = 0$ in the standing wave the magnetic field strength is a maximum at all points as indicated in the third of fig. 6·1 (a).

Because the constituent waves travel in opposite senses, the electric fields are in opposition when the magnetic fields add. Consequently, the electric fields cancel at $t = 0$, and at this instant the field in the standing wave is entirely magnetic. Fig. 6·1 (b) illustrates the situation at time $t = \frac{1}{4}T$, one-quarter cycle later, T being the period of the oscillations. The travelling wave patterns in fig. 6·1 (b) are displaced one-quarter wave-length $\frac{1}{4}\lambda_g$ to the right and left respectively, as can be seen from the displacement of the pair of loops which have been distinguished by a horizontal arrow at the centre. When the patterns are superimposed the magnetic fields cancel but the electric fields add to produce a maximum. Thus at $t = \frac{1}{4}T$ the field in the standing wave is entirely electric. Similarly, from fig. 6·1 (c) we deduce that at $t = \frac{1}{2}T$ the electric field vanishes and that the field is again entirely magnetic but reversed in direction compared with the field at $t = 0$. At $t = \frac{3}{4}T$ we should again find the resultant field to be entirely electric but reversed in direction relative to that at $t = \frac{1}{4}T$. At time $t = T$ the field is again that of fig. 6·1 (a). At other instants in the cycle the electric and the magnetic components are both present; further, the magnetic loops and fixed electric lines, although the fields are no longer at maximum intensity, remain fixed in position and do not progress to right or left as in the constituent progressive H_{01}-waves.

We may note the following features of the electromagnetic field of a stationary wave:

(i) The field patterns of the magnetic and electric fields are individually the same as those in the progressive waves.

(ii) In the progressive wave the field patterns are propagated along the axis of the wave guide at speed v_g, but the field intensities in the moving pattern are unchanged; in the stationary wave the patterns are fixed in position but the field intensities oscillate harmonically between maximum positive and negative values.

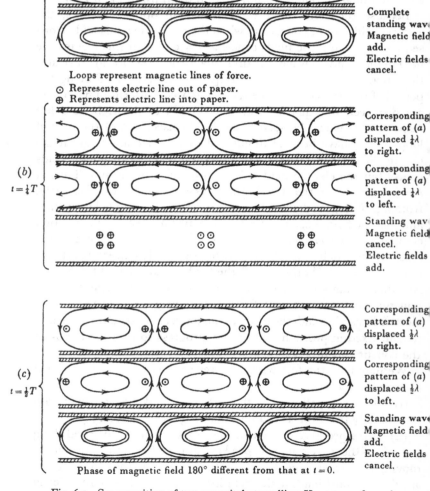

Fig. 6·1. Superposition of two oppositely travelling H_{01}-waves of equal
strength to produce a complete standing wave.

(iii) In the progressive waves the maximum transverse components of the electric and magnetic fields coincide (near the ends of the largest loops), but in the stationary wave the positions at which the transverse components of the magnetic and electric fields have maximum amplitudes are separated by a quarter of a wavelength ($\frac{1}{4}\lambda_g$). (The electric field is concentrated around the centre of a magnetic loop.)

(iv) The resultant electric and magnetic fields within each $\frac{1}{2}\lambda_g$ cell of the pattern oscillate in quadrature. There is no mean flow of power along the axis of the wave guide, but the energy stored in each cell of the pattern is transformed every quarter period from the magnetic to the electric form and back again. The oscillations of the electromagnetic field are therefore entirely analogous to the mechanical vibrations of a pendulum or an escapement wheel in which the energy is transformed alternately from one to the other of the kinetic and potential forms.

6·3. Field patterns in cavity resonators

We have obtained a stationary but oscillating electromagnetic field; we now ask whether this field is one that can exist within a cavity closed by conducting walls. At the walls, the field must satisfy the following conditions: the magnetic field cannot intersect any portion of the conducting surface of the cavity, that is, it may lie tangentially against the surface at some places and vanish at others; the electric field, on the other hand, cannot lie along the boundary, it must either stand at right angles to the surface or vanish at the surface. Evidently, as appears from fig. 6·2, the standing wave pattern of fig. 6·1 can be fitted into a rectangular box, formed by placing conducting partitions across the wave guide a distance apart equal to $p\frac{1}{2}\lambda_g$, where p is an integer. It is then possible to fit p cells of the pattern into the box provided the end-walls touch but do not cut a magnetic loop. A possible disposition is shown in fig. 6·2 for the case $p = 2$. Fig. 6·2 (a) and (b) illustrate respectively the fields at times $t = 0$, when the field is entirely magnetic, and at $t = \frac{1}{4}T$ when the field is entirely electric. If we choose a set of Cartesian axes of reference with the origin O at the near left-hand corner as shown, then with respect to these axes the constituent travelling waves (fig. 6·1) that combine to give the stationary wave are H_{01}-waves. Since $p = 2$, that is, two cells of the pattern are fitted into the resonator, and the mode is derived from progressive H_{01}-waves, it is designated H_{012}.

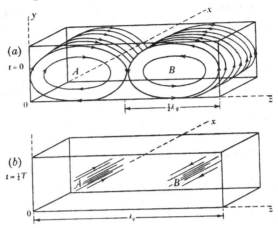

Fig. 6·2. Field patterns of the H_{012}-mode in a rectangular cavity.

It is evident that the same procedure will lead to the field patterns of more general modes of oscillation both in rectangular and cylindrical resonators. First, find the standing wave pattern in the unclosed guide, corresponding to the general E_{mn}- or H_{mn}-wave. The

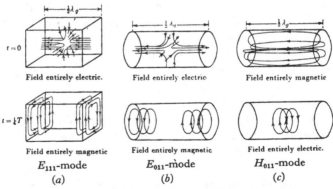

Fig. 6·3. Examples of modes of oscillation in cavities.

guide is then converted to a resonator by the introduction of conducting partitions at positions such as to enclose p-cells of the pattern within the resonator without violation of the conditions imposed on the behaviour of the electromagnetic field at the cavity surface.

The resulting mode is then designated, according to the type of its constituent progressive waves, an E_{mnp}- or an H_{mnp}-mode. Examples are shown in fig. 6·3 (*a*), (*b*) and (*c*), which show respectively the E_{111}-mode in a rectangular cavity, and the E_{011}- and H_{011}-modes in a cylindrical cavity, at times $t = 0$ and $t = \frac{1}{4}T$. We note, again, that the magnetic and electric fields oscillate in quadrature and that they are displaced relatively by $\frac{1}{4}\lambda_g$ in comparison with their positions in the corresponding pattern of the travelling wave.

6·4. Resonant wave-length of a cavity

We have seen that the E_{mnp}- and H_{mnp}-modes have field patterns the same as those of sets of p elementary cells of the corresponding E_{mn} and H_{mn} standing waves in a wave guide.

It follows that the length of the resonator must be $p\frac{1}{2}\lambda_g$, since each cell occupies $\frac{1}{2}\lambda_g$ of the axis of the wave guide. If, therefore, the length of the resonator is d when resonance occurs for an *mnp* mode, then

$$d = p\tfrac{1}{2}\lambda_g, \tag{1}$$

where λ_g is the wave-guide wave-length of the associated E_{mn}- or H_{mn}-wave in the wave guide whose cross-section is identical with that of the resonator.

We know, however, that the wave-length λ_g is related with the free-space wave-length λ and the cut-off wave-length λ_c by the equation

$$\frac{1}{\lambda^2} = \frac{1}{\lambda_g^2} + \frac{1}{\lambda_c^2}. \tag{2}$$

It follows from (1) and (2) that the resonant free-space wave-length of the cavity is given by

$$\frac{1}{\lambda^2} = \left(\frac{p}{2d}\right)^2 + \frac{1}{\lambda_c^2}. \tag{3}$$

We consider in turn rectangular and cylindrical resonators.

Rectangular resonators. Let the linear dimensions of the resonator along OX, OY and OZ (fig. 6·2) be respectively a, b and d.

The cut-off wave-length λ_c of an E_{mn}- and an H_{mn}-wave in a rectangular wave guide whose cross-section has linear dimensions a and b, is given by

$$\frac{1}{\lambda_c^2} = \left(\frac{m}{2a}\right)^2 + \left(\frac{n}{2b}\right)^2. \tag{4}$$

Finally, the resonant free-space wave-length λ_{mnp} of the E_{mnp}- and H_{mnp}-modes, is, from (3) and (4),

$$\frac{1}{\lambda_{mnp}^2} = \left(\frac{m}{2a}\right)^2 + \left(\frac{n}{2b}\right)^2 + \left(\frac{p}{2d}\right)^2. \tag{5}$$

The resonant frequency is

$$f_{mnp} = \frac{v}{\lambda_{mnp}} = \frac{v}{2}\sqrt{\left[\left(\frac{m}{a}\right)^2 + \left(\frac{n}{b}\right)^2 + \left(\frac{p}{d}\right)^2\right]}, \tag{6}$$

where v is the velocity of the TEM-wave. We have throughout assumed the cavity to be empty or air filled.

For instance, the resonant frequency of the H_{011}-mode is

$$f_{011} = \frac{v}{2}\sqrt{\left[\frac{1}{b^2} + \frac{1}{d^2}\right]},$$

and if $b > a$ this is the lowest frequency at which the cavity will resonate. Since there is no E_{0m}- or E_{m0}-wave in a rectangular wave guide the E-mode of the resonator with lowest frequency is the E_{111} whose frequency f_{111} is also that of the H_{111}:

$$f_{111} = \frac{v}{2}\sqrt{\left[\frac{1}{a^2} + \frac{1}{b^2} + \frac{1}{d^2}\right]}.$$

The resonant wave-length of the H_{011}-mode is

$$\frac{1}{\lambda_{011}} = \frac{1}{2}\sqrt{\left[\frac{1}{b^2} + \frac{1}{d^2}\right]}.$$

When the resonator is a cube ($a = b = d$) then

$$\frac{1}{\lambda_{011}} = \frac{1}{2}\frac{\sqrt{2}}{a}$$

or
$$\lambda_{011} = \sqrt{2}\,a = \text{diagonal of a face.}$$

This shows, as mentioned in paragraph 1, that the fundamental wave-length is of the order of magnitude of the linear dimensions of the cavity. The smallest cube therefore that will resonate at a wave-length of 10 cm. (3000 Mc./sec.) has an edge length of 7·07 cm.

Cylindrical resonators. The cut-off wave-lengths λ_c of waves in circular guides are not given by a simple formula such as 2·3 (5) but depend on the roots of Bessel functions. The cut-off wave-length λ_c in a wave guide of radius a, for some of the lower order modes are

Mode	H_{11}	E_{01}	H_{01} and E_{11}
λ_c	3·42a	2·61a	1·64a

The resonant wave-lengths of these modes in a cylindrical cavity of radius a and length d, are therefore given by

$$\frac{1}{\lambda^2_{mnp}} = \begin{cases} \left(\frac{1}{3\cdot42a}\right)^2 + \left(\frac{1}{2d}\right)^2 & ...H_{111}, \\ \left(\frac{1}{2\cdot61a}\right)^2 + \left(\frac{1}{2d}\right)^2 & ...E_{011}, \\ \left(\frac{1}{1\cdot64a}\right)^2 + \left(\frac{1}{2d}\right)^2 & ...H_{011} \text{ and } E_{111}. \end{cases} \qquad (1)$$

6·5. Charges and currents on internal surface of resonator

When electric lines of force begin on one portion of the boundary and end on another, they terminate on electric surface charges of

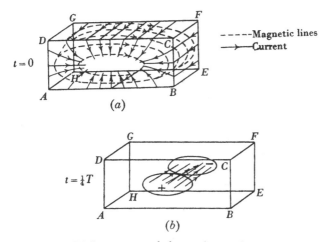

Fig. 6·4. Wall currents and charges in a cavity resonator.

opposite sign. For instance, in the resonator shown in fig. 6·2 (b), positive charge resides on the region around A and negative charge on that around B, with corresponding compensating charges on the opposite wall. When the magnetic field is present skin currents flow on the interior surface everywhere at right angles to the contiguous tangential magnetic field. Where there is no surface field there is no current. We consider the simple case of the H_{011}-mode in a rectangular resonator at the moment when the field is entirely magnetic (fig. 6·4 (a)) and later when it is entirely electric (fig. 6·4 (b)). The lines of current flow are shown running perpendicular to the magnetic field at the surface. The current is shown converging towards the central region of face $ABCD$ (fig. 6·4 (a)) and away from face $EFGH$. These faces become fully charged one-quarter of a cycle later, as shown in fig. 6·4 (b), and electric lines of force run from the positive to the negative charges.

6·6. Method of excitation of a cavity resonator

A cavity resonator may be excited, either by a loop (fig. 6·5 (a)) or a probe (fig. 6·5 (b)). In the former case the loop must be so introduced that lines of magnetic force can thread through it. Fig. 6·5 (a)

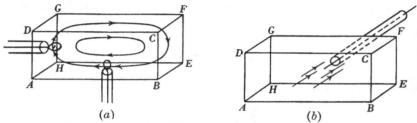

Fig. 6·5. Excitation of H_{01}-mode in a rectangular resonator.

shows two of several possible positions for the loop. As shown, one loop could be an input loop and the other an output loop. The degree of coupling can be controlled by rotation of the plane of the loop. When a probe is used it is introduced into a plane of maximum electric field strength and is set parallel to the electric field. Thus in fig. 6·5 (b) the probe is shown projecting into the cavity from the face $EFGH$.

Excitation to resonance can also be made by means of a slot cut in a face such as to interrupt the flow of current. This face could be made common with the wall of a wave guide from which current could be fed into the cavity.

6·7. The Q-factor of a cavity (quality factor)

Because of the finite conductivity of the walls, power is dissipated as heat in the walls and free oscillations decay exponentially. The Q-factor of the cavity is defined by the expression

$$Q = 2\pi \left(\frac{\text{Energy stored}}{\text{Energy dissipated per cycle}} \right). \tag{1}$$

In practice the walls are made of copper or silver, and the energy dissipated per cycle is only a small fraction of the stored energy; consequently Q is very large.

Let w be the energy stored, then the energy dissipated per cycle is $(dw/dt) T$ and expression (1) may be written

$$Q = \frac{2\pi}{T} \frac{w}{(dw/dt)} = \omega \frac{w}{(dw/dt)}, \tag{2}$$

whence

$$\frac{dw}{dt} = \frac{\omega}{Q} w. \tag{3}$$

Thus

$$w = w_0 \epsilon^{-(\omega/Q)t}, \tag{4}$$

where w_0 is the stored energy at $t = 0$. If, therefore, the resonator is shock excited and left to oscillate freely, the stored energy is reduced to $1/\epsilon$ of its initial value in a time $t' = Q/\omega = QT/2\pi$

or

$$Q = 2\pi \frac{t'}{T}. \tag{5}$$

According to (5) an alternative interpretation of Q is that it is 2π times the number of cycles required for the stored energy to decay to $1/\epsilon$ (approx. one-third) of its initial value.

Since Q values of 10^4 and greater are easily achieved, it is evident that a cavity will ring for a great many cycles before the stored energy is reduced to a small fraction of its initial value.

A useful approximate rule which gives the order of magnitude of Q* in terms of the skin depth δ of the wall currents, and the dimensions of the cavity is

$$Q \doteqdot \frac{\text{Volume of cavity}}{\delta \times \text{surface of cavity}}. \tag{6}$$

A formula for δ in terms of the wave-length and wall conductivity is

$$\delta = 2 \cdot 82 \times 10^{-2} \sqrt{\frac{\lambda}{\sigma}} \text{ m.},$$

where δ and λ are in metres and σ is in mhos per metre cube.

Suppose the resonator to be made of copper for which

$$\sigma = 5 \cdot 8 \times 10^7 \text{ mhos/m.}$$

Then

$$\delta = 3 \cdot 7 \times 10^{-6} \sqrt{\lambda} \text{ m.}$$

---

* The precise expression for Q depends on the mode of oscillation and the geometry of the cavity and requires separate calculation. Some typical examples are given in Sarbacher and Edson, *Hyper- and Ultra-High-Frequency Engineering*, p. 396.

According to equation (6) the Q value of a cubical resonator of edge a m., is

$$Q \doteqdot \frac{a}{6\delta} = \frac{a \times 10^6}{6 \times 3 \cdot 7 \sqrt{\lambda}}. \tag{7}$$

For the H_{011}-mode, $\lambda = a\sqrt{2}$, then

$$Q \doteqdot \frac{10^6 \sqrt{\frac{1}{2}\lambda}}{22 \cdot 2}.$$

For $\lambda = 10$ cm. $(= \frac{1}{10}$ m.$)$, this becomes

$$Q = 1 \cdot 01 \times 10^4.$$

Such a cavity, if shock-excited, would ring for

$$\frac{Q}{2\pi} \doteqdot \frac{10^4}{6} = 1 \cdot 6 \times 10^3$$

periods before the stored energy was reduced to $1/\epsilon$ of its initial value. Each period, $\lambda = 10$ cm., is $\frac{1}{3000}\,\mu$sec., so that the time of ring is about $0 \cdot 5\,\mu$sec.

Because they are highly selective, cavity resonators are used as wave-meters at centimetre wave-lengths. This, and other applications of importance in radar, are described below.

6·8. Applications of cavity resonators

(i) *Wave-meters*

Coaxial line wave-meter. A convenient wave-meter for use at wave-lengths of 9–11 cm. is the coaxial line wave-meter shown in fig. 6·6. It comprises a cylindrical cavity into which a rod can be intruded axially to any desired amount by means of a rack and pinion. The metal block serves as a guide for the rod and as a short circuit to the coaxial transmission-line system formed by the cavity and the rod. The cavity is excited by injecting an e.m.f. into the input loop from the source whose wave-length is required, and the rod is moved by means of the pinion until a position of resonance is indicated by the detecting crystal and microammeter fed from the output loop. The shortest resonant length l of the rod within the cavity is slightly less than $\frac{1}{4}\lambda$, since the transmission-line system is open-circuited but with some small end-capacitance, other resonant positions correspond to $(l + n\frac{1}{2}\lambda)$, where n is an integer, since the end-capacitance remains the same irrespective of the position of the end of the rod. This is because the end of the rod never closely approaches the closed end of the cavity C opposite to B. The rack and pinion carry a scale and vernier from which the displacement of the rod can be measured directly in centimetres. The displacement of the rod between successive resonances is equal to $\frac{1}{2}\lambda$ and gives directly the wave-length on the coaxial line, which is the same as the free-space wave-length of the source. The diameter of the cavity is small enough to ensure that cavity modes of oscillation cannot occur.

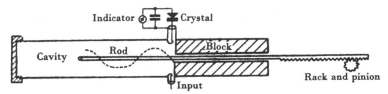

Fig. 6·6. Co-axial line wave-meter.

Resonant cavity wave-meter. A wave-meter suitable for measuring changes in wave-length with great accuracy is the cavity wave-meter shown in fig. 6·7 (*a*). It is a metal cylinder whose length can be adjusted by rotation of the screw head to which the upper end of the cylinder is attached. The resonant mode employed is the H_{011}-mode which is excited by an input loop near the middle of the cavity, as shown, and resonance is indicated by means of an output loop, at the same height but shifted through a quadrant, and a crystal detector and microammeter combination. The magnetic lines of force at resonance are indicated in fig. 6·7 (*a*). To prevent the excitation of the E_{111}-mode, which has the same resonant frequency as the H_{011}-mode, the movable end clears the cylinder wall with a small gap. The currents in the H_{011}-mode flow on the cavity surface in circles about the axis (fig. 6·7 (*b*)) and no flow occurs from the flat ends to the curved surface. The current distribution is therefore indifferent to the presence of the gap between the movable flat end and the curved wall. In all other modes, except H_{0mn}-modes, current is required to flow from the flat ends to the curved wall; consequently the introduction of the gap effectively suppresses these unwanted modes. Wire filters of suitable form can also be used as suppressors, but they are less convenient. In principle the resonant wave-length could be obtained from the

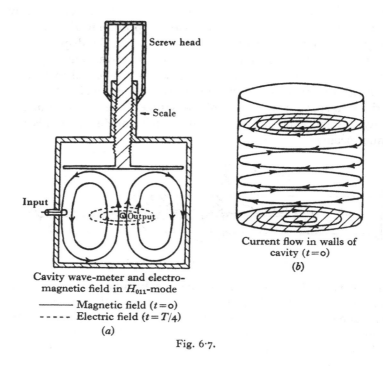

Screw head

← Scale

Input

Output

Cavity wave-meter and electromagnetic field in H_{011}-mode

——— Magnetic field $(t=0)$
- - - - - Electric field $(t=T/4)$

(a)

Current flow in walls of
cavity $(t=0)$

(b)

Fig. 6·7.

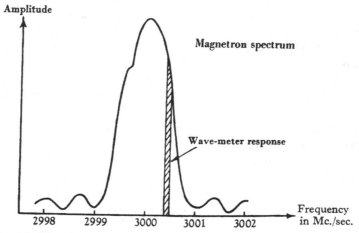

Amplitude

Magnetron spectrum

Wave-meter response

Frequency
in Mc./sec.

2998   2999   3000   3001   3002

Fig. 6·8. Use of wave-meter for examining magnetron frequency spectrum.

dimensions of the cavity at resonance, but in practice it is more accurate to measure displacement of the lid by means of a micrometer screw and scale, as indicated.

This scale is then calibrated against the harmonics of a crystal oscillator, or, more crudely, against the coaxial line wave-meter described above, using a tunable source of e.m.f. The wave-length scale is very open, and a small change in wave-length corresponds to a relatively large displacement of the movable end. The interior of the cavity is usually silvered so that a large Q and sharp resonance results. The wave-meter is then suitable for examining the r.f. spectrum of a magnetron pulse (fig. 6·8).

(ii) *Echo boxes*

It is often necessary to check the overall performance of a centimetre-wave radar equipment in situations where it is difficult to obtain echoes from objects at suitable ranges. For instance, a radar equipment in an aircraft may have its scanner directed downwards so that it is impossible to obtain echoes when the aircraft is on the ground. The echo box is a simple device for checking roughly the overall performance of a set. It is merely a resonant cavity designed to possess a high Q. The cavity is shock-excited by the transmitter pulse and continues to ring and emit a signal which is spread along the time base for an appreciable distance after the cessation of the transmitter pulse. A possible arrangement is shown in fig. 6·9. The echo box is fed via a screened low-loss cable from a pick-up probe fixed near the edge of the mirror. The energy abstracted from the transmitted pulse is stored as a resonant mode in the box and is re-radiated to the receiver as an exponentially decaying signal. For the greater part of its duration the signal saturates the receiver but finally decays to a level at which it no longer does so. The appearance on a type A display is illustrated in the figure. The range at which the echo box response disappears into the 'noise' gives an indication of the overall performance of the set.

Echo boxes are of two types—tuned and untuned. Tuned echo boxes are the same as the wave-meter already described and illustrated in fig. 6·7. They require tuning to the frequency of the transmitter which then shock excites the H_{011}-mode. The output loop is not used. Untuned echo boxes are very large cavity resonators whose lowest modes, H_{111}, E_{111}, etc., correspond to wave-lengths much greater than the wave-length of the radar set. They are usually hollow cubes with copper walls.

The modes that are excited by the radar transmitter are therefore higher order E_{mnp}- and H_{mnp}-modes where m, n and p are relatively large integers.

Consider a cubical resonator whose edge a comprises several wave-lengths λ.

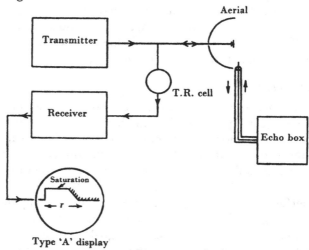

Fig. 6·9. Radar system with echo box.

It can be shown that the spacing of the higher modes is such that the number of modes comprised within the wave-length range λ to $(\lambda + \Delta\lambda)$ or frequency range f to $(f - \Delta f)$ in a rectangular resonator is

$$N \doteq 8\pi \frac{V}{\lambda^3} \left(\frac{\Delta\lambda}{\lambda}\right) = 8\pi \frac{V}{\lambda^3} \left(\frac{\Delta f}{f}\right),$$

where V is the volume of the resonator; thus for a cubical resonator of edge a

$$N \doteq 8\pi \left(\frac{a}{\lambda}\right)^3 \left(\frac{\Delta\lambda}{\lambda}\right) = 8\pi \left(\frac{a}{\lambda}\right)^3 \left(\frac{\Delta f}{f}\right).$$

Let us consider the case of a resonator of edge $a = 1$ m. excited by a magnetron pulse for which $f = 3000$ Mc./sec., $\lambda = 10$ cm. and the band width is 1 Mc./sec. The formula gives, for the number of modes covered by the band width,

$$N \doteq \frac{8\pi \times 10^3}{3 \times 10^3} = \frac{8\pi}{3},$$

i.e. there are eight modes. Thus, the transmitter pulse is able to excite a number of resonant modes whatever its frequency f, and it is not necessary to tune the resonator. In the case discussed, the mean frequency separation of the modes is $\frac{1}{8}$ Mc./sec. = 125 Kc./sec.

Since the ratio of volume to surface of a cubical resonator is proportional to the edge a, these large untuned resonators possess very high Q values and will ring for many microseconds.

(iii) *Cavity resonators in centimetre-wave oscillators*

The usual resonant circuits of radio comprising lumped inductances, capacitances and resistances cannot be constructed in a useful form at wave-lengths of 10 cm. and less, and it is necessary to replace them by other resonant systems such as resonant lengths of coaxial transmission-line or resonant cavities. In the klystron oscillator, whose main application is as a low-power local oscillator in centimetre-wave radar receivers, a resonant cavity is used in which a mode of oscillation is maintained by a 'bunched' electron stream. In order to bring about 'bunching' and to abstract power from the 'bunched' stream, the electrons must pass through the oscillating fields within the cavity against or parallel to the electric field where it is most intense. Further, the time of transit must be small compared with the period of oscillation. It is not possible to accomplish this is any of the resonators described so far because their dimensions are comparable with the wave-length λ, and it would be necessary for the electrons to move at speeds comparable with the velocity of light in order that the transit time should be very much less than the period of oscillation. What is required therefore is á resonant cavity (rhumbatron) in which an intense oscillating electric field is concentrated across a short path so that electrons can travel the whole extent of a line of force in a time short compared with the period.

In one type of reflector klystron, the Sutton tube, an example of which is the CV 67, the resonator (rhumbatron) assumes the form indicated in fig. 6·10 (b). This rhumbatron may itself be regarded as a distorted form of the prototype shown in fig. 6·10 (a). In fig. 6·10 (a), A and B represent a pair of coaxial conducting cones with their tips removed so as to form a small gap between them. It is known that when an alternating e.m.f. is applied across the gap the pair of cones forms a transmission-line system and that a principal (TEM) wave is guided along them. The lines of electric

and magnetic force run as shown in the figure. If a conducting
spherical surface C of radius $\frac{1}{4}\lambda$ concentric with the middle of the

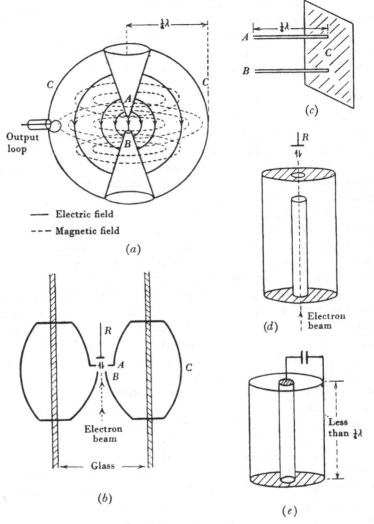

Fig. 6·10. Common types of resonators.

gap is used to close the cones, then the TEM wave is reflected
without distortion and a complete standing wave is produced on
the transmission line, which then forms a resonant system. The

equivalent twin transmission-line system is shown in fig. 6·10 (c). We may, however, regard the system of fig. 6·10 (a) as a hollow spherical cavity resonator with a pair of conical projections. It follows from what has been said that the fundamental resonant wave-length of this resonator is four times the radius of the sphere. A voltage antinode is located at the gap AB where the electric field attains its greatest intensity. It is possible, therefore, to maintain such a rhumbatron in resonance by passing a bunched electron stream across the gap AB. In practice, in the reflector klystron used as a local oscillator, it is necessary to control the resonant frequency by means of external tuning screws; consequently, the rhumbatron is divided in two by a glass tubular envelope which is evacuated and contains the electron gun assembly and reflecting electrodes. The portion of the resonator external to the glass envelope carries the tuning screws. To introduce the glass envelope it is necessary to distort the shape of the rhumbatron to that shown in fig. 6·10 (b). One of the cones is also distorted to bring the reflector close to the gap. Power is abstracted through a loop placed, as shown in fig. 6·10 (a), with its plane parallel to the axis of the cones.

The input impedance to the transmission-line system shown in fig. 6·10 (a) is large at resonance, and it is necessary to drive the rhumbatron from a high-impedance source. Thus the power supplied by the electron stream must be in the form of relatively high-voltage electrons and relatively small current. For instance, in the CV67, the accelerating voltage is 1200 V. and the electron current is 6 mA. The maximum power output at $\lambda = 9$ cm. is 200 mW. which is ample for a local oscillator.

The high-operating voltage is an inconvenience, and a more convenient form of klystron operates on a voltage of 300 and is therefore able to use the same power pack as the radar receiver. The resonator here comprises a short-circuited coaxial transmission line with a gap between the inner and the end-plate of the outer at the top, as shown in fig. 6·10 (d).

This end-plate has a hole in its centre, and both this hole and the end of the inner are covered by a wire gauze through which the electron stream passes into and out of the gap. The length of the transmission line is less than $\frac{1}{4}\lambda$, and it is brought to resonance by the capacitance between the face end of the inner conductor and the end-plate.

The equivalent resonant system is shown in fig. 6·10(e). The shunting impedance of this system is much lower than that of the double-cone system, and the power carried by the electron stream can be supplied at lower voltage and larger current. Typical operating values are: voltage 300, current 30 mA., r.f. output 20 mW.

An important feature of this oscillator, due to the small Q value of the resonator, is that the frequency can be varied considerably by changing the reflector voltage. This makes the valve suitable for incorporation in systems for A.F.C. at centimetre wave-lengths. It should be noted that the Q value of any cavity resonator is greatly reduced by coupling it to a matched output cable. In the case of the CV 67 the Q of the loaded rhumbatron is of the order 300.

6·9. Measurement of power factor of dielectrics

The high-Q properties of cavity resonators have been used in a method for measuring the power factor at microwave-lengths of low-loss dielectric materials such as polythene* ($F = 0.0005$).

When a piece of dielectric material is placed in an alternating electric field some power is wasted as heat in the dielectric. The power-factor F of the dielectric may be defined to be

$$F = \left(\frac{\text{Power lost}}{\omega \times \text{peak energy density of electric field during the cycle}} \right), \quad (1)$$

where $\omega = 2\pi f$. In many substances F is a constant independent of frequency, consequently, in such substances, the wastage of power is proportional to ω. The physical interpretation of this result is that in each cycle of polarization the same fraction of the stored energy is dissipated; thus the rate of dissipation is proportional to the number of cycles per second, that is, to the frequency.

Suppose a specimen of dielectric to be placed inside a resonant cavity. Then, to the loss of power caused by the conductivity of the walls, must be added the power lost in the dielectric specimen. The Q-factor of the cavity as defined in equation 6·7 (1) may also be written

$$Q = \frac{\omega \times \text{energy stored}}{(\text{Power lost})}. \quad (2)$$

* 'Resonant methods of dielectric measurement at centimetre wave-lengths', by F. Horner, T. A. Taylor, R. Dunsmuir, J. Lamb and Willis Jackson, *J. Inst. Elect. Engrs*, 1945, vol. 53, part III, p. 53.

The power lost is the sum of two terms, one P_R, the power lost in the walls, and the other P_D, that lost in the dielectric. Equation (2) may therefore be written

$$\frac{1}{Q} = \frac{P_R}{\omega W} + \frac{P_D}{\omega W} = \frac{1}{Q_R} + \frac{1}{Q_D}, \tag{3}$$

where W is the energy stored in the oscillation.

The quantity Q_R is the Q-factor of the cavity when there is no dielectric loss, and Q_D is

$$Q_D = \frac{\omega W}{P_D}.$$

But from (1), $P/\omega = AWF$

or $\dfrac{1}{Q_D} = AF,$ (4)

where A is a constant, which depends on the size and form of the specimen of dielectric, the mode of oscillation, and the position of the dielectric in the cavity. This constant can be calculated from the conditions of the experiment.

The experimental procedure is to measure Q and Q_R for the cavity, with and without the dielectric respectively, and to obtain Q_D from (3) and F from (4).

The measurement of Q and Q_R is based upon a definition of Q which is essentially equivalent to (2). If the tuning curve is taken of amplitude of oscillation against frequency about the resonant frequency f_0, then the frequencies $(f \pm \Delta f)$ at which the amplitude falls to $1/\sqrt{2}$ of its resonant value serve to determine the Q of the system through the following equation:

$$Q = f_0/2\Delta f. \tag{5}$$

If resonance is indicated by a crystal detector then Δf corresponds to a crystal current one-half maximum, since for small currents the crystal has a square-law rectified response. Two types of cavity were used; the first is a cylindrical cavity in which the E_{010}-mode is excited. This mode is illustrated in fig. 6·11. It differs from the cylindrical modes discussed earlier in this chapter, in that it is not resolvable into a pair of waves travelling in opposite senses along the axis of the cylinder but into a pair of radial waves, one converging on to the axis and the other diverging from it. There is no variation of the pattern in the direction of the axis, nor with azimuth angle θ, but only radially.

The resonant frequency is independent of the length of the cylinder, and the cavity oscillating in this mode cannot therefore be used as a wave-meter tuned by variation of its length. The tuning curve is here obtained by variation of the exciting frequency f.

The dielectric specimen takes the form of a cylindrical rod lying along the axis of the cavity and extending the whole distance between the ends.

A more convenient method is to employ a resonant cavity of the form described in §6·8, in an H_{01p}-mode of oscillation. Here the length of the cavity is variable by known amounts read from a micrometer screw. A tuning curve of crystal response in terms of cavity length near the resonant length can be translated, by a suitable formula, into the equivalent

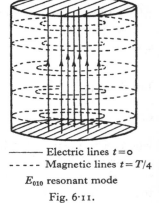

—————— Electric lines $t=0$

- - - - - Magnetic lines $t=T/4$

E_{010} resonant mode

Fig. 6·11.

tuning curve of the cavity with fixed length but with a variable exciting frequency. This method is found to be more convenient than the former which demands a variable frequency oscillator.

These methods also give the dielectric constant of the specimen, from the resonant frequency of the cavity containing the specimen.

A coaxial line resonator is used at longer wave-lengths.

6·10. Q-factor of a dielectric

Instead of describing the electric behaviour of a dielectric in terms of its power factor F, as is the current practice, it is more convenient to employ, as has been suggested by Schelkunoff, a quality factor Q defined for the dielectric as

$$Q = \frac{\omega\epsilon}{g}, \tag{1}$$

where $\epsilon = \epsilon_0 K_e$ and g is the conductance of the dielectric material at the frequency $\omega/2\pi$.

Equation (1) may be written

$$Q = \frac{\omega\epsilon E^2}{g E^2} = \left(\frac{\omega \times \text{peak energy density}}{\text{Mean power loss}}\right). \tag{2}$$

Consequently, from (2) and 6·9 (1),

$$Q = \frac{1}{F}.$$

For instance, if $F = 0\cdot0006$, then $Q = 1\cdot66 \times 10^3$. Thus the term Q_D in equations 6·9 (3) and (4) is proportional to the Q-factor of the dielectric.

This factor is also of value in describing the reflexion of electromagnetic waves from sea water or soil. It is known that at frequencies much less than a certain 'critical frequency' f_c, reflexion from the surface of a partial conductor resembles that from a good conductor and at frequency much greater than f_c, that from a dielectric. It is interesting to note that the critical frequency f_c is that at which the Q-factor of the partial conductor becomes equal to unity. The critical frequency is therefore

$$f_c = \frac{g}{2\pi\epsilon_0 K_e} = 1\cdot8 \times 10^{10} \frac{g}{K_e}. \tag{3}$$

The corresponding 'free-space' wave-length $\lambda_c = \frac{3\cdot10^8}{f_c}$ is

$$\lambda_c = \frac{K_e}{60g} \text{ m}. \tag{4}$$

For instance, for sea water, $g = 5$, $K_e = 78$, $\lambda_c = 0\cdot26$ m. Thus the sea reflects electromagnetic waves whose wave-lengths are less than 26 cm. as though it were a dielectric, but longer waves as though it were a conductor. For dry earth, $g \doteqdot 10^{-4}$, $K_e \doteqdot 6$, $\lambda_c = 10^5$ m. For wave-lengths less than this dry earth or rock behaves as a dielectric.

6·11. Equivalent circuit of a resonator

The input impedance of a resonator excited through a coaxial cable by a loop or probe, as shown for instance in fig. 6·5 (a) and (b), has been investigated theoretically by Hansen, Condon, Slater, Schwinger and other American mathematical physicists (see §§ 7·18·3 and ·4).

It emerges from these investigations, which are based on the theory of the electromagnetic field, that the input impedance to an unloaded cavity excited by a loop is represented by an expression of the form

$$Z = R + \sum_{k=1}^{\infty} \frac{j\omega M_k^2}{\left(\omega_k^2 - \omega^2 + j\frac{\omega_k\omega}{Q_k}\right)}, \tag{1}$$

in which ω is the angular frequency of the exciting current in the loop, ω_k that of resonance of the kth mode and Q_k the Q-factor of the cavity for the kth mode. R is the self-resistance of the loop and may be neglected. M_k is called the coupling coefficient of the kth mode. When ω coincides with one of the resonant values ω_k then the denominator of the term involving the suffix k becomes small and that term predominates. This implies that the corresponding mode is strongly excited. The input impedance then becomes

$$Z \doteqdot M_k^2 Q_k / \omega_k.$$

The equivalent circuit is a chain of rejector L-C circuits in series with each other and the loop resistance. Each rejector circuit is shunted by a resistance and is tuned to resonate at one of the ω_k. The input impedance Z, which is the total impedance across this chain, becomes large whenever ω becomes equal to one of the ω_k.

The coupling coefficients M_k^2 are proportional to the square of the area of the loop and inversely proportional to the fifth power of the linear dimensions of the cavity. The input impedance at resonance is therefore very sensitive to change in cavity dimensions.

In the position of maximum coupling shown in fig. 6·5 (a) the value of M_k^2 for a loop of area S feeding a cubical cavity with side a is

$$M_k^2 = 4 \cdot 46 \times 10^{12} S^2 / a^5$$

(when Z in (1) is expressed in ohms and all lengths are in metres and areas in square metres). For a cylindrical cavity of radius a and length l driven in the H_{011}-mode (fig. 6·3 (c))

$$M_k^2 = 1 \cdot 06 \times 10^{12} S^2 / a^4 l.$$

The corresponding input impedance with probe excitation is

$$Z = -j\omega C + j \sum_{k=1}^{\infty} \frac{\omega_k (M_k')^2}{\left(\omega_k^2 - \omega^2 + j\dfrac{\omega_k \omega}{Q_k}\right)}. \tag{2}$$

The equivalent circuit is the same as before, but with the loop resistance R replaced by the series capacity C—the electrostatic capacity between the probe and the walls of the vessel. These expressions (1) and (2) were first given by Condon.

When ω does not coincide with an ω_k then Z in (1) is

$$Z \doteqdot j \sum_{\omega < \omega_k} \frac{\omega M_k^2}{(\omega_k^2 - \omega^2)} - j \sum_{\omega > \omega_k} \frac{\omega M_k^2}{(\omega^2 - \omega_k^2)}.$$

It is therefore a series combination of inductive and capacitive reactances.

The equivalent result follows from (2) for the probe feed.

6·12. Resonator method for precision measurement of wave-guide discontinuities

When a discontinuity, such as an obstacle, is introduced into a cavity resonator the resonant dimensions of the cavity are changed. This fact was used by W. H. Pickering, D. W. Hagelbarger, C. Y. Meng and L. C. Snowden in a method of measuring with great precision the constituent elements in the circuit equivalent of a wave-guide discontinuity such as an obstacle or sharp bend.

The resonant cavity (fig. 6·12 (a)), which is rectangular, comprises a length of accurately constructed American X-band rectangular wave guide (internal dimensions 0.9×0.4 in.) terminated at each end by choke plungers. The movement of each plunger along the wave guide is controlled by accurate micrometer screws which are mounted against a coupling flange (not shown) that can be clamped to the flange at the end of the wave guide. The cavity is excited to resonance through a small hole in the face of one of the plungers which is made hollow and itself forms a wave guide which is fed from a klystron oscillator through a coaxial cable and a wave guide.

Attenuators are introduced in order to decouple the resonant cavity almost completely from the oscillator whose frequency is virtually unaffected when the resonant cavity is tuned. The klystron output is modulated by a 1000-cycle square wave and resonance of the cavity is indicated by the 1000-cycle output signal from a low-frequency amplifier whose input is fed from a small pick-up probe that projects a very small distance into the cavity as shown.

The micrometer screw readings R_1 and R_2 can be calibrated so that their sum indicates the distance between the faces of the plungers. First, consider the relation between the readings R_1 and R_2 at resonance when the cavity is empty. Let R_1 be given a series of values and let the corresponding values of R_2 for resonance be measured. Then $(R_1 + R_2)$ remains constant unless the wave guide is not uniform in cross-section. This forms a good test of the quality of the wave guide. Suppose that the wave guide satisfies the test for uniformity and that next a simple symmetrical obstacle such as a thin wire, iris or thin post is mounted in it with the axis of the obstacle in the cross-section of the wave guide.

The equivalent circuit of the cavity is now that shown in fig. 6·12 (b). It comprises a pair of short-circuited lengths of transmission line with unit characteristic impedance and respective

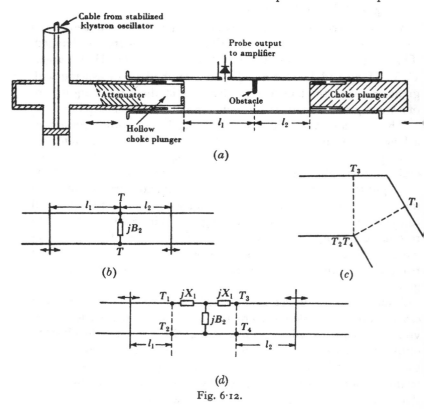

(a)

(b)   (c)

(d)

Fig. 6·12.

lengths l_1 and l_2, and a susceptance jB_2 all connected in shunt across a common pair of terminals TT. The susceptance presented by the lines at TT are respectively

$$jb_1 = -j \cot k l_1 \quad \text{and} \quad jb_2 = -j \cot k l_2,$$

where $k = 2\pi/\lambda_g$.

The condition for resonance is that the total admittance across TT is zero, that is

$$b_1 + b_2 + B_2 = 0, \qquad (1)$$

or

$$B_2 = \cot k l_1 + \cot k l_2. \qquad (2)$$

When a set of pairs of distances l_1 and l_2 of the plungers from the axis of the obstacle has been found at resonance, the same value of B_2 should be obtained from (2) by using any pair of values l_1 and l_2 of the set. When the obstacle is thin this is what is found, but when the obstacle is thick, in the direction of the wave-guide axis, although still symmetrical, the susceptance B_2 is not found to be the same for all pairs of values of l_1 and l_2 when relation (2) is used to calculate it. The reason is that a thick obstacle cannot be adequately represented by a simple-shunt susceptance as in fig. 6·12 (b), but proves, instead, to be accurately represented by a filter section comprising both series and shunt elements. In fig. 6·12 (d), which is the equivalent circuit of a resonator containing a thick obstacle, a symmetrical T-section filter has been chosen, with series reactances X_1 and shunt susceptance B_2.

The terminals T_1, T_2, T_3 and T_4 coincide within the wave guide at the axis of the obstacle, from which the distances l_1 and l_2 of the plungers are supposed to be measured.

The condition for resonance is

$$jB_2 + \frac{1}{\left(jX_1 + \frac{1}{jb_1}\right)} + \frac{1}{\left(jX_1 + \frac{1}{jb_2}\right)} = 0$$

or
$$B_2 + \frac{b_1}{(1 - b_1 X_1)} + \frac{b_2}{(1 - b_2 X_1)} = 0. \qquad (3)$$

Equation (3) is equivalent to

$$\left[b_1 + b_2 - \frac{2}{X_1}\frac{(1 - B_2 X_1)}{(2 - B_2 X_1)}\right]^2 = \left[\frac{2}{X_1(2 - B_2 X_1)}\right]^2 + (b_1 - b_2)^2. \qquad (4)$$

Equation (4) expresses $(b_1 + b_2)$ as a function of $(b_1 - b_2)$, where b_1 and b_2 are any pair of shunting susceptances of the wave-guide arms that bring the obstacle to resonance. The two branches of the hyperbola, of which (4) is the equation, are shown in fig. 6·13, which is the graphical representation of the dependence of $(b_1 + b_2)$ on $(b_1 - b_2)$.

At the points P and Q, one on each branch, where $b_1 - b_2 = 0$, $b_1 + b_2 = 2/X_1$ and $-2B_2/(2 - X_1 B_2)$ respectively, or
$$b_1 = 1/X_1 \quad \text{or} \quad -B_2/(2 - X_1 B_2).$$

Thus, there are two distances $l_1 = L_1$ and $l_1 = L_2$ at which resonance occurs with the subsidiary condition $l_1 = l_2$. Further, at these distances

$$X_1 = \frac{1}{b_1} = -\tan\left(\frac{2\pi L_1}{\lambda_g}\right), \quad \left(X_1 - \frac{2}{B_2}\right) = -\tan\left(\frac{2\pi L_2}{\lambda_g}\right). \qquad (5)$$

If, therefore, these distances can be measured the elements X_1 and B_2 are derived immediately from them by means of equations (5). The advantage of using these special distances lies in the fact that they involve measurement of the total distance only between the plunger faces and not of the distance of each plunger face from the axis of the obstacle.

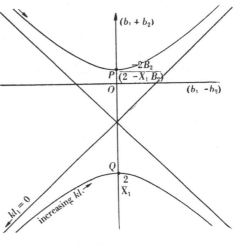

Fig. 6·13.

The positions of the plungers, when $l_1 = l_2$, are found as follows: an extension of the theory shows that if l_1 be increased and l_2 be adjusted to give resonance then a curve in which $(l_1 + l_2)$ (the distance between the piston faces at resonance) is shown as a function of l_1 possesses a maximum and minimum, and that at the maximum $l_1 = L_1$ and at the minimum $l_1 = L_2$. It suffices to plot the sum of the micrometer readings $(R_1 + R_2)$ against R_1, as is done in fig. 6·14 (a).

This curve, which is taken from an unpublished (American) report by Pickering, Snowden and Hagelbarger, represents the experimental results obtained with a metal post of diameter $\frac{1}{16}$ in., ratio length to diameter 0·746, a wave-length $\lambda = 3·4$ cm. and $\lambda_g = 2·0000$ in. When the maxima and minima have been roughly located from this curve, they are next accurately found by taking a great number of measurements around the maximum and minimum and then replotting these portions of the graph on an enlarged scale.

L_1 and L_2 are half the separations of the plungers at the positions of the maximum and minimum respectively.

The equivalent filter section may be specified either in terms of X_1 and $X_2 = -1/B_2$, or of quantities $S = X_1$ and

$$T = (X_1 + 2X_2) = (X_1 - 2/B_2),$$

which are obtained immediately from equations (5).

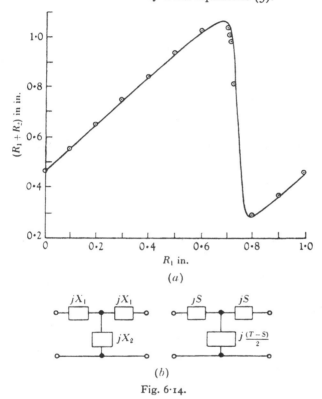

(a)

(b)

Fig. 6·14.

The equivalent T-section is shown in fig. 6·14 (b).

The method was also applied to find the equivalent T-sections of wave-guide corners such as that shown in fig. 6·12 (c). Here the distances l_1 and l_2 of the plungers are measured from the cross-sections T_2, T_3 and T_4, T_1, and the equivalent T-section represents the portion of the corner bounded by these cross-sections. They correspond to the terminals T_1, T_2 and T_3, T_4 in fig. 6·12 (d). The distances L_1 and L_2 are found, as before, but here the micrometers move within lengths of wave guide coupled through the corner by means of choke flanges, the distances l_1 and l_2 being measured from the sections T_1, T_2 and T_3, T_4 respectively.

The following are specimen results of measurements on metal posts:

Metal post: height h, diameter $b = \frac{1}{16}$ in. (flat end),
$\lambda = 3\cdot4$ cm., $\lambda_g = 2\cdot0000$ in.

h/b	0·249	0·497	0·746	0·871
X_1	−0·0045	−0·0099	−0·0143	−0·165
X_2	−6·4809	−1·0147	−1·894	−0·0345
B_2	0·1543	0·9855	5·280	29·99

h/b	0·921	0·934	0·993	1·000
X_1	−0·1080	−0·0183	−0·0195	−0·0195
X_2	0·0160	0·0307	0·1510	0·240
B_2	−62·60	−42·57	−6·623	−4·151

Metal post: height h, diameter $b = \frac{1}{4}$ in. (flat end),
$\lambda = 3\cdot4$ cm., $\lambda_g = 1\cdot9998$ in.

h/b	0·252	0·499	0·760	0·935	1·000
X_1	−0·047	−0·101	−0·174	−0·227	−0·256
X_2	−0·1775	−0·468	−0·166	−0·053	0·026
B_2	0·5633	2·135	6·027	18·80	−37·76

It can be seen that the series reactance of the thin post is negligible in comparison with the shunt reactance, which is capacitive for short posts but becomes positive on passing through resonance. The series reactance is important in the thick post which therefore cannot be represented by a simple-shunt element.

The method gives an accuracy of a different order from that of the standing wave indicator. This is due in part to the sharpness of the resonance, and to the fact that the probe of the resonance detector does not require to be moved. Every care was taken to stabilize the oscillator frequency, the oscillator being water cooled.

The measurements are thought to be accurate to 0·1 %.

REFERENCES

Sarbacher and Edson. *Hyper- and Ultra-High Frequency Engineering.* John Wiley and Chapman and Hall.

M.I.T. Radar School. *Principles of Radar.* McGraw-Hill Book Co.

Chapter 7

MATHEMATICAL TREATMENT
OF SELECTED TOPICS

7·1. Introduction

The preceding chapters form a survey of the subject of wave guides from which all but relatively simple mathematical analysis has been excluded. This final chapter has been added for the sake of those who may prefer a less physical and more formal approach to the subject. The chapter is, therefore, of the nature of a mathematical appendix in which some of the matters mentioned in the survey receive a more systematic mathematical treatment. As there is little reason to repeat in detail the treatments to be found in a number of recent treatises* where the analysis is developed in terms of the field vectors **E** and **H**, and of vector and scalar potentials, the opportunity is taken here of demonstrating the advantages of using special single-component Hertz vectors from which the field components may be derived by the appropriate differential operations.

The method that we adopt was originally introduced by Debye † in 1908, who showed that it is often possible to derive all the components of an electromagnetic field from a single scalar quantity which is the magnitude of a single component vector. We therefore proceed to examine what are the conditions to be satisfied by a system of orthogonal curvilinear coordinates in order that the components in it of an electromagnetic field shall be derivable from a single scalar quantity.

7·2. Maxwell's field equations and Hertz vectors

Maxwell's equations of the electromagnetic field, expressed in m.k.s. units, are:

$$\operatorname{curl} \mathbf{E} = -\dot{\mathbf{B}}, \qquad \operatorname{div} \mathbf{B} = 0, \\ \operatorname{curl} \mathbf{H} = \dot{\mathbf{D}} + \mathbf{J}, \qquad \operatorname{div} \mathbf{D} = \rho, \tag{1}$$

* Ramo and Whinnery, *Fields and Waves in Modern Radio*; Sarbacher and Edson, *Hyper- and Ultra-High Frequency Engineering*.

† Frank and von Mises, *Die Differential und Integral Gleichungen der Mechanik und Physik*, part II, p. 873; Lamont, *Wave Guides*, Methuen Monograph, chapter 2.

where $\mathbf{B} = \mu\mathbf{H}, \quad \mathbf{D} = \epsilon\mathbf{E}, \quad \mu = \mu_0 K_m, \quad \epsilon = \epsilon_0 K_e,$

and K_m and K_e are respectively the magnetic and electric specific inductive capacities, and μ_0 and ϵ_0 the magnetic and electric inductive capacities of a vacuum (§ 1·2).

$$\mathbf{J} = \text{current density in amp./sq.m.}$$

$$\rho = \text{charge density in coulombs/m.}^3$$

We shall define Hertz vectors from which $\mathbf{B}$ and $\mathbf{E}$ may be derived, assuming, however, that in (1) $\mathbf{J}$ and ρ are zero, as always occurs in those applications that will concern us.

We write therefore:

$$\text{curl}\,\mathbf{E} = -\dot{\mathbf{B}}, \tag{2}$$

$$\text{curl}\,\mathbf{H} = \dot{\mathbf{D}}, \tag{3}$$

$$\text{div}\,\mathbf{B} = 0, \tag{4}$$

$$\text{div}\,\mathbf{D} = 0. \tag{5}$$

7·3. The electric Hertz vector Π

To satisfy equation 7·2 (4), define a Hertz vector Π as follows:

$$\mathbf{B} = \mu\epsilon\,\text{curl}\,\dot{\Pi}, \tag{1}$$

whence, from 7·2 (2),

$$\text{curl}\,\mathbf{E} = -\mu\epsilon\,\text{curl}\,\ddot{\Pi},$$

or $$\mathbf{E} = -\mu\epsilon\ddot{\Pi} + \text{grad}\,F, \tag{2}$$

where F is a scalar field function such that $\text{grad}\,F$ is of the nature of a functional constant of integration whose curl is zero, and whose form may be assigned according to convenience.

It follows from equations 7·2 (3) and 7·3 (1) that

$$\text{curl}\,\mathbf{H} = \frac{1}{\mu}\,\text{curl}\,\mathbf{B} = \epsilon\,\text{curl}\,\text{curl}\,\dot{\Pi} = \epsilon\mathbf{E},$$

whence $$\mathbf{E} = \text{curl}\,\text{curl}\,\Pi. \tag{3}$$

Comparison of the two expressions (2) and (3) for $\mathbf{E}$ shows that Π satisfies the following differential equation:

$$\text{curl}\,\text{curl}\,\Pi + \mu\epsilon\,\ddot{\Pi} = \text{grad}\,F. \tag{4}$$

7·4. The magnetic Hertz vector $\mathbf{\Pi}_0$

Since we have assumed $\operatorname{div}\mathbf{D} = \mathrm{o}$, we may introduce a second Hertz vector $\mathbf{\Pi}_0$ as follows:

$$\mathbf{D} = -\mu\epsilon\operatorname{curl}\mathbf{\Pi}_0. \tag{1}$$

It follows from equation 7·2 (3) that

$$\mathbf{H} = -\mu\epsilon\ddot{\mathbf{\Pi}}_0 + \operatorname{grad}G, \tag{2}$$

where $\operatorname{grad}G$ is a functional constant of integration. Further, from 7·2 (2)

$$\mathbf{H} = \operatorname{curl}\operatorname{curl}\mathbf{\Pi}_0. \tag{3}$$

Consequently, from (2) and (3)

$$\operatorname{curl}\operatorname{curl}\mathbf{\Pi}_0 + \mu\epsilon\ddot{\mathbf{\Pi}}_0 = \operatorname{grad}G. \tag{4}$$

7·5. Derivation of an electromagnetic field from a single scalar quantity

We have to consider what restrictions must be placed upon a system of orthogonal curvilinear coordinates in order that the components in it, of the field vectors of an electromagnetic field, may be derived from a single component Hertz vector of the form

$$\mathbf{\Pi} = \mathbf{i}_1 U + \mathbf{i}_2 . \mathrm{o} + \mathbf{i}_3 . \mathrm{o},$$

where U is the scalar magnitude of the vector and $\mathbf{i}_1$, $\mathbf{i}_2$ and $\mathbf{i}_3$ are unit vectors in the orthogonal curvilinear coordinate system;* that is, if $\mathbf{ds}$ is an elementary space interval at an arbitrary point, then

$$\mathbf{ds} = \mathbf{i}_1 h_1 du_1 + \mathbf{i}_2 h_2 du_2 + \mathbf{i}_3 h_3 du_3,$$

where h_1, h_2 and h_3 are the differential multipliers of the system and u_1, u_2, u_3 the variables.

Since $\mathbf{\Pi} = \mathbf{i}_1 U$ satisfies the vector equation 7·3 (4), we deduce

$$\left. \begin{aligned} (\operatorname{curl}\operatorname{curl}\mathbf{\Pi})_1 + \mu\epsilon\ddot{\mathbf{\Pi}}_1 &= (\operatorname{grad}F)_1, \\ (\operatorname{curl}\operatorname{curl}\mathbf{\Pi})_2 &= (\operatorname{grad}F)_2, \\ (\operatorname{curl}\operatorname{curl}\mathbf{\Pi})_3 &= (\operatorname{grad}F)_3. \end{aligned} \right\} \tag{1}$$

---

* Stratton, *Electromagnetic Theory*, pp. 47–59; Ramo and Whinnery, *Fields and Waves in Modern Radio*, pp. 84–91.

F must be chosen to satisfy the second and third of equations (1), since

$$\operatorname{curl} \mathbf{\Pi} = \frac{1}{h_1 h_2 h_3} \begin{vmatrix} h_1 \mathbf{i}_1 & h_2 \mathbf{i}_2 & h_3 \mathbf{i}_3 \\ \dfrac{\partial}{\partial u_1} & \dfrac{\partial}{\partial u_2} & \dfrac{\partial}{\partial u_3} \\ h_1 U & 0 & 0 \end{vmatrix}$$

$$= \frac{1}{h_1 h_2 h_3} \left[\mathbf{i}_2 h_2 \frac{\partial}{\partial u_3}(h_1 U) - \mathbf{i}_3 h_3 \frac{\partial}{\partial u_2}(h_1 U) \right], \qquad (2)$$

and

$$\operatorname{curl} \operatorname{curl} \mathbf{\Pi} = \frac{1}{h_1 h_2 h_3} \begin{vmatrix} h_1 \mathbf{i}_1 & h_2 \mathbf{i}_2 & h_3 \mathbf{i}_3 \\ \dfrac{\partial}{\partial u_1} & \dfrac{\partial}{\partial u_2} & \dfrac{\partial}{\partial u_3} \\ 0 & \dfrac{h_2}{h_1 h_3} \dfrac{\partial}{\partial u_3}(h_1 U) & -\dfrac{h_3}{h_1 h_2} \dfrac{\partial}{\partial u_2}(h_1 U) \end{vmatrix}$$

$$= -\frac{\mathbf{i}_1}{h_2 h_3} \left[\frac{\partial}{\partial u_2} \left\{ \frac{h_3}{h_1 h_2} \frac{\partial}{\partial u_2}(h_1 U) \right\} + \frac{\partial}{\partial u_3} \left\{ \frac{h_2}{h_1 h_3} \frac{\partial}{\partial u_3}(h_1 U) \right\} \right]$$

$$+ \frac{\mathbf{i}_2}{h_1 h_3} \frac{\partial}{\partial u_1} \left\{ \frac{h_3}{h_1 h_2} \frac{\partial}{\partial u_2}(h_1 U) \right\} + \frac{\mathbf{i}_3}{h_1 h_2} \frac{\partial}{\partial u_1} \left\{ \frac{h_2}{h_1 h_3} \frac{\partial}{\partial u_3}(h_1 U) \right\}. \quad (3)$$

It follows from (1) and (2) that F must be so chosen that

$$(\operatorname{grad} F)_2 = \frac{1}{h_2} \frac{\partial F}{\partial u_2} = \frac{1}{h_1 h_3} \frac{\partial}{\partial u_1} \left\{ \frac{h_3}{h_1 h_2} \frac{\partial}{\partial u_2}(h_1 U) \right\}$$

$$\text{and} \qquad (\operatorname{grad} F)_3 = \frac{1}{h_3} \frac{\partial F}{\partial u_3} = \frac{1}{h_1 h_2} \frac{\partial}{\partial u_1} \left\{ \frac{h_2}{h_1 h_3} \frac{\partial}{\partial u_3}(h_1 U) \right\}. \qquad (4)$$

Let the orthogonal curvilinear system of coordinates possess the special property (the restriction on it that in fact we are seeking) that h_1 is independent of u_2 and of u_3 and also that h_3/h_2 is independent of u_1; then equations (4) reduce to

$$\frac{\partial F}{\partial u_2} = \frac{1}{h_1} \frac{\partial^2 U}{\partial u_2 \partial u_1}, \quad \frac{\partial F}{\partial u_3} = \frac{1}{h_1} \frac{\partial^2 U}{\partial u_3 \partial u_1}. \qquad (5)$$

With these restrictions on the coordinate system it follows from (5) that the quantity F that satisfied both equations (4) is

$$F = \frac{1}{h_1} \frac{\partial U}{\partial u_1} = (\operatorname{grad} U)_1. \qquad (6)$$

The differential equation satisfied by U is obtained by inserting this value of F in the first of equations (1), whence

$$\frac{1}{h_1}\frac{\partial}{\partial u_1}\left(\frac{1}{h_1}\frac{\partial U}{\partial u_1}\right)+\frac{1}{h_2 h_3}\left[\frac{\partial}{\partial u_2}\left\{\frac{h_3}{h_1 h_2}\frac{\partial}{\partial u_2}(h_1 U)\right\}\right.$$

$$\left.+\frac{\partial}{\partial u_3}\left\{\frac{h_2}{h_1 h_3}\frac{\partial}{\partial u_3}(h_1 U)\right\}\right]=\mu\epsilon\frac{\partial^2 U}{\partial t^2}. \tag{7}$$

Since the time almost always enters in the factor $e^{j\omega t}$, the term $\mu\epsilon(\partial^2 U/\partial t^2)$ usually becomes $-\omega^2\mu\epsilon U=-k^2 U$.

7·6. Derivation of E and H from U. E-type fields

When a solution U of 7·5 (7) has been found, the components of the electric field with respect to the system of curvilinear coordinates are obtained from the relation 7·3 (3)

$$\mathbf{E}=\operatorname{curl}\operatorname{curl}\mathbf{\Pi}=\operatorname{curl}\operatorname{curl}(\mathbf{i}_1 U), \tag{1}$$

whence
$$E_1=\omega^2\mu\epsilon U+\frac{1}{h_1}\frac{\partial}{\partial u_1}\left(\frac{1}{h_1}\frac{\partial U}{\partial u_1}\right),$$

$$E_2=\frac{1}{h_2}\frac{\partial}{\partial u_2}\left(\frac{1}{h_1}\frac{\partial U}{\partial u_1}\right)=\frac{1}{h_1 h_2}\frac{\partial^2 U}{\partial u_2\,\partial u_1}, \tag{2}$$

$$E_3=\frac{1}{h_3}\frac{\partial}{\partial u_3}\left(\frac{1}{h_1}\frac{\partial U}{\partial u_1}\right)=\frac{1}{h_1 h_3}\frac{\partial^2 U}{\partial u_3\,\partial u_1}.$$

The components of $\mathbf{H}$ follow from equation 7·3 (1),

$$\mathbf{B}=\mu\mathbf{H}=\mu\epsilon\operatorname{curl}\dot{\mathbf{\Pi}}=j\omega\mu\epsilon\operatorname{curl}(\mathbf{i}_1 U),$$

whence
$$H_1=0,$$

$$H_2=j\frac{\omega\epsilon}{h_1 h_3}\frac{\partial}{\partial u_3}(h_1 U)=j\frac{\omega\epsilon}{h_3}\frac{\partial U}{\partial u_3}, \tag{3}$$

$$H_3=-j\frac{\omega\epsilon}{h_1 h_2}\frac{\partial}{\partial u_2}(h_1 U)=-j\frac{\omega\epsilon}{h_1 h_2}\frac{\partial U}{\partial u_2}.$$

In equations (2) and (3) we have taken advantage of the postulate that h_1 is independent of u_2 and u_3 and h_3/h_2 of u_1.

These equations, (2) and (3), show that the Hertz vector $\mathbf{\Pi}=\mathbf{i}_1 U$ leads to electric or E-type fields in which H_1 is zero. It is for this reason that $\mathbf{\Pi}$ is called the electric Hertz vector.

7·7. Boundary conditions for U at the surface of a perfect conductor

In general, to obtain practical solutions of equation 7·5 (7) we require to know how U behaves at the bounding surfaces of the region throughout which the solutions are valid.

The physical bounding surfaces that occur in practice are usually metal walls which we suppose, for simplicity, to be perfectly conducting. The behaviour of U on such a surface is easily derived from the behaviour of $\mathbf{E}$ and $\mathbf{H}$ on perfectly conducting boundaries.

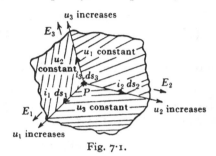

In practice, solutions are usually only obtained in those cases where the bounding surfaces can be represented as surfaces $u_1 = $ constant,

Fig. 7·1.

$u_2 = $ constant, $u_3 = $ constant in an appropriately chosen system of orthogonal curvilinear coordinates, as indicated in fig. 7·1.

The components of $\mathbf{E}$ and $\mathbf{H}$ behave as follows on perfectly conducting boundaries:

$E_1 = 0$ over boundaries $u_2 = $ constant and $u_3 = $ constant,

$E_2 = 0$ over boundaries $u_1 = $ constant and $u_3 = $ constant,

$E_3 = 0$ over boundaries $u_1 = $ constant and $u_2 = $ constant,

$H_2 = 0$ over the boundary $u_2 = $ constant,

$H_3 = 0$ over the boundary $u_3 = $ constant.

Inspection of equations 7·6(2) and (3) show the boundary conditions for U to be

$$\left. \begin{array}{l} U = 0 \\ \dfrac{\partial U}{\partial u_1} = 0 \end{array} \right\} \text{ over the boundaries } u_2 = \text{ constant and } u_3 = \text{ constant,}$$

$$\dfrac{\partial U}{\partial u_2} = 0 \quad \text{ over the boundaries } u_3 = \text{ constant,}$$

$$\dfrac{\partial U}{\partial u_3} = 0 \quad \text{ over the boundaries } u_2 = \text{ constant,}$$

provided the boundaries are perfectly conducting.

7·8. The scalar V—magnetic or H-type fields

As was explained in §7·4, electromagnetic fields may also be derived from a magnetic Hertz vector $\mathbf{\Pi}_0$. We suppose this also to be a single component vector

$$\mathbf{\Pi}_0 = \mathbf{i}_1 V + \mathbf{i}_2 \cdot 0 + \mathbf{i}_3 \cdot 0, \qquad (1)$$

and seek the form of the scalar function G of equation 7·4 (2), and the differential equation satisfied by V.

With the same postulate, that h_1 is independent of u_2 and u_3, and h_3/h_2 of u_1, we find, as in §7·5

$$G = \frac{1}{h_1} \frac{\partial V}{\partial u_1}, \qquad (2)$$

and that V and U satisfy the same differential equation

$$\frac{1}{h_1} \frac{\partial}{\partial u_1} \left(\frac{1}{h_1} \frac{\partial V}{\partial u_1} \right) + \frac{1}{h_2 h_3} \left[\frac{\partial}{\partial u_3} \left\{ \frac{h_3}{h_1 h_2} \frac{\partial}{\partial u_2} (h_1 V) \right\} \right.$$
$$\left. + \frac{\partial}{\partial u_3} \left\{ \frac{h_2}{h_1 h_3} \frac{\partial}{\partial u_3} (h_1 V) \right\} \right] + \omega^2 \mu \epsilon V = 0. \qquad (3)$$

According to 7·4 (3) and (1) the components of $\mathbf{H}$ and $\mathbf{E}$ are

$$\left. \begin{aligned} H_1 &= \omega^2 \mu \epsilon V + \frac{1}{h_1} \frac{\partial}{\partial u_1} \left(\frac{1}{h_1} \frac{\partial V}{\partial u_1} \right), \\ H_2 &= \frac{1}{h_2 h_1} \frac{\partial^2 V}{\partial u_2 \partial u_1}, \\ H_3 &= \frac{1}{h_1 h_3} \frac{\partial^2 V}{\partial u_3 \partial u_1}, \end{aligned} \right\} \qquad (4)$$

$$E_1 = 0, \quad E_2 = -j \frac{\omega \mu}{h_3} \frac{\partial V}{\partial u_3}, \quad E_3 = j \frac{\omega \mu}{h_2} \frac{\partial V}{\partial u_2}. \qquad (5)$$

The magnetic Hertz vector $\mathbf{\Pi}_0 = \mathbf{i}_1 V$ evidently gives magnetic or H-type fields.

7·9. Boundary conditions for V

To the boundary conditions for $\mathbf{E}$ and $\mathbf{H}$ displayed in §7·7 there is the additional condition

$$H_1 = 0 \text{ over the boundary, } u_1 = \text{constant.}$$

From these we deduce the behaviour of V to be the following:

$$\left.\begin{aligned} V &= 0 \\ \frac{\partial V}{\partial u_1} &= 0 \end{aligned}\right\} \quad \begin{aligned} &\text{over a perfectly conducting boundary} \\ &\quad u_1 = \text{constant,} \end{aligned}$$

$$\frac{\partial V}{\partial u_2} = 0 \quad \begin{aligned} &\text{over perfectly conducting boundaries} \\ &\quad u_2 = \text{constant and } u_1 = \text{constant,} \end{aligned}$$

$$\frac{\partial V}{\partial u_3} = 0 \quad \begin{aligned} &\text{over perfectly conducting boundaries} \\ &\quad u_3 = \text{constant and } u_1 = \text{constant.} \end{aligned}$$

7·10. Application of method to specific coordinate systems

7·10·1. *Introduction*

In what follows we apply the general theory of the previous sections to those few systems of curvilinear coordinates that are most commonly employed to study electromagnetic fields in wave guides and resonators. It is everywhere assumed that time enters in the factor $e^{j\omega t}$, and it is verified that the curvilinear system in each case is such that h_1 is independent of u_2 and u_3, and h_3/h_2 of u_1.

To summarize:

Solutions are required of the basic differential equation

$$\frac{1}{h_1}\frac{\partial}{\partial u_1}\left(\frac{1}{h_1}\frac{\partial T}{\partial u_1}\right) + \frac{1}{h_2 h_3}\left[\frac{\partial}{\partial u_2}\left\{\frac{h_3}{h_1 h_2}\frac{\partial}{\partial u_2}(h_1 T)\right\}\right.$$
$$\left. + \frac{\partial}{\partial u_3}\left\{\frac{h_2}{h_1 h_3}\frac{\partial}{\partial u_3}(h_1 T)\right\}\right] + k^2 T = 0, \qquad (1)$$

where T denotes U or V and $k^2 = \omega^2 \mu \epsilon = (2\pi/\lambda)^2$.

U and V are subject respectively to the boundary conditions formulated in §§ 7·7 and 7·9.

In practice, solutions of (1) are usually obtained by use of the method of the separation of the variables and are most commonly obtained in the form

$$T = f_1(u_1) f_2(u_2) f_3(u_3) e^{j\omega t}.$$

The boundary conditions (§§ 7·7 and 7·9) for U and V now simplify as follows:

Electric type oscillations, $T \equiv U$:

$$f_2(u_2) = 0 \text{ on the boundary } u_2 = \text{constant,}$$
$$f_3(u_3) = 0 \text{ on the boundary } u_3 = \text{constant.}$$

Magnetic type oscillations, $T \equiv V$:

$$\frac{d}{du_2} f_2(u_2) = 0 \text{ on the boundary } u_2 = \text{constant},$$

$$\frac{d}{du_3} f_3(u_3) = 0 \text{ on the boundary } u_3 = \text{constant}.$$

The field components are:

Electric type fields, $T \equiv U$:

$$E_1 = k^2 U + \frac{1}{h_1} \frac{\partial}{\partial u_1} \left(\frac{1}{h_1} \frac{\partial U}{\partial u_1} \right), \quad E_2 = \frac{1}{h_1 h_2} \frac{\partial^2 U}{\partial u_2 \partial u_1}, \quad E_3 = \frac{1}{h_1 h_3} \frac{\partial^2 U}{\partial u_3 \partial u_1}, \tag{2}$$

$$H_1 = 0, \quad H_2 = j \frac{\omega \epsilon}{h_3} \frac{\partial U}{\partial u_3}, \quad H_3 = -j \frac{\omega \epsilon}{h_2} \frac{\partial U}{\partial u_2}. \tag{3}$$

Magnetic type fields, $T \equiv V$:

$$H_1 = k^2 V + \frac{1}{h_1} \frac{\partial}{\partial u_1} \left(h_1 \frac{\partial V}{\partial u_1} \right), \quad H_2 = \frac{1}{h_1 h_2} \frac{\partial^2 V}{\partial u_2 \partial u_1}, \quad H_3 = \frac{1}{h_1 h_3} \frac{\partial^2 V}{\partial u_3 \partial u_1}, \tag{4}$$

$$E_1 = 0, \quad E_2 = -j \frac{\omega \mu}{h_3} \frac{\partial V}{\partial u_3}, \quad E_3 = j \frac{\omega \mu}{h_2} \frac{\partial V}{\partial u_2}. \tag{5}$$

7·10·2. *Solutions in Cartesian coordinates*

Here $\quad u_2 = x, \quad u_3 = y, \quad u_1 = z; \quad h_1 = h_2 = h_3 = 1.$

The condition, h_1 to be independent of u_2 and u_3 and h_3/h_2 of u_1, may therefore be satisfied by a unicomponent vector directed along any one of the axes, and we have arbitrarily selected the z-axis as that along which the Hertz vector is directed. The basic differential equation 7·10·1 (1) becomes

$$\frac{\partial^2 T}{\partial x^2} + \frac{\partial^2 T}{\partial y^2} + \frac{\partial^2 T}{\partial z^2} + k^2 T = 0. \tag{1}$$

Let $\qquad T = X(x) \, Y(y) \, Z(z) \, e^{j\omega t},$

where X, Y and Z are functions of x, y and z only. Substitute this expression for T in (1), then

$$\frac{1}{X} \frac{d^2 X}{dx^2} + \frac{1}{Y} \frac{d^2 Y}{dy^2} + \frac{1}{Z} \frac{d^2 Z}{dz^2} + k^2 = 0.$$

This equation resolves into the three equations

$$\frac{d^2X}{dx^2} = -\alpha^2 X, \quad \frac{d^2Y}{dy^2} = -\beta^2 Y, \quad \frac{d^2Z}{dz^2} = -\gamma^2 Z,$$

where $$\alpha^2 + \beta^2 + \gamma^2 = k^2, \tag{2}$$

but are otherwise arbitrary constants.

It follows that
$$\left.\begin{aligned}
X &= \frac{\cos}{\sin}(\alpha x) \text{ or } e^{\pm j\alpha x}, \\
Y &= \frac{\cos}{\sin}(\beta y) \text{ or } e^{\pm j\beta y}, \\
Z &= \frac{\cos}{\sin}(\gamma z) \text{ or } e^{\pm j\gamma z}.
\end{aligned}\right\} \tag{3}$$

Rectangular wave guides

Let Oz be the axis of propagation and let the walls be the planes
$$y = 0 \text{ and } y = b, \quad x = 0 \text{ and } x = a.$$

To obtain the expressions for E-waves put
$$T = U = \sin \alpha x \sin \beta y e^{j(\omega t - \gamma z)},$$

in which U is the z-component Π_z.

The boundary conditions require U to be zero at the walls which are the planes $u_2 = \text{constant}$ and $u_3 = \text{constant}$. That is,
$$\alpha = \frac{m\pi}{a}, \quad \beta = \frac{n\pi}{b}.$$

γ is also determined through (2) as
$$\gamma^2 = k^2 - \left[\left(\frac{m\pi}{a}\right)^2 + \left(\frac{n\pi}{b}\right)^2\right] = \left(\frac{2\pi}{\lambda_g}\right)^2. \tag{4}$$

Whence
$$\frac{1}{\lambda^2} - \frac{1}{\lambda_g^2} = \left[\left(\frac{m}{2a}\right)^2 + \left(\frac{n}{2b}\right)^2\right]$$

$$= \frac{1}{\lambda_c^2}. \tag{5}$$

The cut-off wave-length λ_c is therefore given by
$$\frac{1}{\lambda_c^2} = \left(\frac{m}{2a}\right)^2 + \left(\frac{n}{2b}\right)^2. \tag{6}$$

Thus the component of the electric Hertz vector of the E_{mn}-wave is
$$U = \sin\left(\frac{m\pi x}{a}\right) \sin\left(\frac{n\pi y}{b}\right) e^{j(\omega t - \gamma z)}. \tag{7}$$

Similarly, the component of the magnetic Hertz vector of the H_{mn}-wave is

$$V = \cos\left(\frac{m\pi x}{a}\right)\cos\left(\frac{n\pi y}{b}\right)e^{j(\omega t - \gamma z)}. \tag{8}$$

Field components

E_{mn}-*waves* (equations 7·10·1 (2) and (3)):

$$\left.\begin{aligned}
E_x &= E_2 = \frac{\partial}{\partial x}\left(\frac{\partial U}{\partial z}\right) = -j\gamma\left(\frac{m\pi}{a}\right)\cos\left(\frac{m\pi x}{a}\right)\sin\left(\frac{n\pi y}{b}\right)e^{j(\omega t - \gamma z)}, \\
E_y &= E_3 = \frac{\partial}{\partial y}\left(\frac{\partial U}{\partial z}\right) = -j\gamma\left(\frac{n\pi}{b}\right)\sin\left(\frac{m\pi x}{a}\right)\cos\left(\frac{n\pi y}{b}\right)e^{j(\omega t - \gamma z)}, \\
E_z &= E_1 = k^2 U + \frac{\partial^2 U}{\partial z^2} = (k^2 - \gamma^2)\,U = \left[\left(\frac{n\pi}{a}\right)^2 + \left(\frac{m\pi}{b}\right)^2\right]U, \\
H_z &= H_1 = 0, \\
H_x &= j\omega\epsilon\frac{\partial U}{\partial y} = j\omega\epsilon\left(\frac{n\pi}{b}\right)\sin\left(\frac{m\pi x}{a}\right)\cos\left(\frac{n\pi y}{b}\right)e^{j(\omega t - \gamma z)}, \\
H_y &= -j\omega\epsilon\frac{\partial U}{\partial x} = -j\omega\epsilon\left(\frac{m\pi}{a}\right)\cos\left(\frac{m\pi x}{a}\right)\sin\left(\frac{n\pi y}{b}\right)e^{j(\omega t - \gamma z)}.
\end{aligned}\right\} \tag{9}$$

H_{mn}-*waves* (equations (4), (5) and (6)):

$$\left.\begin{aligned}
H_x &= H_2 = \frac{\partial}{\partial x}\left(\frac{\partial V}{\partial z}\right) = j\gamma\left(\frac{m\pi}{a}\right)\sin\left(\frac{m\pi x}{a}\right)\cos\left(\frac{n\pi y}{b}\right)e^{j(\omega t - \gamma z)}, \\
H_y &= H_3 = \frac{\partial}{\partial y}\left(\frac{\partial V}{\partial z}\right) = j\gamma\left(\frac{n\pi}{b}\right)\cos\left(\frac{m\pi x}{a}\right)\sin\left(\frac{n\pi y}{b}\right)e^{j(\omega t - \gamma z)}, \\
H_z &= H_1 = \frac{\partial^2 V}{\partial z^2} + k^2 V = (k^2 - \gamma^2)\,V = \left[\left(\frac{m\pi}{a}\right)^2 + \left(\frac{n\pi}{b}\right)^2\right]V, \\
E_z &= 0, \\
E_x &= -j\omega\mu\frac{\partial V}{\partial y} = j\omega\mu\left(\frac{n\pi}{b}\right)\cos\left(\frac{m\pi x}{a}\right)\sin\left(\frac{n\pi y}{b}\right)e^{j(\omega t - \gamma z)}, \\
E_y &= j\omega\mu\frac{\partial V}{\partial x} = -j\omega\mu\left(\frac{m\pi}{a}\right)\sin\left(\frac{m\pi x}{a}\right)\cos\left(\frac{n\pi y}{b}\right)e^{j(\omega t - \gamma z)},
\end{aligned}\right\} \tag{10}$$

where γ is given by (4).

These expressions are equivalent to those quoted in § 3·3. Expressions (9) and (10) show that the longitudinal components vibrate in quadrature with the other components.

As explained in § 3·5, evanescent solutions appear when γ, as given by (4), is a mathematically imaginary quantity, the condition being

$$k^2 < \left[\left(\frac{m\pi}{a} \right)^2 + \left(\frac{n\pi}{b} \right)^2 \right],$$

or, from (5),

$$\lambda > \lambda_c.$$

We may then write $\gamma = \pm j\alpha$, where

$$\alpha = \sqrt{\left\{ \left[\left(\frac{m\pi}{a} \right)^2 + \left(\frac{n\pi}{b} \right)^2 \right] - k^2 \right\}} = \sqrt{\left(\frac{1}{\lambda_c^2} - \frac{1}{\lambda^2} \right)}.$$

The terms in (9) and (10) that are prefixed by $j\gamma$ change phase by $90°$ in the evanescent as compared with the progressive solution. The transverse components of $\mathbf{H}$ then oscillate in quadrature with the transverse components of $\mathbf{E}$.

Rectangular resonators

Let the conducting walls of the resonator be the planes

$$x = 0 \text{ and } x = a, \quad y = 0 \text{ and } y = b, \quad z = 0 \text{ and } z = c.$$

E_{mnp}-*modes.* We consider E-type modes first. We select terms (3) which make U vanish over the walls x or $y = $ constant, but not over the remaining pair $z = $ constant.

We write therefore

$$U = \sin \alpha x \sin \beta y \cos \gamma z;$$

$$U = \sin \left(\frac{m\pi x}{a} \right) \sin \left(\frac{n\pi y}{b} \right) \cos \left(\frac{p\pi z}{c} \right),$$

where m, n and p are integers. Further, from (2)

$$\left(\frac{m\pi}{a} \right)^2 + \left(\frac{n\pi}{b} \right)^2 + \left(\frac{p\pi}{c} \right)^2 = k^2 = \left(\frac{2\pi}{\lambda} \right)^2.$$

The resonant wave-length of the E_{mnp}-mode is therefore given by

$$\frac{1}{\lambda_{mnp}^2} = \left(\frac{m}{2a} \right)^2 + \left(\frac{n}{2b} \right)^2 + \left(\frac{p}{2c} \right)^2, \tag{11}$$

in agreement with 6·4 (5).

H_{mnp}-*modes.* To satisfy the boundary conditions, $Z = 0$ on the walls $z =$ constant, $\partial X/\partial x$ and $\partial Y/\partial y$ zero on the remaining walls, put

$$V = \cos\left(\frac{m\pi x}{a}\right)\cos\left(\frac{n\pi y}{b}\right)\sin\left(\frac{p\pi z}{c}\right),$$

with

$$\left(\frac{m\pi}{a}\right)^2 + \left(\frac{n\pi}{b}\right)^2 + \left(\frac{p\pi}{c}\right)^2 = k^2.$$

The resonant wave-length is also given by (11). The field components are obtained directly from these expressions for U and V when they are differentiated according to the procedures indicated in (2), (3), (4) and (5) of the preceding section.

When the expressions for the field components have been obtained it is easy to see that all the components of **E** oscillate in quadrature with those of **H**, as stated in §6·2.

7·10·3. *Electromagnetic waves on cylinders. General discussion*

We consider first a general cylinder whose axis is directed along Oz and whose conducting boundary or boundaries are general cylindrical surfaces $f(x, y) =$ constant.

In 7·5 (1), we put $\mathbf{\Pi} = \mathbf{i}_1 \Pi_z = \mathbf{i}_1 T$, then, since T is a rectilinear component, the term curl curl $\mathbf{\Pi}$ can be replaced by grad div $\mathbf{\Pi} - \nabla^2\mathbf{\Pi}$ and 7·5 (1) becomes

$$\text{grad}\left(\frac{\partial T}{\partial z}\right) - \nabla^2 T - k^2 T = \text{grad}\,F.$$

Thus we may choose $F = \partial T/\partial z$ and obtain T as the solution of

$$\nabla^2 T + k^2 T = 0,$$

where $k^2 = \omega^2\mu\epsilon$.

In practice the boundaries will be surfaces $u_1 =$ constant or $u_2 =$ constant and T will be obtained in the form

$$T = f_1(u_1)\,f_2(u_2)\,Z(z)\,e^{j\omega t}.$$

Cylindrical waves propagated along Oz

In curvilinear coordinates $\nabla^2 T + k^2 T = 0$ becomes

$$\frac{\partial^2 T}{\partial z^2} + \frac{1}{h_2 h_3}\left[\frac{\partial}{\partial u_2}\left(\frac{h_3}{h_2}\frac{\partial T}{\partial u_2}\right) + \frac{\partial}{\partial u_3}\left(\frac{h_2}{h_3}\frac{\partial T}{\partial u_3}\right)\right] + k^2 T = 0. \tag{1}$$

To represent a wave propagated along Oz put

$$T = f_2(u_2)\,f_3(u_3)\,e^{j(\omega t - \gamma z)}, \tag{2}$$

and determine the functions f_2 and f_3 from (1) for the particular coordinate system under discussion.

Suppose the walls to be surfaces $u_2 = $ constant, then to discuss E-modes we put $T = U$ and choose $f_2(u_2)$ so that it vanishes at the wall or walls. For H-modes, put $T = V$ and choose $f_2(u_2)$ so that $d/du_2 f_2(u_2)$ vanishes at the walls.

When U and V have been found, the field components of the E- and H-modes are derived from them by the standard procedure of §7·10·1 (2), (3), (4) and (5).

7·10·4. *TEM-waves*

The condition that a wave should be a TEM-wave propagated along Oz is that $E_z = H_z = 0$ everywhere. This condition gives, from equations 7·10·1 (2) and (4), the following general results ($h_1 = 1$):

$$\frac{\partial^2 T}{\partial z^2} + k^2 T = 0,$$

or,

$$(k^2 - \gamma^2)\, T = 0.$$

Thus $k = \gamma$ whence

$$v = v_g = \frac{1}{\sqrt{(\mu\epsilon)}}, \quad \lambda = \lambda_g.$$

The speed of all TEM-waves is therefore

$$v = \frac{1}{\sqrt{(\mu\epsilon)}}. \tag{1}$$

The field components in the pattern are (from 7·10·1 (2) and (3))

$$E_2 = \frac{1}{h_2}\frac{\partial}{\partial u_2}\left(\frac{\partial T}{\partial z}\right) = -j\gamma\,\frac{1}{h_2}\frac{\partial T}{\partial u_2},$$

$$E_3 = \frac{1}{h_3}\frac{\partial}{\partial u_3}\left(\frac{\partial T}{\partial z}\right) = -j\gamma\,\frac{1}{h_3}\frac{\partial T}{\partial u_3},$$

$$H_2 = \frac{j\omega\epsilon}{h_3}\frac{\partial T}{\partial u_3}, \quad H_3 = -\frac{j\omega\epsilon}{h_2}\frac{\partial T}{\partial u_2}.$$

Thus $\quad \mathbf{E} = -j\gamma\,\mathrm{grad}\,T,$
$\quad \mathbf{H} = \dfrac{j\omega\epsilon}{h_3}\,\mathrm{curl}\,(\mathbf{i}_1 T),\Big\}$ ($\mathbf{i}_1 =$ unit vector along Oz). (2)

These are the conditions that the fields **E** and **H** in the wave front shall be two-dimensional electrostatic and magnetostatic fields. Since a two-dimensional electrostatic field cannot exist within a hollow metal cylinder free of space charge, it is essential for the propagation of TEM-waves that a pair of cylinders be used, so that the electric lines of force may arise on the one and terminate on the other. This is the only means of avoiding a longitudinal component E_z.

The wave impedance of a TEM-wave is

$$Z = \frac{E_2}{H_3} = -\frac{E_3}{H_2} = \frac{\gamma}{\omega\epsilon} = \frac{k}{\omega\epsilon} = \frac{\omega\sqrt{(\mu\epsilon)}}{\omega\epsilon}$$

$$= \sqrt{\frac{\mu}{\epsilon}} \doteq 120\pi \sqrt{\frac{K_m}{K_e}} \text{ ohms}, \qquad (3)$$

since $k = \omega\sqrt{(\mu\epsilon)}$.

The results of this section were assumed in Chapter 1 and are of fundamental importance for the theory of transmission lines.

7·10·5. *Circular cylinders*

The variables are $u_1 = z$, $u_2 = r$, $u_3 = \theta$ and the differential multipliers are $h_1 = 1$, $h_2 = 1$ and $h_3 = r$.

The fundamental differential equation 7·10·1 (1) or 7·10·3 (1) becomes

$$\frac{\partial^2 T}{\partial z^2} + \frac{1}{r}\left[\frac{\partial}{\partial\theta}\left(\frac{1}{r}\frac{\partial T}{\partial\theta}\right) + \frac{\partial}{\partial r}\left(\frac{r\,\partial T}{\partial r}\right)\right] + k^2 T = 0. \qquad (1)$$

Put $$T = R(r)\,F(\theta)\,Z(z)\,e^{j\omega t}.$$

Then (1) reduces to

$$\frac{1}{R}\frac{1}{r}\frac{d}{dr}\left(\frac{r\,dR}{dr}\right) + \frac{1}{F}\frac{1}{r^2}\frac{d^2F}{d\theta^2} + \frac{1}{Z}\frac{d^2Z}{dz^2} + k^2 = 0. \qquad (2)$$

Let

$$\frac{1}{Z}\frac{d^2Z}{dz^2} = -\gamma^2, \quad \frac{1}{F}\frac{d^2F}{d\theta^2} = -m^2, \qquad (3)$$

where m and γ are constants independent of r, θ and z, m being an integer in the present application.

Then, from (3),

$$Z = \frac{\cos}{\sin}(\gamma z) \text{ or } e^{\pm j\gamma z}, \quad F = \frac{\cos}{\sin}(m\theta) \text{ or } e^{\pm jm\theta}. \qquad (4)$$

Replace the terms in (2) by their corresponding values in (3) to obtain the following differential equation for R

$$\frac{1}{r}\frac{d}{dr}\left(\frac{r\,dR}{dr}\right) + \left[(k^2 - \gamma^2) - \frac{m^2}{r^2}\right]R = 0. \tag{5}$$

The solution of (5) is some Bessel function of order m which we shall write $Z_m[\sqrt{(k^2 - \gamma^2)}\,r]$. The type of solution of (1) that we require is therefore

$$T = Z_m[\sqrt{(k^2 - \gamma^2)}\,r] \begin{matrix} \cos \\ \sin \end{matrix} m\theta Z(z). \tag{6}$$

The choice of Bessel function and of the form of $Z(z)$ is determined by the problem in hand.

Wave propagation in circular wave guides

E_{mn}-*modes.* Put

$$T = U, \quad Z_m[\sqrt{(k^2 - \gamma^2)}\,r] = J_m[\sqrt{(k^2 - \gamma^2)}\,r], \quad Z(z) = e^{-j\gamma z},$$

where $J_m(x)$ is a Bessel function of the first kind and of order m. It is chosen because it is finite for $x = 0$ and does not make T infinite on the axis $r = 0$.

Thus $\qquad U = J_m[\sqrt{(k^2 - \gamma^2)}\,r]\cos n\theta e^{j(\omega t - \gamma z)}. \tag{7}$

The conducting boundary is the cylinder $u_2 = r = a$ (a = radius of cylinder) over which we require

$$f_2(u_2) = J_m[\sqrt{(k^2 - \gamma^2)}\,r] \quad \text{to be zero.}$$

Therefore $\qquad \sqrt{(k^2 - \gamma^2)}\,a = \text{a root of } J_m(x).$

Let ρ_{mn} be the nth root of $J_m(x) = 0$, then the value of γ for the E_{mn}-mode is to be obtained from

$$\sqrt{(k^2 - \gamma^2)}\,a = \rho_{mn}$$

or $\qquad k^2 - \gamma^2 = \left(\frac{\rho_{mn}}{a}\right)^2. \tag{8}$

That is, since $k = 2\pi/\lambda$ and $\gamma = 2\pi/\lambda_g$,

$$\frac{1}{\lambda_g^2} = \frac{1}{\lambda^2} - \left(\frac{\rho_{mn}}{2\pi a}\right)^2. \tag{9}$$

The cut-off wave-length of the E_{mn}-mode is therefore

$$\lambda_c = \frac{2\pi a}{\rho_{mn}}. \tag{10}$$

The values of ρ_{mn} for small values of m and n are shown in the following table:

ρ_{mn}

m \ n	1	2	3
0	2·405	5·52	8·65
1	3·85	7·02	10·17
2	5·13	8·42	11·62

whence the cut-off wave-lengths of the E_{01}- and E_{11}-waves are respectively

$$\lambda_c = \frac{2\pi a}{2 \cdot 405} = 2 \cdot 61a$$

and

$$\lambda_c = \frac{2\pi a}{3 \cdot 83} = 1 \cdot 64a.$$

The field components are obtained from (7) in the usual manner through 7·10·1 (2) and (3), with $h_1 = h_3 = 1$ and $h_2 = r$

$$E_z = E_1 = k^2 U + \frac{\partial^2 U}{\partial z^2} = (k^2 - \gamma^2)\,U = \left(\frac{\rho_{mn}}{a}\right)^2 U,$$

$$E_\theta = E_3 = \frac{1}{r}\frac{\partial}{\partial \theta}\left(\frac{\partial U}{\partial z}\right) = j\gamma\frac{m}{r}J_m\left(\frac{\rho_{mn}r}{a}\right)\sin m\theta e^{j(\omega t - \gamma z)},$$

$$E_r = \frac{\partial}{\partial r}\left(\frac{\partial U}{\partial z}\right) = -j\gamma\frac{\partial U}{\partial r} = -j\gamma\left(\frac{\rho_{mn}}{a}\right)J'_m\left(\frac{\rho_{mn}r}{a}\right)\cos m\theta e^{j(\omega t - \gamma z)},$$

$$H_z = 0,$$

$$H_\theta = -j\omega\epsilon\frac{\partial U}{\partial r} = -j\omega\epsilon\left(\frac{\rho_{mn}}{a}\right)J'_m\left(\frac{\rho_{mn}a}{r}\right)\cos m\theta e^{j(\omega t - \gamma z)},$$

$$H_r = +j\frac{\omega\epsilon}{r}\frac{\partial U}{\partial \theta} = -j\frac{\omega\epsilon m}{r}J_m\left(\frac{\rho_{mn}a}{r}\right)\sin m\theta e^{j(\omega t - \gamma z)}.$$

The field patterns of the E_{01}- and E_{11}-waves are indicated in figs. 2·16 and 2·19.

H_{mn}-modes. The boundary condition is

$$\frac{d}{du_2}f_2(u_2) \equiv \frac{d}{dr}J_m[\sqrt{(k^2 - \gamma^2)}\,r] = 0, \quad \text{on the cylinder } r = a.$$

The field is therefore to be derived from

$$V = J_m\left(\frac{\sigma_{mn}}{a}r\right)\cos m\theta e^{j(\omega t - \gamma z)}, \tag{11}$$

where σ_{mn} is the nth root of $J'_m(x) = 0$. The cut-off wave-length is derived from

$$k^2 - \gamma^2 = \left(\frac{\sigma_{mn}}{a}\right)^2,$$

or

$$\frac{1}{\lambda_g^2} = \frac{1}{\lambda^2} - \left(\frac{\sigma_{mn}}{2\pi a}\right)^2,$$

that is

$$\lambda_c = \frac{2\pi a}{\sigma_{mn}}. \tag{12}$$

The following table shows the roots of $J'_m(x) = 0$ for the smaller values of m and n:

σ_{mn}

m \ n	1	2	3
0	3·83	7·02	10·17
1	1·84	5·33	8·54
2	3·05	6·70	9·96

According to (12) the cut-off wave-lengths of the H_{01}- and H_{11}-waves are respectively

$$\lambda_c = \frac{2\pi a}{3\cdot83} = 1\cdot64a,$$

and

$$\lambda_c = \frac{2\pi a}{1\cdot84} = 3\cdot42a.$$

The field components are obtained from 7·10·1 (4) and (5), and (11) above, as

$$H_z = (k^2 - \gamma^2)V = \left(\frac{\sigma_{mn}}{a}\right)^2 V,$$

$$H_\theta = \frac{1}{r}\frac{\partial}{\partial\theta}\left(\frac{dV}{dz}\right) = j\gamma\frac{m}{r}J_m\left(\frac{\sigma_{mn}r}{a}\right)\sin m\theta e^{j(\omega t - \gamma z)},$$

$$H_r = -j\gamma\left(\frac{\sigma_{mn}}{a}\right)J'_m\left(\frac{\sigma_{mn}r}{a}\right)\cos m\theta e^{j(\omega t - \gamma z)},$$

$$E_z = 0,$$

$$E_3 = E_\theta = +j\omega\mu\frac{\partial V}{\partial r} = +j\omega\mu\left(\frac{\sigma_{mn}}{a}\right)J'_m\left(\frac{\sigma_{mn}r}{a}\right)\cos m\theta e^{j(\omega t - kr)},$$

$$E_2 = E_r = -j\frac{\omega\mu}{r}\frac{\partial V}{\partial\theta} = +j\frac{\omega\mu}{r}mJ_m\left(\frac{\sigma_{mn}r}{a}\right)\sin m\theta e^{j(\omega t - \gamma r)}.$$

The patterns of the H_{01}- and H_{11}-waves are indicated in figs. 2·15 and 2·18.

Cylindrical resonators (hollow)

The cylinder is bounded by the surfaces

$$r = u_2 = a, \quad u_1 = z = 0 \text{ or } d.$$

E_{mnp}-*modes*. Put $T = U = f_1(u_1) f_2(u_1) f_3(u_3) e^{j\omega t}$. The boundary conditions require

$$f_1(u_1) + 0 \text{ over the ends } u_1 = z = 0 \text{ or } d,$$

$$f_2(u_2) = 0 \text{ on the surface } u_2 = r = a.$$

From (6) select the following expression for U:

$$U = J_m[\sqrt{(k^2 - \gamma^2)}\, r] \cos(m\theta) \cos(\gamma z)\, e^{j\omega t}. \tag{13}$$

The boundary conditions require

$$\sqrt{(k^2 - \gamma^2)} = \frac{\rho_{mn}}{a} \quad \text{and} \quad \gamma = \frac{p\pi}{d} \quad (p \text{ an integer})$$

and

$$U = J_m\left(\frac{\rho_{mn} r}{a}\right) \cos m\theta \cos\left(\frac{p\pi z}{d}\right) e^{j\omega t}.$$

Whence

$$\left(\frac{2\pi}{\lambda_{mnp}}\right)^2 = \left(\frac{\rho_{mn}}{a}\right)^2 + \left(\frac{p\pi}{d}\right)^2,$$

or

$$\frac{1}{\lambda_{mnp}^2} = \left(\frac{\rho_{mn}}{2\pi a}\right)^2 + \left(\frac{p}{2d}\right)^2 = \left(\frac{1}{\lambda_c}\right)^2 + \left(\frac{p}{2d}\right)^2. \tag{14}$$

This is the formula for the resonant wave-lengths given in § 6·4. Similarly, for the H_{mnp}-modes

$$\left(\frac{2\pi}{\lambda_{mnp}}\right)^2 = \left(\frac{\sigma_{mn}}{a}\right)^2 + \left(\frac{p\pi}{d}\right)^2,$$

which leads again to formula (14).

For these modes

$$V = J_m\left(\frac{\sigma_{mn}}{a} r\right) \cos m\theta \sin\left(\frac{p\pi z}{d}\right) e^{j\omega t}. \tag{15}$$

The fields of the E_{mnp} and H_{mnp} modes may be derived from these expressions for U and V by the usual method.

We next consider briefly cases where choices from the possible functions shown in (4) and (6), different from those we have made in (7), (11), (14) and (15), are required. These are shown in fig. 7·2.

Fig. 7·2 (a) shows a pair of coaxial cylinders of radii a and b. For E-type waves or oscillations we require the factor $Z_m[\sqrt{(k^2-\gamma^2)}\,r]$ in (6) to vanish both at $r = a$ and $r = b$, and for H-type modes $Z'_m[\sqrt{(k^2-\gamma^2)}\,r]$ to vanish on these surfaces.

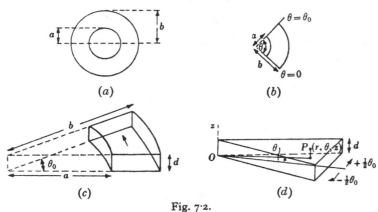

Fig. 7·2.

Z_m is therefore made the sum of mth order Bessel functions of the first and second kinds with suitable constant coefficients.

Thus $\quad Z_m = A J_m[\sqrt{(k^2-\gamma^2)}\,r] + B N_m[\sqrt{(k^2-\gamma^2)}\,r]$.

The boundary conditions for E-modes require

$$Z_m = 0 \quad \text{at} \quad r = a \quad \text{and} \quad r = b,$$

and for H-modes,

$$Z'_m = 0 \quad \text{at} \quad r = a \quad \text{and} \quad r = b.$$

The value of $(k^2 - \gamma^2)$ for E-modes is determined therefore from the following transcendental equation:[*]

$$\frac{J_m[\sqrt{(k^2-\gamma^2)}\,a]}{N_m[\sqrt{(k^2-\gamma^2)}\,a]} = \frac{J_m[\sqrt{(k^2-\gamma^2)}\,b]}{N_m[\sqrt{(k^2-\gamma^2)}\,b]}, \tag{16}$$

and for H-modes from

$$\frac{J'_m[\sqrt{(k^2-\gamma^2)}\,a]}{N'_m[\sqrt{(k^2-\gamma^2)}\,a]} = \frac{J'_m[\sqrt{(k^2-\gamma^2)}\,b]}{N'_m[\sqrt{(k^2-\gamma^2)}\,b]}. \tag{17}$$

The ratio of the constants is

$$\frac{A}{B} = -\frac{N_m[\sqrt{(k^2-\gamma^2)}\,a]}{J_m[\sqrt{(k^2-\gamma^2)}\,a]}.$$

[*] Stratton, *Electromagnetic Theory*, p. 548.

These modes are the supplementary modes mentioned in §2·7. When $k = \gamma$ and $m = 0$ the corresponding field is that of a TEM-mode.

Equation (5) becomes

$$\frac{d}{dr}\left(r\frac{dR}{dr}\right) = 0,$$

or
$$R = A \log r + B,$$

and
$$T = (A \log r + B)\, e^{j(\omega t - kz)}. \tag{18}$$

From 7·10·1 (2) and (3)

$$E_z = E_\theta = 0, \quad E_r = -j\frac{kA}{r}\, e^{j(\omega t - kr)},$$

$$H_z = H_r = 0, \quad H_\theta = -j\frac{\omega\epsilon}{r}\, A e^{j(\omega t - kr)}.$$

Whence
$$Z_{\text{TEM}} = \frac{E_r}{H_\theta} = \frac{k}{\omega\epsilon} = \frac{\omega\sqrt{(\mu\epsilon)}}{\omega\epsilon} = \sqrt{\frac{\mu}{\epsilon}},$$

in agreement with 7·10·4 (3).

The configuration shown in fig. 7·2 (b) is a coaxial system with conducting septa placed radially with their planes meeting on Oz at an angle θ.

To find the E-modes we require Z_m to vanish on the cylindrical surfaces and $\genfrac{}{}{0pt}{}{\sin}{\cos} m\theta$ to vanish on the septa at $\theta = 0$ and $\theta = \theta_0$. Thus $m\theta_0 = s\pi$ and the sine term becomes $\sin\left(\dfrac{s\pi}{\theta_0}\theta\right)$.

The order m of the Bessel function is no longer integral. The fields of E-waves are to be derived from

$$U = Z_{(s\pi/\theta_0)}[\sqrt{(k^2 - \gamma^2)}\, r]\sin\left(\frac{s\pi}{\theta_0}\theta\right) e^{j(\omega t - \gamma z)},$$

where s is an integer. The cut-off wave-lengths are to be obtained from (16) with $m = s\pi/\theta_0$.

The H-modes are derived from

$$V = Z_{(s\pi/\theta_0)}[\sqrt{(k^2 - \gamma^2)}\, r]\cos\left(\frac{s\pi}{\theta_0}\theta\right) e^{j(\omega t - \gamma z)}$$

and their cut-off wave-lengths from (17).

When $\theta_0 = 2\pi$ the coaxial system of fig. 7·2 (b) transforms to the septate coaxial system of fig. 2·17 where dominant H mode is indicated. The order of the Bessel functions is now $s/2$ with $s = 1$ for the dominant H-mode.

Fig. 7·2 (c) shows the portion of a bend in a rectangular wave guide regarded as a coaxial system closed by planes at $z = 0$ and $z = d$, with the end sections of the wave guide lying in the planes $\theta = 0$ and $\theta = \theta_0$.

We are here concerned with a wave propagated in the direction of increasing θ as shown by the arrow in the figure.

The expressions for U and V, chosen from (4) and (6) to satisfy the boundary conditions, are

$$U = \{AJ_l[\sqrt{(k^2 - \gamma^2)}\,r] + BN_l[\sqrt{(k^2 - \gamma^2)}\,r]\} \cos\frac{p\pi z}{d}\, e^{j(\omega t - l\theta)},$$

where $\gamma = (p\pi/d)$ and p is an integer, and

$$V = \{CJ_l[\sqrt{(k^2 - \gamma^2)}\,r] + DN_l[\sqrt{(k^2 - \gamma^2)}\,r]\} \sin\frac{p\pi z}{d}\, e^{j(\omega t - l\theta)}.$$

Since $k^2 - \gamma^2 = k^2 - (p\pi/d)^2$ is now given equations (16) and (17) are now used to find the value l of the order of the Bessel functions and of the effective propagation constant.

The fields derived from U have H_z zero and those from V, E_z zero. They correspond to what are termed longitudinal section waves in rectangular straight wave guides* (§ 7·10·8). When $p = 0$, $V = 0$ and in the wave derived from U, it can be seen from 7·10·1 (2) and (3) —with $u_1 = z_1$, $u_2 = r$, $u_3 = \theta$ as before—that all components of the wave vanish except E_z and H_r. The wave therefore resembles an H_{10}-wave in a straight wave guide and is the one that is readily excited in an H-plane bend attached to a straight section of wave guide. It would be anticipated from this that an H-plane bend would give smaller reflexion than an E-plane bend.

Finally, in fig. 7·2 (d), we are shown a wedge bounded by conducting walls $z = 0$ and d, and $\theta = \pm\theta_0/2$. We require the field components of wave propagated radially outwards from the origin. The problem clearly relates to wave propagation in horns. We here require a different type of Bessel function from those employed above, namely Hankel functions.†

These functions are of two types that are related to the Bessel functions $J_m(x)$ and $N_m(x)$ as follows:

$$H_m^{(1)}(x) = J_m(x) + jN_m(x), \quad H_m^{(2)}(x) = J_m(x) - jN_m(x). \quad (19)$$

---

* Lamont, *Wave Guides*, p. 58, Methuen Monograph.
† Janke and Emde, *Tables of Functions*, p. 134, 3rd ed.

There is more than a formal similarity between relations (17) and the well-known trigonometrical relations

$$e^{jx} = \cos x + j \sin x, \quad e^{-jx} = \cos x - j \sin x. \tag{20}$$

Just as two-plane oppositely travelling waves

$$e^{j(\omega t + \gamma x)} \quad \text{and} \quad e^{j(\omega t - \gamma x)}$$

can be superimposed to produce a standing wave $2e^{j\omega t} \cos \gamma x$, so it proves that the functions

$$e^{j\omega t} H^{(1)}[\sqrt{(k^2 - \gamma^2)} \, r] \quad \text{and} \quad e^{j\omega t} H^{(2)}[\sqrt{(k^2 - \gamma^2)} \, r]$$

represent respectively inwards and outwards travelling cylindrical waves, which according to (19) may be superimposed to produce a cylindrical standing wave

$$2e^{j\omega t} J_m[\sqrt{(k^2 - \gamma^2)} \, r] = e^{j\omega t}\{H^{(1)}[\sqrt{(k^2 - \gamma^2)} \, r] + H^{(2)}[\sqrt{(k^2 - \gamma^2)} \, r]\}.$$

Thus the Bessel function of the first kind which appears in (13) and (15) may be regarded as a standing wave formed from cylindrical waves incident upon and reflected from the cylindrical boundary. After these remarks it should be clear that the form for U appropriate to outwardly travelling waves in the wedge of fig. 7·2 (d) is

$$U = H_m^{(2)}[\sqrt{(k^2 - \gamma^2)} \, r] \cos m\theta \, \cos \gamma z e^{j\omega t},$$

where

$$\gamma = \left(\frac{p\pi}{d}\right) \quad \text{and} \quad m = \frac{s\pi}{\theta_0},$$

with p and s integers. The solutions derived from U were used by Barrow and Chu* to discuss wave propagation in horns.

7·10·6. *Intrinsic (or wave) impedance of waves along cylinders*

The intrinsic (or wave) impedance of a wave propagated along Oz is

$$Z_0 = \frac{E_2}{H_3} = -\frac{E_3}{H_2}.$$

For E-waves, we put $Z_0 = Z_E$ and use 7·10·1·(2) and (3) to obtain ($h_1 = 1$ and $u_1 = z$)

$$Z_E = \left(-j\gamma \frac{1}{h_2} \frac{\partial U}{\partial u_2}\right) \bigg/ \left(-j\frac{\omega\epsilon}{h_2} \frac{\partial U}{\partial u_2}\right)$$

$$= \gamma/\omega\epsilon = \frac{2\pi}{\lambda_g \, \omega \sqrt{(\mu\epsilon)}} \sqrt{\left(\frac{\mu}{\epsilon}\right)} = \frac{1}{\lambda_g} \frac{v}{f} \sqrt{\left(\frac{\mu}{\epsilon}\right)}$$

$$= \sqrt{\left(\frac{\mu}{\epsilon}\right)} \frac{\lambda}{\lambda_g}.$$

* W. L. Barrow and J. L. Chu, *Proc. Inst. Radio Engrs*, vol. 27, p. 51 (1939).

Similarly, from 7·10·1 (4) and (5) for H-waves $Z_0 = Z_H$

$$Z_H = \sqrt{\left/\left(\frac{\mu}{\epsilon}\right)\right.} \frac{\lambda_g}{\lambda}.$$

These results were stated in § 3·4·1.

7·10·7. *Solutions in spherical polar coordinates*

The variables are: the radial distance $r = u_1$, co-latitude $\theta = u_2$ and longitude $\phi = u_3$, and the differential multipliers are, $h_1 = 1$, $h_2 = r$ and $h_3 = r\sin\theta$. The basic differential equation 7·10·1 (1) therefore assumes the form

$$\frac{\partial^2 T}{\partial r^2} + \frac{1}{r^2 \sin\theta}\left[\frac{\partial}{\partial\theta}\left(\sin\theta\frac{\partial T}{\partial\theta}\right) + \frac{\partial}{\partial\phi}\left(\frac{1}{\sin\theta}\frac{\partial T}{\partial\phi}\right)\right] + k^2 T = 0, \quad (1)$$

where $k = \omega\sqrt{(\mu\epsilon)} = 2\pi/\lambda$. Put

$$T = R(r)\,S(\theta,\phi)\,e^{j\omega t}, \tag{2}$$

where $R(r)$ is a function of r only and $S(\theta,\phi)$ a function of θ and ϕ.

When this expression for T is substituted in (1) the variables may be separated in the usual manner to give the following equations, in which n is the separation constant:

$$\frac{1}{\sin\theta}\frac{\partial}{\partial\theta}\left(\sin\theta\frac{\partial S}{\partial\theta}\right) + \frac{1}{\sin^2\theta}\frac{\partial^2 S}{\partial\phi^2} + n(n+1)S = 0, \tag{3}$$

and

$$\frac{d^2 R}{dr^2} + \left[k^2 - \frac{n(n+1)}{r^2}\right]R = 0. \tag{4}$$

Equation (3) is Legendre's equation, and $S(\theta,\phi)$ is therefore a surface harmonic.*

Equation (4) may be thrown into a more familiar form by replacing n in it by $(p - \frac{1}{2})$. That is, put $(n+1) = (p + \frac{1}{2})$.

Equation (4) then transforms to

$$\frac{d^2 R}{dr^2} + \left[k^2 + \frac{(\frac{1}{4} - p^2)}{r^2}\right]R = 0. \tag{5}$$

It is known* that the solution of

$$\frac{d^2 y}{dx^2} + \left(\frac{1 - 2\alpha}{x}\right)\frac{dy}{dx} + \left(k^2 + \frac{\alpha^2 - p^2}{x^2}\right)y = 0, \tag{6}$$

---

* Janke and Emde, *Tables of Functions*, pp. 107–25. Loc. cit. p. 146, § 7 (equation (4)).

in which α, k and p are constants, is

$$y = x^\alpha Z_p(kx), \qquad (7)$$

where $Z_p(kx)$ is a Bessel function of order p.

Equation (5) is a special case of (6) with $\alpha = \frac{1}{2}$ and its solution is therefore $R = r^{\frac{1}{2}} Z_p(kr)$. Since $p = (n+\frac{1}{2})$, the solution of (4) is therefore,

$$R = r^{\frac{1}{2}} Z_{(n+\frac{1}{2})}(kr). \qquad (8)$$

Since the solution of (3) will be of the general form

$$S(\theta, \phi) = S_n^m(\cos\theta) \cos m\phi,$$

where $S_n^m(\cos\theta)$ is a spherical harmonic, the solution of (1) is of the form

$$T = r^{\frac{1}{2}} Z_{(n+\frac{1}{2})}(kr) S_n^m(\cos\theta) \cos m\phi e^{j\omega t}. \qquad (9)$$

E-modes. $T = U$; from equations 7·10·1 (2) and (3).

$$E_r = E_1 = k^2 U + \frac{\partial^2 U}{\partial r^2} = \frac{n(n+1)\,U}{r^2} \quad \text{(from (4))},$$

$$\left.\begin{aligned}
E_\theta = E_2 &= \frac{1}{r}\frac{\partial^2 U}{\partial\theta\,\partial r}, \\
E_\phi = E_3 &= \frac{1}{r\sin\theta}\frac{\partial^2 U}{\partial\phi\,\partial r}, \\
H_\theta = H_2 &= j\frac{\omega\epsilon}{r\sin\theta}\frac{\partial U}{\partial\phi}, \\
H_\phi = H_3 &= -j\frac{\omega\epsilon}{r}\frac{\partial U}{\partial\theta}.
\end{aligned}\right\} \qquad (10)$$

H-modes. Put $T = V$ and apply equations 7·10·1 (4) and (5).

$$\left.\begin{aligned}
H_r = H_1 &= k^2 V + \frac{\partial^2 V}{\partial r^2} = \frac{n(n+1)\,V}{r^2}, \\
H_\theta = H_2 &= \frac{1}{r}\frac{\partial^2 V}{\partial\theta\,\partial r}, \\
H_\phi = H_3 &= \frac{1}{r\sin\theta}\frac{\partial^2 V}{\partial\phi\,\partial r}, \\
E_\theta = E_2 &= -j\frac{\omega\mu}{r\sin\theta}\frac{\partial V}{\partial\phi}, \\
E_\phi = E_3 &= j\frac{\omega\mu}{r}\frac{\partial V}{\partial\theta}.
\end{aligned}\right\} \qquad (11)$$

Spherical waves

To obtain the field components of general spherical diverging waves, give T in (9) the form

$$T = r^{\frac{1}{2}}H^{(2)}_{(n+\frac{1}{2})}(kr)\, S^m_n(\cos\theta)\, \cos m\phi e^{j\omega t}, \qquad (12)$$

and operate on T according to (10) and (11) to obtain E and H waves. The simplest cases correspond to $m = 0$, $n = 0$ and 1.

Case $m = 0$, $n = 1$:

$$T = r^{\frac{1}{2}}H^{(2)}_{\frac{3}{2}}(kr)\cos\theta e^{j\omega t}. \qquad (13)$$

Since

$$H^{(2)}_p(x) = j^{(p+\frac{1}{2})}\sqrt{\left(\frac{2}{\pi x}\right)}e^{-jx}\, S_p(j2x)$$

and

$$H^{(1)}_p(x) = j^{-(p+\frac{1}{2})}\sqrt{\left(\frac{2}{\pi x}\right)}e^{+jx}\, S_p(-j2x),$$

$$\qquad (14)$$

where

$$S_p(y) = 1 + \frac{(4p^2-1)}{1!\,4y} + \frac{(4p^2-1)(4p^2-9)}{2!\,(4y)^2}$$

$$+ \frac{(4p^2-1)(4p^2-9)(4p^2-25)}{3!\,(4y)^3} + \cdots,$$

so that

$$S_p(\pm j2x) = P_p(2x) \mp jQ_p(2x),$$

where

$$P_p(x) = 1 - \frac{(4p^2-1)(4p^2-9)}{2!\,(8x)^2}$$

$$+ \frac{(4p^2-1)(4p^2-9)(4p^2-25)(4p^2-36)}{4!\,(8x)^4} + \cdots$$

and

$$Q_p(x) = \frac{(4p^2-1)}{1!\,(8x)} - \frac{(4p^2-1)(4p^2-9)(4p^2-25)}{3!\,(8x)^3} + \cdots,$$

it follows that

$$H^{(2)}_{\frac{3}{2}}(kr) = -\sqrt{\left(\frac{2}{\pi kr}\right)}e^{-jkr}\left[1 - \frac{j}{kr}\right],$$

so that, in (13),

$$T = -\sqrt{\left(\frac{2}{\pi k}\right)}\left[1 - \frac{j}{kr}\right]e^{j(\omega t - kr)}\cos\theta$$

$$= \sqrt{\left(\frac{2}{\pi k}\right)}\left[\frac{j}{kr} - 1\right]e^{j(\omega t - kr)}\cos\theta. \qquad (15)$$

E-wave. Put $T = U$, and apply (10), with $n = 1$,

$$E_r = \frac{n(n+1)}{r^2} U = 2 \sqrt{\left(\frac{2}{\pi k}\right)} \left[\frac{j}{kr^3} - \frac{1}{r^2}\right] \cos \theta e^{j(\omega t - kr)},$$

$$E_\theta = \sqrt{\left(\frac{2}{\pi k}\right)} \sin \theta \left[-\frac{jk}{r} - \frac{1}{r^2} + \frac{j}{kr^3}\right],$$

$$E_\phi = 0,$$

$$H_\theta = 0,$$

$$H_\phi = j\omega\epsilon \sin \theta \sqrt{\left(\frac{2}{\pi k}\right)} \left[\frac{j}{kr^2} - \frac{1}{r}\right] e^{j(\omega t - kr)}.$$

This is the radiation field of an electric Hertz doublet.

The radiation field due to an oscillating magnetic doublet is similarly found by putting $T = V$ in (15) and by using (11).

The case $n = 0$, $m = 0$ which is degenerate, is not covered by (12), and it is necessary to return to the differential equations (3) and (4), with $n = 0$, and $\partial S / \partial \phi = 0$ $(m = 0)$.

Whence, from (3), $\qquad \sin \theta \dfrac{dS}{d\theta} = A$

or $\qquad\qquad\qquad S = A \displaystyle\int \dfrac{d\theta}{\sin \theta} + B,$

where A and B are integration constants.

Thus $\qquad\qquad\qquad S = A \log \cot \tfrac{1}{2}\theta + B.$

From (4), with $n = 0$, $\qquad \dfrac{d^2 R}{dr^2} = -k^2 R.$

Whence $\qquad\qquad\qquad R = Ce^{\pm jkr}.$

Thus, for a diverging wave,

$$T = [A \log (\cot \tfrac{1}{2}\theta) + B] e^{j(\omega t - kr)}. \qquad (16)$$

To find the *E*-wave, put $T = U$ and apply (10). We find

$$E_r = \frac{n(n+1)}{r^2} U = 0 \quad \text{(since } n = 0),$$

$$E_\theta = \frac{1}{r} \frac{\partial^2 U}{\partial \theta \, \partial r} = A \frac{(-jk)}{r \sin \theta} e^{j(\omega t - kr)},$$

$$E_\phi = 0 = H_\theta,$$

$$H_\phi = -j\frac{\omega\epsilon}{r} \frac{\partial U}{\partial \theta} = A \frac{(-j\omega\epsilon)}{r \sin \theta} e^{j(\omega t - kr)}.$$

The wave is therefore a TEM-wave and requires a pair of conductors to guide it. Further, since T and the field components become infinite for $\theta = 0$ or π, this axis must be excluded from the region. Also since E_r is zero everywhere and $E = E_0$, the appropriate conductors are a pair of coaxial cones with the common axis $\theta = 0$ or π, such, for instance, as the pair of cones shown in fig. 6·10 (a). Let the angles of the conical conductors be θ_1 and θ_2. The voltage in the TEM-wave guided by them is

$$v = \int_{\theta_1}^{\theta_2} E_\theta r\, d\theta = \int_{\theta_1}^{\theta_2} d\left[\frac{\partial U}{\partial r}\right]$$

$$= \left[\frac{\partial U}{\partial r}\right]_{\theta_1}^{\theta_2} = A(-jk)\,e^{j(\omega t - kr)}\,[\log \cot \tfrac{1}{2}\phi]_{\theta_1}^{\theta_2}.$$

The current in the conductor θ_1 at distance r is

$$i = 2\pi r \sin\theta_1 H_\phi = -j\omega\epsilon 2\pi A e^{-j(\omega t - kr)}.$$

The characteristic impedance of the double-cone transmission line is

$$Z_0 = \frac{v}{i} = \frac{k}{2\pi\omega\epsilon}\,[\log\cot\tfrac{1}{2}\phi]_{\theta_1}^{\theta_2}.$$

But $k = \omega\sqrt{(\mu\epsilon)}$. Therefore

$$Z_0 = \frac{1}{2\pi}\sqrt{\left(\frac{\mu}{\epsilon}\right)}\,[\log\cot\tfrac{1}{2}\phi]_{\theta_1}^{\theta_2}.$$

When $\theta_1 = -\theta_2 = \theta_0$, $\quad Z_0 = \frac{1}{\pi}\sqrt{\left(\frac{\mu}{\epsilon}\right)}\log(\cot\tfrac{1}{2}\theta_0)$.

Spherical resonators

We shall suppose the sphere to be hollow and with radius a. We give T in (9) the form

$$T = r^{\frac{1}{2}}J_{(n+\frac{1}{2})}(kr)\,P_n^m(\cos\theta)\cos m\phi e^{j\omega t}.$$

The cases of greatest interest are those with axial symmetry, that is, with $m = 0$.

The functions $J_{n+\frac{1}{2}}(kr)$ are easily obtained from the Hankel functions in (14) with $p = (n+\frac{1}{2})$ as

$$J_{n+\frac{1}{2}}(kr) = \tfrac{1}{2}[H^{(1)}_{(n+\frac{1}{2})}(kr) + H^{(2)}_{(n+\frac{1}{2})}(kr)]$$

$$= \sqrt{\left(\frac{1}{2\pi kr}\right)}j^{-(n+1)}[e^{jkr}S_{(n+\frac{1}{2})}(-2jkr) + j^{2(n+1)}e^{-jkr}S_{(n+\frac{1}{2})}(2jkr)]$$

$$= j^{-(n+1)}\frac{1}{\sqrt{(2\pi kr)}}[e^{jkr}S_{n+\frac{1}{2}}(-2jkr) - e^{-jkr}S_{n+\frac{1}{2}}(2jkr)].$$

From this it follows that

$$J_{\frac{1}{2}}(kr) = \sqrt{\left(\frac{2}{\pi kr}\right)} \sin kr,$$

$$J_{\frac{3}{2}}(kr) = \sqrt{\left(\frac{2}{\pi kr}\right)} \left[\frac{\sin kr}{kr} - \cos kr\right],$$

$$J_{\frac{5}{2}}(kr) = \sqrt{\left(\frac{2}{\pi kr}\right)} \left[\left(\frac{3}{(kr)^2} - 1\right) \sin kr - \frac{3}{kr} \cos kr\right], \text{ etc.}$$

The natural frequencies are found from the conditions $V = 0$ and $U \neq 0$ at $r = a$. Thus for H-modes: $J_{(n+\frac{1}{2})}(ka) = 0$, $k = \rho_{nq}/a$; for E-modes: $\frac{d}{d(ka)} [\sqrt{(ka)} J(ka)] = 0$, $k = \sigma_{nq}/a$.

Here ρ_{nq} and σ_{nq} are the qth roots respectively of $J(x) = 0$ and $\frac{d}{dx}[x^{\frac{1}{2}} J(x)] = 0$. Some values are: $\rho_{11} = 5\cdot8$, $\rho_{12} = 7\cdot64$, $\sigma_{11} = 2\cdot75$.

These are the (nqm) H- and E-modes of a spherical resonator.

7·10·8. *Longitudinal section waves*

If the expressions §7·10·2 (9) and (10) for the field components of E_{mn}- and H_{mn}-waves in rectangular wave guides be examined it will be noted that the components E_y are, apart from constant coefficients, the same functions of x, y and z, since both types of wave travel at the same speed. If therefore the wave fields derived from

$$\omega\mu\left(\frac{m\pi}{a}\right) U \quad \text{and} \quad \gamma\left(\frac{n\pi}{b}\right) V$$

are subtracted, the component E_y disappears and a new wave pattern results in which E has no transverse component. This wave is called a longitudinal section wave.

Similarly, by superimposing the fields of

$$\gamma\left(\frac{n\pi}{b}\right) U \quad \text{and} \quad -\omega\epsilon\left(\frac{m\pi}{a}\right) V,$$

the resulting longitudinal section wave is one in which H_y is absent.

The lowest order longitudinal wave is that for which $m = n = 1$.

7·11. Example of a method for calculating the susceptance of an obstacle

Although in §§5·4 and 5·5 we have discussed the interpretation of the term self-susceptance of an obstacle in a wave guide, it is useful to illustrate what was said, by demonstrating how the theoretical expression for the susceptance is actually obtained in a specific instance. The object chosen for investigation is a thin wire stretched across the centre of the cross-section of the wave guide parallel to the electric field of the H_{10}-wave. This example is well adapted to our purpose, since the analysis is relatively simple. The principle of the method is that employed by Macfarlane.

Consider first the magnetostatic field excited by the plane grid of parallel infinitely long and evenly spaced fine wires, a normal section of which is shown in fig. 7·3. The currents are equal but flow in opposite senses in adjacent wires. Let the planes of the grid and of the paper be respectively XOY and XOZ, with the origin placed on one of the wires. The figure indicates roughly the configuration of the magnetic field.

Let the separation of adjacent wires be equal to a and the current in each wire I. It is known that the two-dimensional magnetic field is to be derived from the following complex potential:*

$$w = u + jv = j\frac{I}{2\pi}\log\tanh\left[\frac{\pi}{2a}(z+jx)\right]$$

$$= j\frac{I}{2\pi}\log\left(\frac{1-e^{-\pi(z+jx)/a}}{1+e^{-\pi(z+jx)/a}}\right). \tag{1}$$

The magnetic field is $\mathbf{H} = -\operatorname{grad} u$, that is

$$H_z = -\frac{\partial u}{\partial z}, \quad H_x = -\frac{\partial u}{\partial x}. \tag{2}$$

According to the Cauchy-Riemann equations

$$\frac{\partial u}{\partial z} = \frac{\partial v}{\partial x}, \quad \frac{\partial u}{\partial x} = -\frac{\partial v}{\partial z},$$

consequently $\mathbf{H}$ is also obtained from v, as follows:

$$H_z = -\frac{\partial v}{\partial x}, \quad H_x = \frac{\partial v}{\partial z}. \tag{3}$$

Since $\mathbf{H}$ is curl $\mathbf{A}$, where $\mathbf{A}$ is the electric vector potential of the field (§7·3), it follows from (3) that $\mathbf{A}$ is directed along OY (parallel

* For instance, S. L. Green, *Hydro and Aerodynamics*, Pitman, p. 83.

to the wires) and that its single component A_y is equal to v. Thus u is the magnetostatic potential and v the vector potential.

We proceed to express relation (1) in a form more significant for our present purpose.

This expression is equivalent to

$$w = j\,\frac{I}{2\pi}\left[\log\left(1 - e^{-\pi(z+jx)/a}\right) - \log\left(1 + e^{-\pi(z+jx)/a}\right)\right]$$

$$= j\,\frac{I}{2\pi}\left[-\left(e^{-\pi(z+jx)/a} + \frac{e^{-2\pi(z+jx)/a}}{2} + \frac{e^{-3\pi(z+jx)/a}}{3} + \cdots\right)\right.$$

$$\left.-\left(e^{-\pi(z+jx)/a} - \frac{e^{-2\pi(z+jx)/a}}{2} + \frac{e^{-3\pi(z+jx)/a}}{3} - \cdots\right)\right]$$

$$= -j\,\frac{I}{\pi}\left[e^{-\pi z/a}\left(\cos\frac{\pi x}{a} - j\sin\frac{\pi x}{a}\right) + \frac{e^{-3\pi z/a}}{3}\left(\cos\frac{3\pi z}{a} - j\sin\frac{3\pi z}{a}\right)\right.$$

$$\left.+ \frac{e^{-5\pi z/a}}{5}\left(\cos\frac{5\pi z}{a} - j\sin\frac{5\pi z}{a}\right) + \cdots\right].$$

Whence
$$\left.\begin{aligned} u &= -\frac{I}{\pi}\sum_{m=1}^{\infty}\sin\left(\frac{m\pi x}{a}\right)\frac{e^{-m\pi z/a}}{m}, \\ v &= -\frac{I}{\pi}\sum_{m=1}^{\infty}\cos\left(\frac{m\pi x}{a}\right)\frac{e^{-m\pi z/a}}{m}, \end{aligned}\right\} \tag{4}$$

with m an odd integer.

Next suppose the currents in the wires to oscillate at low frequency $(\lambda \gg a)$ so that (4) becomes

$$\left.\begin{aligned} u &= -\frac{I}{\pi}\sum_{1}^{\infty}\sin\left(\frac{m\pi x}{a}\right)\frac{e^{-(m\pi/a)z}}{m}e^{j\omega t}, \\ v &= -\frac{I}{\pi}\sum_{1}^{\infty}\cos\left(\frac{m\pi x}{a}\right)\frac{e^{-(m\pi/a)z}}{m}e^{j\omega t}, \end{aligned}\right\} \tag{5}$$

(m an odd integer).

We next seek to remove the restriction that ω shall be small $(\lambda \gg a)$ and thus obtain an expression for the field potentials valid at any frequency. This expression must reduce to (5) at low and zero frequencies.

We recall that in a Cartesian coordinate system an electromagnetic field with x- and z-components may be derived from a scalar quantity V which is the z-component of a Hertz vector. The components of the H-type electromagnetic field derived from V are:

$$
\left.
\begin{aligned}
H_x &= \frac{\partial}{\partial x}\left(\frac{\partial V}{\partial z}\right), \quad E_x = -j\omega\mu\frac{\partial V}{\partial y}, \\[2mm]
H_y &= \frac{\partial}{\partial y}\left(\frac{\partial V}{\partial z}\right), \quad E_y = j\omega\mu\frac{\partial V}{\partial x}, \\[2mm]
H_z &= k^2 V + \frac{\partial}{\partial z}\left(\frac{\partial V}{\partial z}\right), \quad k^2 = \omega^2\mu\epsilon.
\end{aligned}
\right\}
\tag{6}
$$

When ω is zero or small, then the field derived from (6) is static or quasi-static, with components

$$
H_x = \frac{\partial}{\partial x}\left(\frac{\partial V}{\partial z}\right), \quad H_y = \frac{\partial}{\partial y}\left(\frac{\partial V}{\partial z}\right), \quad H_z = \frac{\partial}{\partial z}\left(\frac{\partial V}{\partial z}\right).
$$

When in addition this static field is two-dimensional, with H_y (and therefore $\partial V/\partial y$) zero, it is evident from (2) that $-(\partial V/\partial z)$ becomes the magnetostatic scalar potential u.

Thus, at low frequencies,

$$
\frac{\partial V}{\partial z} \to \frac{I}{\pi}\Sigma\sin\left(\frac{m\pi x}{a}\right)\frac{e^{-(m\pi/a)z}}{m}\,e^{j\omega t}.
\tag{7}
$$

But we know, from 7·10·2 (10), that the H_{mn}-field is derived from

$$
V_{mn} = A_{mn}\sin\left(\frac{m\pi x}{a}\right)\cos\left(\frac{n\pi y}{b}\right)e^{-j\gamma_{mn}z}\,e^{j\omega t},
\tag{8}
$$

in which a sine function appears here because of the choice of origin. Put $n = 0$ in (8) and differentiate with respect to z:

$$
\frac{\partial V_m}{\partial z} = -j\gamma_m A_m\sin\left(\frac{m\pi x}{a}\right)e^{-j\gamma_m z}\,e^{j\omega t},
\tag{9}
$$

and the general field independent of y is to be derived from $V = \Sigma V_m$.

We require $\Sigma(\partial V_m/\partial z)$ from (9) to transform into (7) at low frequencies.

Since, according to 7·10·2 (4),

$$
\gamma_m^2 = k^2 - \left(\frac{m\pi}{a}\right)^2,
\tag{10}
$$

at low frequencies this gives

$$
\gamma_m = \pm j\sqrt{\left\{\left(\frac{m\pi}{a}\right)^2 - k^2\right\}} \to \pm j\left(\frac{m\pi}{a}\right),
$$

or

$$
j\gamma_m = \mp\sqrt{\left\{\left(\frac{m\pi}{a}\right)^2 - k^2\right\}}.
$$

Thus the exponential terms in (9) become those in (7) at low frequencies. We also require $-j\gamma_m A_m = I/\pi m$ or

$$A_m = \frac{I}{m\pi\sqrt{\{(m\pi/a)^2 - k^2\}}} = \frac{I}{-j\gamma_m\pi m}.$$

The quantity V which represents the field due to currents in the grid oscillating at any frequency $(\omega/2\pi)$ is,

$$\frac{I}{\pi}\Sigma\frac{\sin(m\pi x/a)}{m\sqrt{[(m\pi/a)^2 - k^2]}}e^{-\{\sqrt{[(m\pi/a)^2-k^2]z}\}}e^{j\omega t} \tag{11}$$

$$(m \text{ an odd integer}).$$

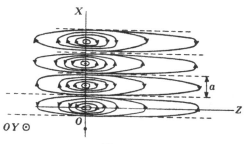

Fig. 7·3.

This expression makes V, and therefore H_z, a maximum on the planes $x = \pm(2p+1)a/2$, where p is an integer. These planes are shown dotted in fig. 7·3. The oscillating field differs from the truly magnetostatic field (4) in an important respect, it possesses an electric component

$$E_y = j\omega\mu\frac{\partial V}{\partial x}$$

$$= j\omega\mu\frac{I}{a}\Sigma\frac{\cos(m\pi x/a)}{\sqrt{[(m\pi/a)^2 - k^2]}}e^{-\{\sqrt{[(m\pi/a)^2-k^2]\}z}}e^{j\omega t}, \tag{12}$$

which vanishes at the planes $x = \pm(2p+1)a/2$.

We may therefore isolate a strip of the field by placing flat plates with infinite conductivity to coincide with the planes $x = \pm\frac{1}{2}a$, and the field between, which is still to be derived from (1), corresponds to that excited by an oscillating current in a thin straight wire placed parallel to and midway between planes with infinite conductivity.

Expression (11) has been written in a form which represents all the terms in the summation as evanescent modes and it serves to illustrate, what has been remarked in §3·8, that the storage fields of evanescent modes are quasi-static.

Suppose the frequency to be increased from a small value to one for which the first term ($m = 1$) in (11) ceases to be evanescent, that is $k > (\pi/a)$, but other terms remain evanescent. In short

$$\frac{3\pi}{a} > k > \frac{\pi}{a}.$$

The first term in (11) may now be written

$$v_1 = \frac{I}{\pi} \frac{\sin(\pi x/a)}{j\gamma} e^{j(\omega t - \gamma z)}, \tag{13}$$

where $\gamma = \sqrt{\{k^2 - (\pi/a)^2\}}$. The electric field of this wave is

$$E_{1y} = j\omega\mu \frac{I}{a} \frac{\cos(\pi x/a)}{j\gamma} e^{j(\omega t - \gamma z)}. \tag{14}$$

This represents a progressive H_{10}-wave between the plates.

We shall now suppose the current in the wire to be excited by the electric field E_{0y} of an H_{10}-wave

$$E_{0y} = E_0 \cos\left(\frac{\pi x}{a}\right) e^{j(\omega t - \gamma z)}. \tag{15}$$

The wave in (14) then represents the scattered wave, and the remaining terms in (12) the storage field. To obtain the susceptance of the wire it is necessary to relate the scattered wave (14) to the incident wave (15).

This we may do by noting that at the surface of the wire (radius r) the tangential components of the total electric field is zero. The total electric field comprises:

The incident field E_{0y} of (15).

The scattered field E_{1y} of (14).

The storage field, obtained from (12).

$$E_{sy} = j\omega\mu \frac{I}{a} \sum_3^\infty \frac{\cos(m\pi x/a)}{\sqrt{\{(m\pi/a)^2 - k^2\}}} e^{-\{\sqrt{[(m\pi/a)^2 - k^2]}z\}} e^{j\omega t}. \tag{16}$$

The condition is $\qquad E_0 + E_{1y} + E_{sy} = 0 \tag{17}$

at the surface of the wire.

The scattering coefficient h is E_{1y}/E_0, so that (17) becomes

$$E_{1y}\left(1 + \frac{1}{h}\right) = -E_{sy}, \quad \text{or} \quad \frac{h}{(1+h)} = -\frac{E_{1y}}{E_{sy}}.$$

The admittance of the obstacle y_1 is, from 5·4 (5),

$$y_1 = \frac{-2h}{(1+h)} = \frac{2E_{1y}}{E_{sy}}, \tag{18}$$

the fields being evaluated at the surface of the wire.

From (14), at the wire $\dfrac{E_{1y}}{j\omega\mu} = \dfrac{I}{j\gamma a} e^{j\omega t} = \dfrac{\lambda_g I}{j2\pi a} e^{j\omega t}$, since the wire, assumed thin, is at $x = 0$.

We require E_{sy} near the wire. Since, in (16), $m \geqslant 3$, $(m\pi/a)^2 \gg k^2$, and we make little error in neglecting k^2 in the z-exponent. Expression (16) may now be written

$$\frac{e^{-j\omega t} E_{sy}}{j\omega\mu} = \frac{I}{\pi} \sum_3^\infty \frac{\cos(m\pi x/a)}{m} e^{-m\pi z/a}$$
$$+ \frac{I}{\pi} \sum_3^\infty \cos\left(\frac{m\pi x}{a}\right) \left[\frac{1}{\sqrt{[1-(2a/m\lambda)^2]}} - 1 \right] \frac{e^{-m\pi z/a}}{m}, \tag{19}$$

in which m is an odd integer.

On referring to (4) we see that the first term in (19) is

$$-\left(v + \frac{I}{\pi} \cos\left(\frac{\pi x}{a}\right) e^{-\pi z/a} \right). \tag{20}$$

To obtain v near the wire return to (1) and write $(z+jx) = \rho e^{j\theta}$ and let $\rho/a \ll 1$.

Then
$$w = u+jv \rightarrow \frac{jI}{2\pi} \log\left(\frac{\pi}{2a}\rho e^{j\theta}\right)$$
$$= \frac{jI}{2\pi}\left[\log\left(\frac{\pi\rho}{2a}\right) + j\theta\right].$$

Whence
$$v = \frac{I}{2\pi} \log\left(\frac{\pi\rho}{2a}\right). \tag{21}$$

The quantity (20) is therefore approximately

$$-\frac{I}{2\pi}\left[\log\left(\frac{\pi\rho}{2a}\right) + 2\right], \tag{22}$$

since x and z are small when ρ is small.

The second term in (19) when (x/a) is small is very nearly equal to

$$\frac{I}{\pi} \sum_3^\infty \left[\frac{1}{\sqrt{[1-(2a/m\lambda)^2]}} - 1 \right] \frac{1}{m}.$$

The susceptance of the wire is therefore, from (18), given by ($\rho = r$ = radius of wire)

$$\frac{1}{y_1} = \frac{E_{sy}}{2E_{1y}} = \frac{-ja}{2\lambda_g}\left[\log\left(\frac{\pi r}{2a}\right) + 2 - 2\sum_3^\infty\left(\frac{1}{\sqrt{[1-(2a/m\lambda)^2]}} - 1\right)\frac{1}{m}\right]$$

$$= \frac{ja}{2\lambda_g}\left[\log\left(\frac{2a}{\pi r}\right) - 2 + 2\sum_3^\infty\left(\frac{1}{\sqrt{[1-(2a/m\lambda)^2]}} - 1\right)\frac{1}{m}\right], \qquad (23)$$

where m is odd. The pair of parallel plates may be converted to a wave guide by introducing a second pair of walls in the planes $y = 0$ and b.

Expression (23) is that given in fig. 5·16(10). This example illustrates how, by proper choice of an electrostatic problem whose solution is known, it is sometimes possible to solve an associated problem where the fields are oscillatory.

7·12. The field energy

We return to §7·10·3 where the propagation of waves along cylinders was discussed in a general manner.

It was shown that it was necessary to obtain solutions of

$$\nabla^2 T + k^2 T = 0 \qquad (1)$$

($k^2 = \omega^2\mu\epsilon$), subject to the boundary conditions $T = 0$ at the boundary when $T = U$ (E-modes) the normal component of $T = 0$ at the boundaries when $T = V$ (H-modes).

Field energy in E-waves. Put $T = U = F(x,y)\sin(\omega t - \gamma z)$. Then, from (1),

$$\frac{\partial^2 F}{\partial x^2} + \frac{\partial^2 F}{\partial y^2} + (k^2 - \gamma^2)F = 0. \qquad (2)$$

The field components are:

$$E_x = \frac{\partial}{\partial x}\left(\frac{\partial U}{\partial z}\right) = -\gamma\frac{\partial F}{\partial x}\cos(\omega t - \gamma z),$$

$$E_y = \frac{\partial}{\partial y}\left(\frac{\partial U}{\partial z}\right) = -\gamma\frac{\partial F}{\partial y}\cos(\omega t - \gamma z),$$

$$E_z = (k^2 - \gamma^2)F(x,y)\sin(\omega t - \gamma z),$$

$$H_x = \omega\epsilon\cos(\omega t - \gamma z)\frac{\partial F}{\partial y},$$

$$H_y = -\omega\epsilon\cos(\omega t - \gamma z)\frac{\partial F}{\partial x}.$$

The mean electric energy per length $\lambda_g = 2\pi/\gamma$ of the wave guide is

$$W_E = \frac{\gamma^2 \epsilon}{4} \int \left[\left(\frac{\partial F}{\partial x}\right)^2 + \left(\frac{\partial F}{\partial y}\right)^2 \right] dx\,dy + \frac{(k^2 - \gamma^2)^2 \epsilon}{4} \int F^2 dx\,dy.$$

The mean magnetic energy per wave-length is

$$W_M = \frac{\omega^2 \epsilon^2 \mu}{4} \int \left[\left(\frac{\partial F}{\partial x}\right)^2 + \left(\frac{\partial F}{\partial y}\right)^2 \right] dx\,dy$$

$$= \frac{k^2 \epsilon}{4} \int \left[\left(\frac{\partial F}{\partial x}\right)^2 + \left(\frac{\partial F}{\partial y}\right)^2 \right] dx\,dy.$$

To relate W_E and W_M we use the two-dimensional form of one of Green's theorems:

$$\int_S \left[\phi \left(\frac{\partial^2 \psi}{\partial x^2} + \frac{\partial^2 \psi}{\partial y^2}\right) + \frac{\partial \phi}{\partial x} \frac{\partial \psi}{\partial x} + \frac{\partial \phi}{\partial y} \frac{\partial \psi}{\partial y} \right] dx\,dy = \int \phi \frac{\partial \psi}{\partial n}\,ds,$$

where ϕ and ψ are scalars, S a surface and $\int ds$ its periphery.

Put $\phi = \psi = F(x, y)$, then using (2)

$$\int_S \left[-(k^2 - \gamma^2) F^2 + \left(\frac{\partial F}{\partial x}\right)^2 + \left(\frac{\partial F}{\partial y}\right)^2 \right] dx\,dy = 0,$$

since F vanishes at the boundary.

Thus
$$\int \left[\left(\frac{\partial F}{\partial x}\right)^2 + \left(\frac{\partial F}{\partial y}\right)^2 \right] dx\,dy = (k^2 - \gamma^2) \int F^2 dx\,dy.$$

It follows that $W_E = W_M$. Thus in a progressive E-wave the mean electric and magnetic energies per wave-length are equal. The same result is valid for H-waves.

When, however, the mode is evanescent a similar investigation shows that in E-modes the electric energy exceeds the magnetic energy and conversely in H-modes.

7·13. Babinet's principle

Babinet's principle, which is described in a number of treatises on physical optics,* relates the diffraction pattern produced by a screen containing one or more apertures with that produced by its complementary screen. If we denote the screen by S, then the

---

* T. Preston, *Theory of Light*, Macmillan; R. W. Wood, *Physical Optics*, Macmillan; P. Drude, *Theory of Optics*, Longmans, Green and Co.

screen C complementary to S is obtained from S by interchanging the opaque and open regions of S. As formulated in text-books of optics, Babinet's principle is derived from the Kirchhoff theory of diffraction, which treats light as a scalar wave phenomenon, consequently this formulation is not strictly applicable to the diffraction of electromagnetic waves. Several investigators* have discussed the correct form of the electromagnetic analogue of Babinet's principle, the first of whom appears to be P. Epstein (1915).[†] The practical importance of the electromagnetic analogue of Babinet's principle in the development of microwave techniques (mainly in the subject of aerials and radiation, but indirectly in wave guides) was stressed by Booker.

Because of its importance in the theory and practice of microwaves and because it affords a good illustration of the usefulness of the U and V functions in general discussions, we give a derivation of the electromagnetic analogue of Babinet's principle valid for infinitely thin perfectly conducting complementary screens (the case of imperfectly conducting screens is also discussed by Sommerfeld—see reference).

We begin with a statement of the electromagnetic analogue of Babinet's principle for a simple case. Consider a pair of plane perfectly conducting and complementary screens S and C of infinite extent, so constructed that the apertures S correspond to the conducting portions of C and conversely, so that if the two screens were superimposed they would fit together to form a conducting plane without apertures. That is, if the conducting regions of S are denoted by σ_1, and the open regions by σ_2, then in C, σ_1 denotes the open regions and σ_2 the closed. Suppose the screen S to occupy the plane $z = 0$ of a Cartesian coordinate system and that a plane-polarized electromagnetic wave $(\mathbf{E}_0, \mathbf{H}_0)$ is incident normally on S, its direction of propagation being along OZ.

Let the electric vector $\mathbf{E}_0$ of the wave oscillate parallel to OY.

The incident wave drives oscillating currents and produces oscillating charge distributions on the conducting regions σ_1, of S,

* See footnote on previous page.

† P. Epstein, *Enzyl. d. Math. Wissensch.* Bd. 5, Art. 24, SS. 510–11, Leipzig, 1915; F. Jentzsch, *Handbuch der Physik*, Bd. 21, Licht und Materil, Kap. 20, SS. 914–15; W. V. Ignatowsky, *Ann. Phys.* Bd. 77, SS. 622, 1925; W. Fischer, *Math. Ann.* Bd. 101, 1929; A. Sommerfeld, *Die Differential und Integralgleichungen der Mechanik und Physik*, Frank und v. Mises, Zweiter Teil, SS. 811–16.

which excite a supplementary electromagnetic field $(\mathbf{E_1}, \mathbf{H_1})$. Consequently, the total electromagnetic field $(\mathbf{E}, \mathbf{H})$ behind the screen is the vector sum of the incident and supplementary fields; that is, $\mathbf{E} = \mathbf{E_0} + \mathbf{E_1}$ and in particular $E_y = E_0 + E_{1y}$.

Suppose the conducting regions σ_1 and the apertures σ_2 of S to be interchanged so that S is transformed to the complementary screen C. Let the incident electromagnetic wave, propagated along OZ, become $(\mathbf{E_0'}, \mathbf{H_0'})$. This wave is chosen to be related to the original wave $(\mathbf{E_0}, \mathbf{H_0})$ as follows:

$\mathbf{H_0'}$ oscillates parallel to $\pm OY$.

The amplitude of $\mathbf{H_0'}$ is the same as $\mathbf{E_0}$, i.e. (ignoring physical dimensions)
$$\mathbf{H_0'} = \mathbf{E_0}.$$

In other words, the second incident wave is polarized at right angles to the first.

Let the supplementary electromagnetic field associated with C under the influence of $(\mathbf{E_0'}, \mathbf{H_0'})$ be $(\mathbf{E_2}, \mathbf{H_2})$. The y-component of the total magnetic field behind C is

$$H_y = H_0' + H_{2y}.$$

The electromagnetic analogue of Babinet's principle asserts that, under the respective influences of the two incident waves $(\mathbf{E_0}, \mathbf{H_0})$ and $(\mathbf{E_0'}, \mathbf{H_0'})$ with $\mathbf{H_0'} = \mathbf{E_0}$ and parallel to $\mathbf{E_0}$, the total magnetic field behind C is equal to the y-component of the supplementary electric field behind S but oppositely directed if the incident waves oscillate in phase.

In symbols, this statement is equivalent to (disregarding physical dimensions)

$$H_y = H_0' + H_{2y} \quad (y\text{-component of total } H\text{-field behind } C),$$
$$= -E_{1y} \quad (y\text{-component of supplementary } E\text{-field behind } S).$$

Evidently, when the supplementary (diffracted) field $\mathbf{E}$, of a screen S, had been found by any means, that of the complementary screen C may be immediately deduced for the case in which the wave incident normally on C is polarized at right angles to that incident normally on S.

We proceed to use the U and V functions to obtain a more general formulation of the electromagnetic analogue of Babinet's principle than that we have just considered.

286 PRINCIPLES AND PRACTICE OF WAVE GUIDES

Let an arbitrary electromagnetic E-type field derived from a Hertz vector $T = U_0$, impinge on screen S, and an electromagnetic H-type field derived from the same function $T = V_0$ be incident on C. That is, the generating function T is common to both fields which are obtained by differentiating T by the respective methods required to give E- and H-fields. For instance, a simple plane wave incident normally on S may be derived from

$$T = U_0 = E_0 y e^{j(\omega t - kz)},$$

and that incident on C from $T = V_0 = H_0' y e^{j(\omega t - kz)}$, with $H_0' = E_0$.

We now consider incident fields of a more general type. The field U_0 incident on S induces surface charges and currents on its conducting portions σ_1, which excite supplementary electromagnetic fields which are superimposed on the exciting field. These supplementary fields comprise, in general, both E- and H-type fields which we suppose to be derivable from potential functions U_1 and V_1 respectively. Similarly, the field V_0 excites a supplementary field U_2, V_2 when it falls on C. The resulting field components in both cases satisfy the following conditions at the surfaces of the screens. Over the conducting portions the tangential components of the total electric field and the normal component of the magnetic field vanish. Over the open portions (in the plane of the screens), since the screens are infinitely thin so that the induced currents and charges are distributed in sheets, the normal component of the supplementary electric field and the tangential component of the supplementary magnetic field, vanish.

Denote the total field behind (on the side away from the source) S by $(\mathbf{E}_S, \mathbf{H}_S)$, and the supplementary field from the screen S by $(\mathbf{E}_{1S}, \mathbf{H}_{1S})$, the corresponding fields for C being $(\mathbf{E}_C, \mathbf{H}_C)$ and $(\mathbf{E}_{1C}, \mathbf{H}_{1C})$. Draw up the comparisons shown in tables 1 and 2 of the field components and their behaviour at the surface of S and C. Suppose the problem of the supplementary fields of S, excited by U_0 has been solved; that is, a solution $(U_1 + V_1)$ satisfying the basic differential equation $\nabla^2 T + k^2 T = 0$ and the boundary conditions over S, has been found. The problem is to find the fields associated with C under the influence of V_0, where the H-type wave V_0 and E-type wave U_0 have the same functional form T.* An inspection

---

* This is a convenient method of postulating that electric and magnetic fields are interchanged in the two waves with a reversal of sign of one of them.

of the table of components and boundary conditions shows what is the solution of the problem for the screen C.

Table 1

Supplementary field behind S $(\mathbf{E}_{1S}, \mathbf{H}_{1S}) = (\mathbf{E}_S, \mathbf{H}_S) - (\mathbf{E}_0, \mathbf{H}_0)$	Behaviour on S	
	σ_1 closed	σ_2 open
$(E_{1S})_x = \dfrac{\partial}{\partial x}\left[\dfrac{\partial}{\partial z}(U_1)\right] - j\omega\mu\,\dfrac{\partial V_1}{\partial y}$	$-\dfrac{\partial}{\partial x}\left(\dfrac{\partial U_0}{\partial z}\right)$	$\neq o$
$(E_{1S})_y = \dfrac{\partial}{\partial y}\left[\dfrac{\partial}{\partial z}(U_1)\right] + j\omega\mu\,\dfrac{\partial V_1}{\partial x}$	$-\dfrac{\partial}{\partial y}\left(\dfrac{\partial U_0}{dz}\right)$	$\neq o$
$(E_{1S})_z = \left(k^2 + \dfrac{\partial^2}{\partial z^2}\right) U_1$	$\neq o$	o
$(H_{1S})_x = \dfrac{\partial}{\partial x}\left(\dfrac{\partial V_1}{\partial z}\right) + j\omega\epsilon\,\dfrac{\partial U_1}{\partial y}$	$\neq o$	o
$(H_{1S})_y = \dfrac{\partial}{\partial y}\left(\dfrac{\partial V_1}{\partial z}\right) - j\omega\epsilon\,\dfrac{\partial U_1}{\partial x}$	$\neq o$	o
$(H_{1S})_z = \left(k^2 + \dfrac{\partial^2}{\partial z^2}\right) V_1$	o	$\neq o$

Table 2

Total field $(\mathbf{E}_c, \mathbf{H}_c)$ behind C	Behaviour on C	
	σ_1 open	σ_2 closed
$(H_c)_x = \dfrac{\partial}{\partial x}\left[\dfrac{\partial}{\partial z}(V_2 + V_0)\right] + j\omega\epsilon\,\dfrac{\partial U_2}{\partial y}$	$\dfrac{\partial}{\partial x}\left(\dfrac{\partial V_0}{\partial z}\right)$	$\neq o$
$(H_c)_y = \dfrac{\partial}{\partial y}\left[\dfrac{\partial}{\partial z}(V_2 + V_0)\right] - j\omega\epsilon\,\dfrac{\partial U_2}{\partial x}$	$\dfrac{\partial}{\partial x}\left(\dfrac{\partial V_0}{\partial z}\right)$	$\neq o$
$(H_c)_z = \left(k^2 + \dfrac{\partial^2}{\partial z^2}\right)(V_2 + V_0)$	$\neq o$	o
$(E_c)_x = \dfrac{\partial}{\partial x}\left(\dfrac{\partial U_2}{\partial z}\right) - j\omega\mu\,\dfrac{\partial}{\partial y}(V_2 + V_0)$	$\neq o$	o
$(E_c)_y = \dfrac{\partial}{\partial y}\left(\dfrac{\partial U_2}{\partial z}\right) + j\omega\mu\,\dfrac{\partial}{\partial x}(V_2 + V_0)$	$\neq o$	o
$(E_c)_z = \left(k^2 + \dfrac{\partial^2}{\partial z^2}\right) U_2$	o	$\neq o$

Put $\quad V_C = V_2 + V_0 = -U_1 \quad$ and $\quad U_C = U_2 = +\dfrac{\mu}{\epsilon}V_1,$ $\qquad$ (1)

where V_C and U_C are the potentials of the total field behind C.

Expression (1) is the solution we require because every component $(H_C)_x$, etc., of the C field automatically satisfies the boundary conditions over the screen C. For instance, consider $(H_C)_x$:

According to (1) and table 2

$$(H_C)_x = \frac{\partial}{\partial x}\left[\frac{\partial}{\partial z}(V_2 + V_0)\right] + j\omega\epsilon\frac{\partial U_2}{\partial y}$$

$$= -\frac{\partial}{\partial x}\left[\frac{\partial}{\partial z}U_1\right] + j\omega\mu\frac{\partial V_1}{\partial y} = -(E_{1S})_x.$$

And the boundary conditions to be satisfied by $(H_x)_C$ are

$$\text{over } \sigma_1: \quad (H_C)_x = \frac{\partial}{\partial x}\left(\frac{\partial V_0}{\partial z}\right) = \frac{\partial}{\partial x}\left(\frac{\partial U_0}{\partial z}\right),$$

$$\text{over } \sigma_2: \quad (H_C)_x = 0;$$

and these are the boundary conditions satisfied by $-(E_{1S})_x$ which we assume to be a known solution.

On the other hand,

$$(E_C)_x = \frac{\partial}{\partial x}\left(\frac{\partial U_2}{\partial z}\right) - j\omega\mu\frac{\partial}{\partial y}(V_2 + V_0)$$

$$= \frac{\mu}{\epsilon}\frac{\partial}{\partial x}\left(\frac{\partial V_1}{\partial z}\right) + j\omega\mu\frac{\partial}{\partial y}U_1 = \frac{\mu}{\epsilon}(H_{1S})_x.$$

We may summarize as follows:

The magnetic field of the total field behind C is identical both in strength and configuration with the electric portion of the supplementary field behind S but reversed in phase everywhere.

The electric portion of the total field behind C is (μ/ϵ) times the magnetic portion of the supplementary field behind S.

These results can be expressed as vector equations:

$$\mathbf{H}_C = -\mathbf{E}_{1S}, \quad \mathbf{E}_C = \frac{\mu}{\epsilon}\mathbf{H}_{1S}, \tag{2}$$

but they are implicit in (1). It should be noted that in equations such as (2) we are comparing magnitudes of vectors and that the equations are not of necessity dimensionally correct.

It is again stressed that in obtaining (2) we have postulated that the exciting fields U_0 and V_0 applied respectively to S and C are E-type and H-type fields derived from a common single component (Π_0) vector T of arbitrary form. (The reason why Babinet's principle is not rigorously applicable to the study of Irises in Wave Guides is that we cannot change the H_{10}-wave into an E_{10}-wave which does not exist in a rectangular wave guide.)

Since $$\mathbf{H}_C = \mathbf{H}_{1C} + \mathbf{H}_0', \quad \mathbf{E}_C = \mathbf{E}_{1C} + \mathbf{E}_0',$$

where $(\mathbf{H}_{01}', \mathbf{E}_0')$ is the electromagnetic field that excites C, it follows from (2) that

$$\mathbf{H}_{1C} = -(\mathbf{E}_{1S} + \mathbf{H}_0) = -(\mathbf{E}_{1S} + \mathbf{E}_0) = -\mathbf{E}_S, \\ \mathbf{E}_{1C} = \frac{\mu}{\epsilon}\mathbf{H}_{1S} - \mathbf{E}_0' = \frac{\mu}{\epsilon}[\mathbf{H}_{1S} + \mathbf{H}_0] = \frac{\mu}{\epsilon}\mathbf{H}_S. \quad (3)$$

The magnetic portion of the supplementary field behind C is equal and opposite to the electric portion of the total field behind S. The electric portion of the supplementary field behind C is μ/ϵ times the total magnetic field behind S.

We could, of course, let an E-type field excite C and an H-type field excite S and relate the resultant fields as we have done above. Finally, we suppose S to be excited by a field of very general type $(U_0 + V_0')$, and C by $(U_0' + V_0)$, such that U_0 and V_0 are the same T function and V_0' and U_0' another function T'.

Equation (1) is then modified to

$$V_C = V_2 + V_0 = -U_1, \quad U_C = U_2 + U_0' = \frac{\mu}{\epsilon}V_1, \quad (4)$$

and reciprocally, from the appropriately modified tables 1 and 2

$$U_S = U_1 + U_0 = -V_2, \quad V_S = V_1 + V_0' = \frac{\epsilon}{\mu}U_2, \quad (5)$$

where U_S and V_S are the potentials of the total field behind S when driven by $(U_0 + V_0)$, and V_2 and U_2 the potentials of the supplementary field behind C when it is excited by a field $(U_0' + V_0)$, such that

$$U_0' \equiv T' = V_0', \quad U_0 \equiv T = V_0.$$

The sign of identity indicates functional identity leading to different fields according to the mode of differentiation, whether E-type or H-type.

Expressions (4) and (5) are very general statements of Babinet's principle, for the case where the screens have perfectly conducting regions.

The supplementary fields (U_{1F}, V_{1F}), (U_{2F}, V_{2F}) in *front* of S and C respectively are related to those *behind* S and C, as follows:

$$\begin{Bmatrix} U_{1F} = -U_1 \\ V_{1F} = V_1 \end{Bmatrix} \quad \text{and} \quad \begin{Bmatrix} U_{2F} = -U_2 \\ V_{2F} = V_2 \end{Bmatrix}.$$

The supplementary field in front of S is related as follows to the total field behind C:

$$U_{1F} + V_C, \quad V_{1F} = \frac{\epsilon}{\mu} U_C.$$

Similarly, from (5),

$$U_{2F} = \frac{\mu}{\epsilon} V_S, \quad V_{2F} = U_S.$$

7·14. Skin effect and equivalent surface resistance

We return to Maxwell's equations 7·2(1)–(4), and suppose the medium to possess a conductivity σ so that the current density $\mathbf{J}$ is $\sigma\mathbf{E}$. Let all components contain the factor $e^{j\omega t}$, and put $\rho = 0$.

The equations become

$$\mathbf{\nabla} \times \mathbf{E} = -j\omega\mu\mathbf{H}, \quad \mathbf{\nabla} \times \mathbf{H} = j\omega\epsilon\mathbf{E} + \sigma\mathbf{E}, \quad \operatorname{div}\mathbf{D} = \operatorname{div}\mathbf{B} = 0. \quad (1)$$

Let the coordinate system be Cartesian and form

$$\mathbf{\nabla} \times \mathbf{\nabla} \times \mathbf{H} = \mathbf{\nabla}(\mathbf{\nabla}.\mathbf{H}) - \nabla^2\mathbf{H}$$

$$= -\nabla^2\mathbf{H} = j\omega\left(\epsilon + \frac{\sigma}{j\omega}\right)\mathbf{\nabla} \times \mathbf{E}.$$

Put $\epsilon' = (\epsilon + \sigma/j\omega)$ and use the first of equations (1), then

$$\left.\begin{aligned} \nabla^2\mathbf{H} = \omega^2\mu\epsilon'\mathbf{H} = k^2\mathbf{H}. \\ \text{Similarly,} \qquad \nabla^2\mathbf{E} = \omega^2\mu\epsilon'\mathbf{E} = k^2\mathbf{E}. \end{aligned}\right\} \quad (2)$$

We require a solution representing a plane wave with $\mathbf{E}$ and $\mathbf{H}$ in the ZOY plane, and direction of propagation along OX.

The following solution of (2) serves our purpose:

$$H_z = H_0 e^{j(\omega t - kx)}, \quad E_y = E_0 e^{j(\omega t - kx)}, \quad \sqrt{(\mu)}\,H_0 = \sqrt{(\epsilon')}\,E_0. \quad (3)$$

We consider the case of propagation in a metal where σ/ω at microwave frequencies ($\sigma/\omega \doteqdot 3 \times 10^{-3}$) is large compared with ϵ ($\sigma/\omega \doteqdot \epsilon_0 = 10^{-9}/36\pi$).

The effective inductive capacity ϵ' may therefore be written

$$\epsilon' = \sigma/j\omega,$$

and the propagation constant k becomes

$$k = \omega\sqrt{(\mu\epsilon')} = \omega\sqrt{\frac{\mu\sigma}{j\omega}}$$

$$= (1-j)\sqrt{\tfrac{1}{2}(\mu\sigma\omega)}. \tag{4}$$

The expression (3) for the waves become

$$H_z = H_0 e^{\sqrt{\frac{1}{2}(\mu\sigma\omega)}x}\, e^{j(\omega t - \sqrt{\frac{1}{2}(\mu\sigma\omega)}x)} = \sqrt{\left(\frac{\epsilon'}{\mu}\right)} E_y. \tag{5}$$

Thus the wave is attenuated in the direction of travel with an attenuation coefficient $\alpha = \sqrt{\tfrac{1}{2}(\mu\sigma\omega)} = \sqrt{(\pi\mu\sigma f)}$ nepers per metre and the amplitude of the wave is reduced by $1/e$ in a distance, in the direction of propagation,

$$\delta = \sqrt{\left(\frac{2}{\mu\sigma\omega}\right)}. \tag{6}$$

The effective propagation constant is

$$\left(\frac{2\pi}{\lambda_m}\right) = \sqrt{\left(\frac{\mu\sigma\omega}{2}\right)} = \frac{1}{\delta}. \tag{7}$$

That is, the wave-length in the metal is

$$\lambda_m = 2\pi\delta. \tag{8}$$

Consider the case of copper and a frequency.

$$f = 3000\,\text{Mc./sec.}, \qquad\qquad \mu = \mu_0 = 4\pi \times 10^{-7},$$
$$\sigma = 5\cdot82 \times 10^7\,\text{mhos/m.}, \qquad \omega = 6\pi \times 10^9.$$

From (6) $\qquad\qquad \delta = 1\cdot2 \times 10^{-6}\,\text{m.},$

from (8) $\qquad\qquad \lambda_m = 7\cdot4 \times 10^{-6}\,\text{m.},$

whereas $\qquad\qquad \lambda_{\text{air}} = 10^{-1}\,\text{m.}$

The refractive index of the metal is

$$\nu = \frac{\lambda_{\text{air}}}{\lambda_m} = \frac{1}{f\sqrt{(\mu_0\epsilon_0)}\,(2\pi\delta)} = \frac{1}{\omega\delta\sqrt{(\mu_0\epsilon_0)}}$$

$$= \sqrt{\frac{\mu\sigma}{2\omega\mu_0\epsilon_0}} = \sqrt{\frac{\sigma}{2\omega\epsilon_0}}, \tag{9}$$

since $\mu \doteqdot \mu_0$.

For copper, $\nu = 1\cdot3 \times 10^4$.

The ratio of the amplitudes of the electric and magnetic fields in the wave are

$$\frac{E_0}{H_0} = \sqrt{\frac{\mu}{\epsilon'}} = \sqrt{\frac{j\omega\mu}{\sigma}} = (1+j)\sqrt{\frac{\omega\mu}{2\sigma}}$$

$$= \sqrt{\frac{\omega\mu}{\sigma}}(\cos\tfrac{1}{4}\pi + j\sin\tfrac{1}{4}\pi)$$

$$= \sqrt{\frac{\omega\mu}{\sigma}}\,e^{\frac{1}{4}j\pi}. \tag{10}$$

Thus E oscillates with a 45° advance in phase with respect to H. Its amplitude, however, is very small compared with that of H. For instance, in copper, according to (10),

$$\frac{|E_0|}{|H_0|} = 2 \times 10^{-2}.$$

Consider the case of a thick metal plate with a plane face above which is a dielectric medium (air) as in fig. 1·3 (c), and let an electromagnetic wave in the dielectric be incident on the metal at an angle of incidence θ_1. A transmitted and a reflected wave are produced at the surface, the former proceeding into the metal at an angle of refraction θ_2 with the normal. According to (9) the refractive index

$$\nu = \frac{\sin\theta_1}{\sin\theta_2} = \sqrt{\frac{\sigma}{2\omega\epsilon_0}};$$

consequently, $\sin\theta_2 = \dfrac{\sin\theta_1}{\nu}$.

As the example of copper shows θ_2 will be a very small angle even when θ_1 approaches closely to 90°. This means that the critical angle in the metal is small.

Put $\sin\theta_2 = 1$; then the upper limit to $\sin\theta_1$ is $1/\nu$ which for copper is $1/1·3 \times 10^4$, or $\theta_1 \approx 7·7 \times 10^{-5}$ radian $= 4·4 \times 10^{-3}$ degree.

We conclude therefore that whatever the angle of incidence, the refracted wave virtually travels into the metal in the direction of the normal to the surface. Since an electromagnetic field of arbitrary form can be resolved into travelling waves it follows that an arbitrary electromagnetic field in the dielectric produces a single plane wave in the metal travelling along the normal which we take as OZ. This wave is highly attenuated and the amplitudes of the magnetic field H_z, of the electric field E_y and of the current density $J_y = \sigma E_y$, all decay exponentially with depth, falling to $e^{-1} = 1/2 \cdot 72$ of their surface values in the distance δ, which is called the skin depth. As already pointed out, this distance is very small in a good conductor like copper in which it equals $1 \cdot 2 \times 10^{-6}$ m. at $\lambda = 10$ cm. Because the electric field is small compared with the magnetic field within the metal and at the surface, the reflexion coefficient of the surface for any wave incident on it within the dielectric is high, since the tangential electric field is the difference in the fields of the incident and reflected waves and the magnetic field the sum of the two magnetic fields.

Thus, when ω is large ($> 10^6$), all the electromagnetic fields vanish inside the metal except within a thin surface layer. In passing from a relatively small depth within the metal, to the surface the magnetic field increases from a negligibly small value to its surface amplitude H_0. Thus H_0 is equal to the total current flowing in the surface layer per unit length normal to the current (§ 1·4).

Equivalent surface resistance. Let the tangential magnetic field at the surface be

$$H = H_0 \cos \omega t.$$

The tangential electric field is therefore, according to (10),

$$E = \sqrt{\left(\frac{\omega \mu}{\sigma} \right)} H_0 \cos \left(\omega t + \tfrac{1}{4} \pi \right).$$

The Poynting flux of energy into the metal is

$$P = EH = \sqrt{\left(\frac{\omega \mu}{\sigma} \right)} H_0^2 \cos \omega t \cos \left(\omega t + \tfrac{1}{4} \pi \right),$$

watts per square metre, and the mean loss of power is

$$\overline{P} = \sqrt{\left(\frac{\mu \omega}{\sigma} \right)} H_0^2 \int_0^{2\pi/\omega} \cos \omega t \cos \left(\omega t + \tfrac{1}{4} \pi \right) dt$$

$$= \frac{1}{2} \sqrt{\left(\frac{\mu \omega}{2\sigma} \right)} H_0^2 = \frac{1}{2} \frac{I^2}{\sigma \delta} \ \text{W./sq.m.,} \tag{11}$$

where $I = H_0$ is the total surface current per metre. This is precisely the power that would be dissipated in heat were the current I distributed uniformly in a layer of thickness δ instead of exponentially through an indefinite thickness. It is for this reason that δ is called the skin depth.

Although we have postulated a flat metal surface, yet even with a curved surface, unless the radius of curvature of the surface is of the order of magnitude of δ (say less than 10δ), the surface resistance will still be given accurately by (11). We call

$$\frac{1}{\sigma\delta} = R = \sqrt{\frac{\mu\omega}{2\sigma}}, \tag{12}$$

the equivalent surface resistance in ohms per square (any units).

7·15. General formula for attenuation coefficient of an empty wave guide

It was shown in §3·9 that the attenuation coefficient in nepers was

$$\alpha = \frac{A}{2W}, \tag{1}$$

where A is the power dissipated in ohmic heating of the walls per unit length; and W is the mean flux of power carried by wave across the cross-section of the wave guide.

We require general expressions for A and W. The loss A per unit length is the loss per unit surface integrated around the perimeter of the wave guide. Thus

$$A = \frac{R}{2} \oint_s H_0^2 \, ds.$$

For E-waves, with OZ the wave-guide axis ($h_1 = 1$), the tangential magnetic field is either H_2 or H_3 according as the portion of the boundary is $u_3 = $ constant or $u_2 = $ constant. In either case, according to 7·10·1 (3) the tangential magnetic field is $(\partial U/\partial n)$ whence, if we omit the exponential factor $e^{j\omega t}$ from U,

$$A = \frac{\omega^2 \epsilon^2 R}{2} \oint_s \left(\frac{\partial U}{\partial n} \right)^2 ds,$$

where $\oint ds$ indicates an integration around the periphery of the cross-section of the wave guide. The element of surface of the cross-section is

$$h_2 h_3 \, du_2 \, du_3.$$

The mean Poynting flux is

$$P = \frac{(E_2 H_3 - E_3 H_2)}{2}.$$

From 7·10·1 (2) and (3) this is ($h_1 = 1$)

$$P = \frac{\gamma \omega \epsilon}{2} \left[\left(\frac{1}{h_2} \frac{\partial U}{\partial u_2} \right)^2 + \left(\frac{1}{h_3} \frac{\partial U}{\partial u_3} \right)^2 \right],$$

$$W = \frac{\gamma \omega \epsilon}{2} \int_S \left[\left(\frac{1}{h_2} \frac{\partial U}{\partial u_2} \right)^2 + \left(\frac{1}{h_3} \frac{\partial U}{\partial u_3} \right)^2 \right] h_2 h_3 \, du_2 \, du_3,$$

$$\alpha = \frac{R \omega \epsilon}{2\gamma} \frac{\oint_s \left(\frac{\partial U}{\partial n} \right)^2 ds}{\int_S \left[\left(\frac{1}{h_2} \frac{\partial U}{\partial u_2} \right)^2 + \left(\frac{1}{h_3} \frac{\partial U}{\partial u_3} \right)^2 \right] h_2 h_3 \, du_2 \, du_3}.$$

Alternatively,

$$\alpha = \frac{R}{2Z_E} \frac{\oint_s \left(\frac{\partial U}{\partial n} \right) ds}{\int_S \left[\left(\frac{1}{h_2} \frac{\partial U}{\partial u_2} \right)^2 + \left(\frac{1}{h_3} \frac{\partial U}{\partial u_3} \right)^2 \right] h_2 h_3 \, du_2 \, du_3}, \tag{2}$$

where $Z_E = \lambda/\lambda_g \sqrt{(\mu/\epsilon)}$ = intrinsic impedance of E-waves.

For H-waves: from 7·10·1 (4)

$$A = \gamma^2 \frac{R}{2} \oint_s \left[\left(\frac{1}{h_3^2} \frac{\partial V}{\partial u_3^2} \right)^2 + \left(\frac{k^2}{\gamma^2} - 1 \right)^2 V^2 \right] ds,$$

where $\frac{2}{3}$ indicates that the subscript 2 is to be used on boundaries u_3 = constant and the subscript 3 on boundaries u_2 = constant,

$$W = \frac{\omega \gamma^2 \mu}{2\gamma} \int_S \left[\left(\frac{1}{h_3} \frac{\partial V}{\partial u_3} \right)^2 + \left(\frac{1}{h_2} \frac{\partial V}{\partial u_2} \right)^2 \right] h_2 h_3 \, du_2 \, du_3,$$

whence

$$\alpha = \frac{R}{2Z_H} \frac{\oint_s \left[\left(\frac{1}{h_3^2} \frac{\partial V}{\partial u_3^2} \right)^2 + \left(\frac{\lambda_g^2}{\lambda_c^2} V \right)^2 \right] ds}{\int_S \left[\left(\frac{1}{h_3} \frac{\partial V}{\partial u_3} \right)^2 + \left(\frac{1}{h_2} \frac{\partial V}{\partial u_2} \right)^2 \right] h_2 h_3 \, du_2 \, du_3},$$

where Z_H is the intrinsic impedance of H-waves and λ_c is the cut-off wave-length.

7·16. The Q-factor of a resonator

The following definition of the Q-factor of an air-filled resonator was given in §6·7,

$$Q = \omega \frac{\text{Energy stored}}{\text{Mean dissipation of power}}. \tag{1}$$

We shall obtain a general formula for calculating Q.

The energy stored is either the maximum electric or the maximum magnetic energy.

From 7·10·1 (3), we find for E-modes

$$\text{Energy stored} = \frac{\mu}{2} \int_{\text{volume}} H^2 d\tau = \frac{\omega^2 \epsilon^2 \mu}{2} \int_{\tau} \left\{ \left[\frac{1}{h_3} \frac{\partial}{\partial u_3} (U) \right]^2 \right.$$
$$\left. + \left[\frac{1}{h_2} \frac{\partial}{\partial u_2} (U) \right]^2 \right\} h_1 h_2 h_3 \, du_1 \, du_2 \, du_3,$$

Mean power dissipated in watts is (surface resistance R)

$$\frac{\omega^2 \epsilon^2 R}{2} \left[\int\!\!\int_{\substack{\text{Surfaces } u_1 \\ =\text{constant}}} \left\{ \left[\frac{1}{h_3} \frac{\partial}{\partial u_3} (U) \right]^2 \right. \right.$$
$$\left. + \left[\frac{1}{h_2} \frac{\partial}{\partial u_2} (U) \right]^2 \right\} h_2 h_3 \, du_2 \, du_3$$
$$+ \int_{\substack{\text{Surfaces } u_2 \\ =\text{constant}}} \left[\frac{1}{h_3} \frac{\partial}{\partial u_3} (U) \right]^2 h_1 h_3 \, du_1 \, du_3$$
$$\left. + \int_{\substack{\text{Surfaces } u_3 \\ =\text{constant}}} \left[\frac{1}{h_2} \frac{\partial}{\partial u_2} (U) \right]^2 h_1 h_2 \, du_1 \, du_2 \right].$$

Thus, from (1) and 7·14 (12)

$$Q = 2R\sigma \frac{\displaystyle \int H^2 d\tau}{\displaystyle \Sigma \int_{\text{surface}} H^2 dS} = 2 \frac{\displaystyle \int H^2 d\tau}{\displaystyle \delta \Sigma \int H^2 ds}, \qquad (2)$$

where δ is the skin depth.

The same procedure is employed with H-modes, but H^2 in the volume integral in (2) is replaced by $(H_1^2 + H_2^2 + H_3^2)$ whose value is given by 7·10·1 (4). Also the surface integrals are to be taken over whatever portions of the boundary $u_1 = \text{constant}$, $u_2 = \text{constant}$, $u_3 = \text{constant}$ exist, with $H^2 = (H_2^2 + H_3^2)$ over $u_1 = \text{constant}$; $H^2 = (H_1^2 + H_3^2)$ over $u_2 = \text{constant}$; $H^2 = (H_1^2 + H_2^2)$ over $u_3 = \text{constant}$. Evidently, the expressions for Q in particular cases will be complicated.

An approximate simple form of (2) may, however, be obtained by assuming that the mean value of H^2 over the surface is double the mean value throughout the volume, since the variations of amplitude are quasi-sinusoidal.

Then
$$Q \doteqdot \frac{\text{Volume}}{\delta \times \text{surface}}. \tag{3}$$

This formula was used in § 6·7.

The mean energy stored in the metal is

$$\frac{\mu}{2} \int \left[\int_0^\infty H_0^2 e^{-2x/\delta}\, dx \right] dS = \frac{\delta\mu}{4} \int H_0^2\, dS$$

(neglecting the electric energy which is relatively small), where H_0 is the amplitude of $\mathbf{H}$ at the surface. The integral $\int H_0^2\, dS$ is the same as the surface integral in (2). Consequently another interpretation of Q is

$$Q = \frac{\text{Energy stored in cavity}}{\text{Energy stored in walls}}. \tag{4}$$

7·17. Lorentz's reciprocal theorem* and equivalent networks—Thévenin's theorem

We consider in this section some reciprocal properties of electromagnetic fields that can exist in a given region and their significance for the interpretation of field phenomena in terms of equivalent circuits.

For this purpose we employ a valuable theorem due to Lorentz which is applicable to simple harmonic fields; Lorentz's reciprocity theorem states that:

If $(\mathbf{E}, \mathbf{H})$ and $(\mathbf{E}', \mathbf{H}')$ are simple harmonic electromagnetic fields that can exist in the same region, then at all points in the region

$$\operatorname{div}[\mathbf{E} \times \mathbf{H}' - \mathbf{E}' \times \mathbf{H}] = 0, \tag{1}$$

where $\times$ represents a vector product as usual.

To establish this result consider the first pair of Maxwell's equations
$$\nabla \times \mathbf{E} = -\mu\dot{\mathbf{H}}, \quad \nabla \times \mathbf{H} = \epsilon\dot{\mathbf{E}} + \sigma\mathbf{E}, \tag{2}$$

where $\mathbf{J}$ has been replaced by $\sigma\mathbf{E}$, the medium being assumed to possess conductivity σ.

The same equations are satisfied by a field $(\mathbf{E}', \mathbf{H}')$.

---

* H. A. Lorentz, *Proc. Acad. Sci. Amst.* vol. 4, p. 176 (1895–6); W. Dallenbach, *Arch. Electrotechn.* vol. 36, no. 3, p. 153 (1942) (this paper gives references to earlier work on reciprocity).

We may regroup the equations for both fields as follows:

$$\mathbf{\nabla} \times \mathbf{E} = -\mu\dot{\mathbf{H}}, \quad \mathbf{\nabla} \times \mathbf{H}' = \epsilon\dot{\mathbf{E}}' + \sigma\mathbf{E}', \tag{3}$$

$$\mathbf{\nabla} \times \mathbf{E}' = -\mu\dot{\mathbf{H}}', \quad \mathbf{\nabla} \times \mathbf{H} = \epsilon\dot{\mathbf{E}} + \sigma\mathbf{E}. \tag{4}$$

Scalarly multiply the first of equations (3) by $\mathbf{H}'$ and the second by $\mathbf{E}$, subtract the second from the first and use the vector theorem

$$\operatorname{div}(\mathbf{P} \times \mathbf{Q}) = \mathbf{Q}.\mathbf{\nabla} \times \mathbf{P} - \mathbf{P}.\mathbf{\nabla} \times \mathbf{Q}$$

to obtain

$$\operatorname{div}(\mathbf{E} \times \mathbf{H}') = -[\mu\mathbf{H}'.\dot{\mathbf{H}} + \epsilon\dot{\mathbf{E}}'.\mathbf{E} + \sigma\dot{\mathbf{E}}'.\mathbf{E}]. \tag{5}$$

In the same way we obtain from (4), or more simply by inter-changing the dashed and undashed symbols in (5),

$$\operatorname{div}(\mathbf{E}' \times \mathbf{H}) = -[\mu\dot{\mathbf{H}}'.\mathbf{H} + \epsilon\mathbf{E}.\dot{\mathbf{E}}' + \sigma\mathbf{E}.\mathbf{E}']. \tag{6}$$

Subtract (6) from (5), then

$$\operatorname{div}[\mathbf{E} \times \mathbf{H}' - \mathbf{E}' \times \mathbf{H}] = -[\mu(\mathbf{H}.\dot{\mathbf{H}}' - \mathbf{H}'.\dot{\mathbf{H}}) + \epsilon(\dot{\mathbf{E}}.\mathbf{E}' - \dot{\mathbf{E}}'.\mathbf{E})]. \tag{7}$$

When $(\mathbf{E}, \mathbf{H})$ and $(\mathbf{E}', \mathbf{H}')$ both oscillate at the same frequency, so that a factor $e^{j\omega t}$ enters into the expressions for all components, then

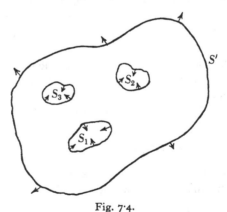

Fig. 7·4.

the terms in the round brackets vanish on the right of equation (7) and Lorentz's theorem (1) is established. It is of very general validity and only demands that the local current density $\mathbf{J}$ shall be a linear function of the local field $\mathbf{E}$ at all points. This excludes the ionosphere, a gas discharge and a thermionic valve, although σ may be a symmetrical tensor (excludes ionosphere where the conductivity is skew-symmetric due to the earth's magnetic field); ϵ and μ may be symmetric tensors or simple constants.

Lorentz's theorem (1) is, however, more usefully expressed in another form as follows:

Let the region, throughout which (1) is valid, be bounded by surfaces S_1, S_2, etc., which are closed, and by a single large outer surface S' which encloses the whole region. These surfaces may be the surfaces of bodies or merely geometrical surfaces, according to convenience. Apply Gauss's theorem to the region, with the normals n to the surfaces S considered positive when directed away from the region bounded by these surfaces as shown in fig. 7·4. We find

$$\int \mathrm{div}\,[\mathbf{E} \times \mathbf{H}' - \mathbf{E}' \times \mathbf{H}]\,d\tau = \sum_{k=1}^{n} \int_{S_R} (\mathbf{E} \times \mathbf{H}' - \mathbf{E}' \times \mathbf{H}).\,d\mathbf{S}$$

$$+ \int_{S'} (\mathbf{E} \times \mathbf{H}' - \mathbf{E}' \times \mathbf{H}).\,d\mathbf{S} = 0. \quad (8)$$

Suppose the sources of the fields $(\mathbf{E}, \mathbf{H})$ and $(\mathbf{E}', \mathbf{H}')$ to be excluded from the region and to be enclosed by one or more of S_1, S_2, etc. Let the surface S' expand to infinity. The fields $(\mathbf{E}, \mathbf{H})$ and $(\mathbf{E}', \mathbf{H}')$ then become spherical waves of large radius and hence TEM-waves such that $\mathbf{H}' = \sqrt{(\epsilon/\mu)}\,\mathbf{E}'$ and $\mathbf{H} = \sqrt{(\epsilon/\mu)}\,\mathbf{E}$. Consequently,

$$(\mathbf{E} \times \mathbf{H}' - \mathbf{E}' \times \mathbf{H}) = \sqrt{\left(\frac{\epsilon}{\mu}\right)}(\mathbf{E} \times \mathbf{E}' - \mathbf{E}' \times \mathbf{E}) = 2\sqrt{\left(\frac{\epsilon}{\mu}\right)}\mathbf{E} \times \mathbf{E}'.$$

If $\mathbf{E}'$ and $\mathbf{E}$ are parallel at infinity or if the medium is slightly absorbing the vector product vanishes on the infinite surface S'. The integral over S' vanishes therefore and we have finally

$$\sum_{k=1}^{n} \int_{S_R} [\mathbf{E} \times \mathbf{H}' - \mathbf{E}' \times \mathbf{H}]\,d\mathbf{S} = 0, \quad (9)$$

provided the region external to S_1, S_2, ..., S_n extends to infinity, or alternatively the integral over S' vanishes even when S' is finite. Otherwise (8) must be used.

Let a surface S_k be that of a perfect conductor, then $\mathbf{E}$ and $\mathbf{E}'$ are normal to S_k and $\mathbf{H}$ and $\mathbf{H}'$ tangential to it. Thus both $(\mathbf{E}' \times \mathbf{H}')$ and $(\mathbf{E}' \times \mathbf{H})$ are tangential to S_k. The surface integral therefore vanishes over a closed conductor. As an example of the application of the Lorentz theorem consider the system shown in fig. 7·5 which comprises a cavity bounded by a conducting surface S through which coaxial transmission lines and wave guides are coupled into the cavity. For the sake of generality we suppose the cavity to contain conducting objects S_1, S_2, etc.

The transmission lines and wave guides either feed power into the cavity or abstract power from it. To form an enclosing surface S' we employ cross-sections of the transmission lines and wave guides, at σ_1, σ_2, etc., which with the surface of the inner and outer conductors of the coaxials, of the wave guides and of the cavity S

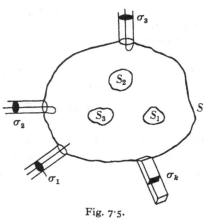

Fig. 7·5.

form a closed surface S' as shown in fig. 7·5. Suppose different fields $(\mathbf{E}, \mathbf{H})$ and $(\mathbf{E}', \mathbf{H}')$ to be excited within the cavity and apply theorem (8) to this system of surfaces. The integrals over the surfaces of the conducting objects S_1, S_2, etc., vanish so that we are left with the integral over S'. Hence

$$\int_{S'} [\mathbf{E} \times \mathbf{H}' - \mathbf{E}' \times \mathbf{H}].d\mathbf{S} = 0$$

or
$$\int_{S'} \mathbf{E} \times \mathbf{H}'.d\mathbf{S} = \int_{S'} \mathbf{E}' \times \mathbf{H}.d\mathbf{S}. \tag{10}$$

These integrals again reduce to those over σ_1, σ_2, etc., which are not occupied by conducting matter, the remainder of S contributing nothing since it comprises conducting surfaces. Thus, (10) breaks up on either side into a sum of integrals over the cross-sections. Suppose these sections to be so placed in the transmission lines that the field is a TEM-field and in the wave guide an H_{10}-field, that of the only progressive mode.

Consider the integrals taken over σ_1. Both fields $(\mathbf{E}, \mathbf{H})$ and $(\mathbf{E}', \mathbf{H}')$ are entirely transverse at σ_1 and, moreover, the E-lines of force run from the inner to the outer and the H-lines form closed loops cutting the E-lines at right angles.

Consequently, $\mathbf{E} \times \mathbf{H}'$ is directed along the normal (outwards from the cavity) when the central conductor is positively charged and the current i' flows out of the cavity at σ_1. As in § 1·5, let dl be an element of length measured along an electric line of force and ds along a magnetic line. The element of surface dS or σ_1 is $dS = dl\,ds$.

The integrals to be evaluated over σ_1 are

$$\int (\mathbf{E} \times \mathbf{H}') . d\mathbf{S} = \int\int |E|\,|H'|\,dl\,ds$$

and

$$\int (\mathbf{E}' \times \mathbf{H}) . d\mathbf{S} = \int\int |E'|\,|H|\,dl\,ds.$$

The former, as in § 1·5, reduces to $V_1 i_1'$ and the latter to $V_1' i_1$, where V_1 and V_1' are the voltages at σ_1 in the two fields and i_1 and i_1' the corresponding currents. The other transmission lines provide similar terms ($V_2 i_2'$ and $V_2' i_2$), etc.

Consider σ_k to be a wave-guide section. The field of the H_{10}-mode (E, H) is

$$E_y = E_0 \sin \frac{\pi x}{a}, \quad H_x = \frac{E_0}{Z_H} \sin \frac{\pi x}{a},$$

with similar terms for $(\mathbf{E}'\mathbf{H}')$. H_z is omitted because it contributes nothing to the integral.

The integrals over σ_k are therefore

$$\int_0^a \int_0^b E_y H_x' dx\, dy = \int_0^b dy \int_0^a E_0 H_0' \sin^2 \frac{\pi x}{a} dx = \frac{ab E_0 H_0'}{2} = \frac{a E_0}{\sqrt{2}} \times \frac{b H_0'}{\sqrt{2}}.$$

$a E_0'$ is the maximum voltage across the section considered as product of field strength and length of field. $a E_0/\sqrt{2}$ is the root mean square voltage averaged across the dimension b. Similarly, $b H_0'/\sqrt{2}$ is the spatial root mean square longitudinal current.

To bring the integral over σ_k into conformity with those over the transmission-line sections it is convenient to define the wave-guide equivalent voltage and current as

$$V_k = \frac{a E_0}{\sqrt{2}} \quad \text{and} \quad i_k' = \frac{b H_0'}{\sqrt{2}}.$$

The integral therefore is equal to $V_k i'_k$. Similarly, the other integral is $V'_k i_k$.

Equation (10) therefore becomes

$$\sum_{k=1}^{n} V_k i'_k = \sum_{k=1}^{n} V'_k i_k, \qquad (11)$$

in which the summation includes contributions from all the cross-sections both of transmission lines and of wave guides. Equation (11) is identical in form with a well-known reciprocal theorem for networks, first formulated by Helmholtz for d.c. networks of a very general character.* It shows that provided the voltages and currents are those of the dominant modes in the feeders and are referred to the same sections σ_k in all cases, the cavity will behave like a $2n$-terminal network, and it is apparent why it is possible to find exact circuit representations of wave-guide systems.

Owing to the linear character of Maxwell's equations the V's are linear functions of the i's. We may therefore write

$$\left. \begin{aligned} V_1 &= Z_{11}i_1 + Z_{12}i_2 + \ldots + Z_{1n}i_n, \\ V_2 &= Z_{21}i_1 + Z_{22}i_2 + \ldots + Z_{2n}i_n, \\ V_k &= Z_{k1}i_1 + \ldots + Z_{kj}i_j + \ldots + Z_{kn}i_n, \\ V_n &= Z_{n1}i_1 + Z_{n2}i_2 + \ldots + Z_{nn}i_n. \end{aligned} \right\} \qquad (12)$$

Consider two distributions (V_k, i_k) and (V'_j, i'_j) in which all currents are zero except i_k in the first, and all except i'_j in the second. Apply the reciprocal relation (11), then

$$V_j i'_j = V'_k i_k$$

or

$$Z_{jk} i_k i'_j = Z_{kj} i'_j i_k,$$

that is

$$Z_{jk} = Z_{kj}. \qquad (13)$$

The impedance matrix is symmetrical.

Equations (12) obtained by this method are applicable to any electromagnetic system including networks as a special case.

We have, corresponding to (12),

$$\left. \begin{aligned} i_1 &= Y_{11}V_1 + Y_{12}V_2 + \ldots + Y_{1n}V_n, \\ i_2 &= Y_{21}V_1, \text{ etc.,} \\ i_n &= Y_{n1}V_1 + \ldots + Y_{nn}V_n, \end{aligned} \right\} \qquad (14)$$

with

$$Y_{jk} = Y_{kj}. \qquad (15)$$

* F. B. Pidduck, *Lectures on the Mathematical Theory of Electricity*, p. 35, Oxford; S. A. Schelkunoff, *Electromagnetic Waves*, p. 104.

Some convention is required about the sign of the currents in relation to the direction of flow. In order to make the input impedances of a passive network positive we define a positive current as one that flows into the network when the central conductor of the coaxial is at a positive potential (which is equivalent to reversing the positive direction of the normal to the areas σ_k in the integrals $\int (\mathbf{E} \times \mathbf{H}') \, d\mathbf{S}$, etc.) and a negative current in (14) as one flowing out of the central conductor when it is at a positive voltage. The analysis also suggests a suitable definition of voltage, current and total impedance of a wave guide to conform with (11), (12) and (14). From what has preceded these are defined to be:

Amplitudes of equivalent voltage and current in H_{10}-wave

$$E = E_0 \sin\left(\frac{\pi x}{a}\right), \quad H = H_0 \sin\left(\frac{\pi x}{a}\right)$$

are respectively

$$V_{\text{max.}} = \left(\frac{E_0 a}{\sqrt{2}}\right), \quad \text{and} \quad i_{\text{max.}} = \left(\frac{H_0 b}{\sqrt{2}}\right). \qquad (16)$$

The total characteristic impedance of the wave guide, defined as $Z_{0g} = V/i$ in a progressive H_{10}-wave, is

$$Z_{0g} = \frac{a}{b} \frac{E_0}{H_0} = \frac{a}{b} Z_H, \qquad (17)$$

where Z_H is the intrinsic impedance of the H_{10}-wave.

According to Poynting's theorem the maximum longitudinal flux of power over the section is

$$W = \iint E_0 H_0 \sin^2 \frac{\pi x}{a} \, dx \, dy = \frac{ab E_0 H_0}{2} = V_{\text{max.}} i_{\text{max.}}.$$

The mean power is one-half this amount.

To apply the Lorentz theorem to aerial systems we consider a system in which generators and receivers are totally enclosed in a conducting envelope through which they communicate to outer space by means of coaxial transmission lines or wave guides feeding aerial systems. The outer enclosing surface S' is expanded to infinity and the surface integral over it thus made to vanish.

It can then be seen that the reciprocal relation (11), applied as before to suitably chosen sections σ_k of the feeder systems, is valid for the system of aerials which can therefore be represented in terms of an equivalent network.

Equivalent four-terminal network. Relations (11), (12) and (14), when applied to a system comprising two feeder systems ($n = 2$) only, are evidently represented by a four-terminal network. Examples would be a resonant cavity (wave-meter), with input and output loops joined to coaxial feeder systems, or fed from wave guides through holes or slots, and a system of two aerials each fed from a pair of coaxial feeders. We proceed to show that Thévenin's theorem is valid for such a system. Refer to fig. 7·6, which represents the equivalent four-terminal network.

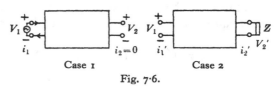

Case 1   Case 2

Fig. 7·6.

We suppose a generator with zero internal impedance and e.m.f. V_1' to be applied to terminals 1. When terminals 2 are open-circuited ($i_2 = 0$) let the voltage between them be V_2 (fig. 7·6, case 1).

Since $$V_1 = Z_{11}i_1 + Z_{12}i_2, \quad V_2 = Z_{12}i_1 + Z_{22}i_2, \tag{18}$$

it follows that on open circuit ($i_2 = 0$)

$$V_1 = Z_{11}i_1 \quad \text{and} \quad V_2 = Z_{12}i_1 = \frac{Z_{12}}{Z_{22}}V_1. \tag{19}$$

Let terminals 2 be closed by an impedance Z (fig. 7·6, case 2). If the voltage and current at these terminals now become V_2' and i_2', those at terminals 1 being V_1 and i_1', it follows that

$V_2' = -Zi_2'$ (since current flows outwards at the positive terminal)

$= Z_{12}i_1' + Z_{22}i_2',$

whence $$i_2' = \frac{-Z_{12}}{(Z + Z_{22})}i_1'. \tag{20}$$

Also $$V_1 = Z_{11}i_1' + Z_{12}i_2' = \left[Z_{11} - \frac{Z_{12}^2}{(Z + Z_{22})}\right]i_1'. \tag{21}$$

Thus, from (20) and (21),

$$i_2' = \frac{-Z_{12}V_1}{Z_{11}[Z + (Z_{22} - Z_{12}^2/Z_{11})]}. \tag{22}$$

It follows from (22) that the current i_2'' through the impedance Z is the same as if Z were connected across the terminals of a generator with e.m.f. $Z_{12}V_1/Z_{11}$ and internal impedance $(Z_{22} - Z_{12}^2/Z_{11})$.

To interpret this equivalent internal impedance in terms of the electrical properties of the four-terminal network suppose the generator to be removed from terminals 1 which are then short-circuited. Let an e.m.f. V_2'' be applied at terminals 2 and suppose the currents at terminals 2 and 1 to be i_2''' and i_1'''.

Since $V_1'' = 0$ it follows that

$$0 = Z_{11}i_1''' + Z_{12}i_2''', \quad V_2'' = Z_{12}i_1''' + Z_{22}i_2''' = (Z_{22} - Z_{12}^2/Z_{11})i_2'''.$$

The input impedance at terminals 2 is, therefore,

$$Z_{in} = \frac{V_2''}{i_2'''} = \left(Z_{22} - \frac{Z_{12}^2}{Z_{11}}\right). \tag{23}$$

The internal impedance of the equivalent generator is, therefore, according to (22) and (23), the same as the input impedance at terminals 2 when terminals 1 are short-circuited.

We have, therefore, established that Thévenin's theorem is implicit in equations (11) and (12).

We may regard an aerial system comprising a transmitter and a receiver, or the resonator mentioned above, as a four-terminal network. According to Thévenin's theorem if a voltage V_2 appears across the open-circuited receiver terminals then when these are closed by an impedance Z the current through Z is $i = V_2/(Z_e + Z_i)$, where Z_e is the input impedance of the receiving aerial used as a transmitter with the terminals of the other aerial short-circuited. Corresponding statements apply to the resonator.

By terminals we mean the two conductors of the coaxial feeder systems or the broad faces of the wave guides at the chosen sections σ_k which in practice are placed at the input terminals of the aerials. In the case of aerials, the mutual coupling between transmitter and receiver is frequently so small that the input impedance Z_e is influenced only to a negligible extent in changing from an open to a short circuit at the other aerial. This, however, would not be generally true for resonators.

The Lorentz reciprocity theorem is of fundamental importance for aerial theory, and through it simple, but general, proofs of the identity of the polar diagrams of an arbitrary aerial system in transmission and in reception, and of the identity of the equivalent transmitting and receiving areas of aerials, can be obtained.

As an example of a simple well-known result of resonator, aerial and network theory, we formulate (11) for a four-terminal network.

Thus
$$V_1 i_1' + V_2 i_2' = V_1' i_1 + V_2' i_2. \tag{24}$$

Let $V_2 = V_1' = 0$ so that terminals 2 are short-circuited in the first distribution and terminals 1 in the second. Equation (24) reduces to

$$V_1 i_1' = V_2' i_2.$$

If therefore $V_1 = V_2'$, then $i_1' = i_2$. That is, if the applied voltage and the short circuit are interchanged, the current through the short circuit remains unaltered.

Enough has been given to indicate the fundamental character of the basic relations (11), (12) and (14) for the study of electromagnetic systems of widely differing characters. These results are more fundamental than the Kirchhoff circuited relations from which they are usually derived, the latter being a species of 'geometrical optics of electromagnetism'.

7·18. Spatial and functional properties of the electromagnetic fields of the characteristic modes

7·18·1. *Spatial orthogonality of the electric and magnetic fields of a mode*

It is easy to see that the **E** and **H** fields belonging to the same mode in a wave guide or resonator are everywhere at right angles. The condition for orthogonality, in terms of curvilinear coordinates, is

$$\mathbf{E}.\mathbf{H} = E_1 H_1 + E_2 H_2 + E_3 H_3 = 0. \tag{1}$$

Consider E-modes, whose components are derived from the Hertz vector $\mathbf{\Pi} = i_1 U$ by use of 7·10·1 (2) and (3). It follows immediately from these equations that $E_1 H_1 + E_2 H_2 + E_3 H_3$ is zero; consequently the fields **E** and **H** are at right angles. Further, since $E_2 H_2 + E_3 H_3 = 0$, the components of **E** and **H** in the surface $u_1 = $ constant are at right angles. For instance, in a wave guide $u_1 = $ constant is a cross-section, and it follows that the transverse components of **E** and **H** intersect everywhere at right angles.

Similarly, it can be seen from equations 7·10·1 (3) and (4) that the **E** and **H** fields of an H-mode also possess these properties, as do all the modes of a twin transmission line.

7·18·2. *Some properties of transmission-line modes*

Suppose the twin system to be quasi-coaxial, that is, the two conductors are general cylinders $u_2 = \text{constant} = c_1$, and $u_2 = c_2$, with their generators parallel to OZ. Suppose also $u_2 = c_2$ to enclose $u_2 = c_1$, so that u_3 is a cyclic coordinate.

Equations 7·10·1 (2) and (3) here become, with $U = f(u_1, u_2)\, e^{j(\omega t - \gamma z)}$,

$$E_3 = E_1 = (k^2 - \gamma^2)\, U, \quad H_3 = H_1 = 0,$$

$$\left.\begin{array}{ll} E_2 = -j\gamma \dfrac{1}{h_2}\dfrac{\partial U}{\partial u_2}, & H_2 = j\omega\epsilon\dfrac{1}{h_3}\dfrac{\partial U}{\partial u_3}, \\[2mm] E_3 = -j\gamma \dfrac{1}{h_3}\dfrac{\partial U}{\partial u_3}, & H_3 = -j\omega\epsilon\dfrac{1}{h_2}\dfrac{\partial U}{\partial u_2}. \end{array}\right\} \tag{1}$$

We define the potential difference v between the conductors at a given position to be the line integral of $\mathbf{E}$ taken along any path between the conductors $u_2 = c_1$ and $u_2 = c_2$ that lies in the cross-section at that position. Thus

$$\begin{aligned} v &= \int_{u_2=c_1}^{u_2=c_2} \mathbf{E}\, ds = \int_{u_2=c_1}^{u_2=c_2} (E_2 h_2\, du_2 + E_3 h_3\, du_3) \\[2mm] &= -j\gamma \int_s \left[\left(\frac{\partial U}{\partial u_2}\right) du_2 + \left(\frac{\partial U}{\partial u_3}\right) du_3 \right] \\[2mm] &= -j\gamma [U_2 - U_1], \end{aligned} \tag{2}$$

where U_2 and U_1 are the values of U at the end and beginning of the path s in the cross-section $z = \text{constant}$.

But at the conductors $u_2 = c_2$ and $u_2 = c_1$, $E_3 = H_2 = 0$, whence from (1)

$$\frac{\partial U}{\partial u_3} = 0, \quad \text{over} \quad u_2 = c_2 \quad \text{and} \quad u_2 = c_1.$$

Consequently, around each boundary where the cross-section $z = \text{constant}$ intersects the cylinders $u_2 = c_2$ and $u_2 = c_1$, U is a constant independent of the remaining variable u_3.

These constant values differ in the two classes formed by the principal mode, and all other modes respectively.

Consider, first, the principal mode in which

$$k = \gamma \quad \text{and} \quad E_3 = H_3 = 0.$$

We cannot make U zero over both $u_2 = c_1$ and $u_2 = c_2$, since the total field $\mathbf{E} = i_1 E_1 + i_2 E_2 + i_3 E_3$ is here a two-dimensional electrostatic field derivable from a potential $-j\gamma U$. Consequently we put $U = 0$ at the boundary of the cross-section $u_2 = c_1$ and $U = -U_2$ on the outer cylinder. The potential difference between the cylinders at the section $z = $ constant is therefore, from (2),

$$v = j\gamma U_2,$$

and is independent of the path between the cylinders provided it lies in the section $z = $ constant. With all other modes, however, $k \neq \gamma$ and E_3 or H_3 do not vanish everywhere.

We require $E_3 = 0$ at the cylinders, and therefore $U = 0$ at both $u_2 = c_1$ and $u_2 = c_2$. Consequently, from (2), $v = 0$. Thus, the electric fields of E-modes contribute nothing to the potential difference between the conductors.

The fields of the H-modes are:

$$\left.\begin{array}{ll} H_z = H_1 = (k^2 - \gamma^2)\,V, & E_z = E_1 = 0, \\[2mm] H_2 = -j\gamma\,\dfrac{1}{h_2}\dfrac{\partial V}{\partial u_2}, & E_2 = -j\omega\mu\,\dfrac{1}{h_3}\dfrac{\partial V}{\partial u_3}, \\[2mm] H_3 = -j\gamma\,\dfrac{1}{h_3}\dfrac{\partial V}{\partial u_3}, & E_3 = j\omega\mu\,\dfrac{1}{h_2}\dfrac{\partial V}{\partial u_2}. \end{array}\right\} \quad (3)$$

The line integral of $\mathbf{E}$ between the conductors is

$$\int(E_2 h_2\,du_2 + E_3 h_3\,du_3) = -j\omega\mu\int\left[\frac{h_2}{h_3}\left(\frac{\partial V}{\partial u_3}\right)du_2 - \frac{h_3}{h_2}\left(\frac{\partial V}{\partial u_3}\right)du_3\right],$$

and this in general is neither equal to zero nor independent of the path between the conductors. Thus in the presence of H-modes the only definable potential is that due to the principal mode. This is because the longitudinal component H_z makes $\operatorname{curl}\mathbf{E}$ not zero in the section and $\mathbf{E}$ not derivable from a potential.

The behaviour of the magnetic field is complementary to that of the electric field. Form the integral of $\operatorname{curl}\mathbf{H}$ across an annulus in the cross-section $z = $ constant, bounded by any two curves in the section. From 7·8 (7)

$$\int(\operatorname{curl}\mathbf{H})_1\,h_2 h_3\,du_2\,du_3 = \int\left[\frac{\partial}{\partial u_2}(h_3 H_3) - \frac{\partial}{\partial u_3}(h_2 H_2)\right]du_2\,du_3.$$

In the case of the TEM- (principal) and the H-modes it can be seen from (1) and (3) that the surface integral vanishes, but, from (2), it does not vanish for E-modes. From Stokes's theorem it follows that the line integral around any closed curve in the 1-ply connected region enclosing the central conductor is the same in the case of TEM- and H-modes and is equal to the total surface current crossing the section on one of the conductors. The line integral around a closed curve in the section with E-modes depends on the closed path that is selected. Consequently the total longitudinal currents on the inner and outer conductors at the cross-section are not the same. This is because the presence of E_z requires (curl H)$_z$ to exist over the cross-section.

7·18·3. *Functional orthogonality of the characteristic modes of a resonator*

In addition to the spatial orthogonality of the $\mathbf{E}$ and $\mathbf{H}$ fields of the same mode, discussed in 7·18·1, there exists a different type of orthogonality, namely, functional orthogonality of the fields of the different characteristic modes of natural oscillation of a given resonator.

To understand what is meant by functional orthogonality in this context, consider two field vectors $\mathbf{A}$ and $\mathbf{B}$ that exist throughout a bounded volume $\int d\tau$.

Let the scalar product $\mathbf{A}.\mathbf{B}$ of these vectors be integrated through-out the volume. If this integral $\int \mathbf{A}.\mathbf{B}\, d\tau$ is zero then the vectors $\mathbf{A}$ and $\mathbf{B}$ are said to be functionally orthogonal. Similarly, two field scalars ϕ and ψ are functionally orthogonal if $\int \phi\psi\, d\tau$ vanishes, when taken over the specified volume.

In earlier sections of this chapter we have derived expressions for the field components of the natural modes of oscillation of resonators with miscellaneous simple geometrical shapes and obtained them from Hertz vectors $\mathbf{\Pi} = i_1 U$ and $\mathbf{\Pi}_0 = i_1 V$. These vectors were obtained as vector solutions of the fundamental differential equation 7·7 (5) for the case of simple harmonic oscilla-tions and $\rho = 0$, namely,

$$\text{curl curl } \mathbf{\Pi} - k^2\mathbf{\Pi} = \text{grad } F, \tag{1}$$

where F was a field scalar whose value is assigned according to convenience.

It was found that solutions were only obtained for a given resonator when $k^2 = \omega^2\mu\epsilon = (2\pi/\lambda)^2$ possessed specific values called the characteristic values of the problem.

For instance, with E_{mnp}- and H_{mnp}-modes in a rectangular resonator with sides a, b and c, the appropriate values of k^2 are $(7\cdot10\cdot2\,(11))$

$$k^2 = \left(\frac{2\pi}{\lambda}\right)^2 = \left[\left(\frac{m\pi}{a}\right)^2 + \left(\frac{n\pi}{b}\right)^2 + \left(\frac{p\pi}{c}\right)^2\right].$$

The E_{mnp}- and H_{mnp}-modes in cylinders, however, require different values of k even when m, n and p are individually the same: the respective characteristic values are here $(7\cdot10\cdot5\,(14))$,

$$k^2 = \left(\frac{\rho_{mn}}{a}\right)^2 + \left(\frac{p\pi}{d}\right)^2 \quad \text{and} \quad k^2 = \left(\frac{\sigma_{mn}}{a}\right)^2 + \left(\frac{p\pi}{d}\right)^2.$$

We first show that if $\mathbf{E}_a$ and $\mathbf{E}_b$ are the electric fields of two characteristic modes of the same resonator, then $\int \mathbf{E}_a . \mathbf{E}_b \, d\tau = 0$.

Since we are concerned with simple harmonic oscillations Maxwell's equations may be written,

$$\operatorname{curl}\mathbf{E} = -j\omega\mu\mathbf{H}, \quad \operatorname{curl}\mathbf{H} = j\omega\epsilon\mathbf{E} + \mathbf{J}, \quad \operatorname{div}\mathbf{B} = 0, \quad \operatorname{div}\mathbf{D} = \rho, \quad (2)$$

whence
$$\operatorname{curl}\operatorname{curl}\mathbf{E} = -j\omega\mu\operatorname{curl}\mathbf{H}$$
$$= \omega^2\mu\epsilon\mathbf{E} - j\omega\mu\mathbf{J}$$
$$= k^2\mathbf{E} - j\omega\mu\mathbf{J}. \quad (3)$$

Similarly,
$$\operatorname{curl}\operatorname{curl}\mathbf{H} = k^2\mathbf{H} + \operatorname{curl}\mathbf{J}. \quad (4)$$

The fields $(\mathbf{E}_a, \mathbf{H}_a)$, $(\mathbf{E}_b, \mathbf{H}_b)$ of different characteristic modes satisfy equations (3) and (4) with $\mathbf{J} = 0$ and with k equal respectively to the corresponding characteristic values k_a and k_b.

Thus,
$$\operatorname{curl}\operatorname{curl}\mathbf{E}_a = k_a^2\mathbf{E}_a, \quad \operatorname{curl}\operatorname{curl}\mathbf{E}_b = k_b^2\mathbf{E}_b, \quad (5)$$

with similar equations for the magnetic fields.

Form the scalar product of both sides of the first of equations (5), with $\mathbf{E}_b$, and integrate over the volume of the resonator

$$k_a^2 \int \mathbf{E}_b . \mathbf{E}_a \, d\tau = \int \mathbf{E}_b . \operatorname{curl}\operatorname{curl}\mathbf{E}_a \, d\tau. \quad (6)$$

Similarly, on multiplying the second of equations (5) by $\mathbf{E}_a$ and integrating, we find

$$k_b^2 \int \mathbf{E}_a . \mathbf{E}_b \, d\tau = \int \mathbf{E}_a . \operatorname{curl}\operatorname{curl}\mathbf{E}_b \, d\tau. \quad (7)$$

Subtract (7) and (6), then

$$(k_a^2 - k_b^2) \int \mathbf{E}_a . \mathbf{E}_b \, d\tau = \int [\mathbf{E}_b . \operatorname{curl} \operatorname{curl} \mathbf{E}_a - \mathbf{E}_a . \operatorname{curl} \operatorname{curl} \mathbf{E}_b] \, d\tau. \quad (8)$$

The right-hand side of this equation may be transformed to a surface integral over the boundary of the resonator by means of the vector analogue of the second of Green's theorems.* This states that if $\mathbf{A}$ and $\mathbf{B}$ are two continuous field vectors, then

$$\int_{\text{volume}} [\mathbf{A} . (\operatorname{curl} \operatorname{curl} \mathbf{B}) - \mathbf{B} . (\operatorname{curl} \operatorname{curl} \mathbf{A})] \, d\tau$$

$$= \int_{\text{surface}} [\mathbf{A} \times (\operatorname{curl} \mathbf{B}) - \mathbf{B} \times (\operatorname{curl} \mathbf{A})] . d\mathbf{S}, \quad (9)$$

where the surface integral is taken over the boundary surfaces of $\int d\tau$.

The right-hand side of (8) may therefore be written as a surface integral

$$\int [\mathbf{E}_b \times \operatorname{curl} \mathbf{E}_a - \mathbf{E}_a \times \operatorname{curl} \mathbf{E}_b] . d\mathbf{S}.$$

According to (2) (with $\mathbf{J} = 0$) this is the same as

$$-j\omega\mu \int [\mathbf{E}_b \times \mu\mathbf{H}_a - \mathbf{E}_a \times \mu\mathbf{H}_b] . d\mathbf{S}.$$

At the boundary, $\mathbf{E}_a$ and $\mathbf{E}_b$ are perpendicular to, and $\mathbf{H}_a$ and $\mathbf{H}_b$ tangential to the plane of $d\mathbf{S}$. Consequently $(\mathbf{E}_b \times \mu\mathbf{H}_a)$ and $(\mathbf{E}_a \times \mu\mathbf{H}_b)$ both lie in the plane of $d\mathbf{S}$ and their scalar product with the vector $d\mathbf{S}$, whose direction is along the normal, is zero. Thus

$$(k_a^2 - k_b^2) \int \mathbf{E}_a . \mathbf{E}_b \, d\tau = 0.$$

If therefore k_a is not equal to k_b, then

$$\int \mathbf{E}_a . \mathbf{E}_b \, d\tau = 0, \quad (10)$$

and the vector fields are functionally orthogonal. When $\mathbf{E}_a$ and $\mathbf{E}_b$ are the same field then $k_a = k_b$ and the integrand is essentially positive and equal to $|E_a|^2$.

The rectangular resonator is exceptional in that the E_{mnp}- and H_{mnp}-modes where the suffixes are identical have the same

* Stratton, *Electromagnetic Theory*, p. 250.

characteristic value $k_{mn\nu}$. It may be shown here, by direct reference to the expressions for the field components, that

$$\int \mathbf{E}_a . \mathbf{E}_b \, d\tau = 0,$$

unless $\mathbf{E}_a$ and $\mathbf{E}_b$ are the electric field of the same mode. Similarly,

$$\int \mathbf{H}_a . \mathbf{H}_b \, d\tau = \int \mathbf{E}_a . \mathbf{H}_b \, d\tau = \int \mathbf{E}_b . \mathbf{H}_a \, d\tau = 0.$$

For instance,

$$(k_a^2 - k_b^2) \int \mathbf{E}_a . \mathbf{H}_b \, d\tau = \int [\mathbf{H}_b . (\text{curl curl } \mathbf{E}_a) - \mathbf{E}_a . (\text{curl curl } \mathbf{H}_b)] \, d\tau$$

$$= \int [\mathbf{H}_b \times \text{curl } \mathbf{E}_a - \mathbf{E}_a \times \text{curl } \mathbf{H}_b] \, d\tau$$

$$= -j\omega \int [\mathbf{H}_b \times \mu \mathbf{H}_a + \mathbf{E}_a \times \epsilon \mathbf{E}_b] . d\mathbf{S} = 0,$$

since at the surface $\mathbf{H}_b$ is parallel to $\mathbf{H}_a$, and $\mathbf{E}_a$ to $\mathbf{E}_b$, with the result that the vector products vanish.

Thus, to summarize, if $\mathbf{A}$ and $\mathbf{B}$ represent any of the vector fields $\mathbf{E}_a$, $\mathbf{E}_b$, $\mathbf{H}_a$, $\mathbf{H}_b$, then $\int \mathbf{A}.\mathbf{B} \, d\tau = 0$, unless $\mathbf{A}$ and $\mathbf{B}$ are identical or $\mathbf{A} = C\mathbf{B}$, where C is a scalar constant.

It is usual to multiply the field components by constants, so that the integrals which do not vanish, namely, $\int \mathbf{E}_a . \mathbf{E}_a \, d\tau$, $\int \mathbf{H}_a . \mathbf{H}_b \, d\tau$ and $\int \mathbf{E}_b . \mathbf{E}_b \, d\tau$, etc., are equal to unity. The vector functions $\mathbf{E}_a$, $\mathbf{E}_b$, $\mathbf{H}_a$, $\mathbf{H}_b$ are then said to be normalized.

The vectors $\mathbf{E}_a$ and $\mathbf{H}_a$ are now proportional to the corresponding vectors in the mode from which they are derived but do not themselves constitute in general an E-H pair satisfying Maxwell's equations.

An arbitrary electromagnetic field within a resonator within which the charge density ρ and current density $\mathbf{J}$ are everywhere zero may now be represented as a linear vector function of the normalized vectors $\mathbf{E}_a$, $\mathbf{E}_b$, etc., $\mathbf{H}_a$, $\mathbf{H}_b$, etc. Let the field vectors of the arbitrary electromagnetic field be $\mathbf{E}$ and $\mathbf{H}$. We write

$$\mathbf{E} = \sum_a^\infty K_a \mathbf{E}_a + \sum_a^\infty K_a^0 \mathbf{E}_a^0, \quad \mathbf{H} = \sum_a^\infty M_a \mathbf{H}_a + \sum_a^\infty M_a^0 \mathbf{H}_a^0, \quad (11)$$

in which the summations over the $\mathbf{E}_a$ and the $\mathbf{H}_a$ embrace all the possible E-modes, and those over the $\mathbf{E}_a^0$ and $\mathbf{H}_a^0$ all the H-modes. Equations (11) therefore represent quasi-Fourier expansions of the vectors $\mathbf{E}$ and $\mathbf{H}$ in terms of the normalized mode-vectors.

To obtain an expression for any one of the coefficients, say K_b, form the scalar produce of $\mathbf{E}_b$ with both sides of the first of equations (11), integrate and make use of the orthogonal properties (10) of the mode vectors; thus

$$\int \mathbf{E} . \mathbf{E}_b \, d\tau = K_b \int \mathbf{E}_b . \mathbf{E}_b \, d\tau = K_b,$$

that is
$$K_b = \int \mathbf{E} . \mathbf{E}_b \, d\tau.$$

Similarly

$$K_b^0 = \int \mathbf{E} . \mathbf{E}_b^0 \, d\tau, \quad M_b = \int \mathbf{H} . \mathbf{H}_b \, d\tau, \quad M_b^0 = \int \mathbf{H} . \mathbf{H}_b^0 \, d\tau.$$

The coefficients K_a, etc., will in general be complex quantities since the constituent modes will not in general vibrate in phase. As an example of the use of normalized mode vectors we obtain a well-known result relating to the total field energy within the resonator. The maximum electric energy in a mode whose electric field is $K_a \mathbf{E}_a$ is

$$\frac{\epsilon}{2} K_a \overline{K}_a \int \mathbf{E}_a . \mathbf{E}_a \, d\tau = \frac{\epsilon}{2} K_a . \overline{K}_a,$$

and this is equal to the maximum magnetic energy, and to the total stored energy in this mode. ($\overline{K}_a$ is the complex conjugate of K_a.) The maximum total electric energy of an arbitrary oscillatory field $(\mathbf{E}, \mathbf{H})$ in the resonator is

$$\frac{\epsilon}{2} \int \mathbf{E} . \overline{\mathbf{E}} \, d\tau = \frac{\epsilon}{2} \int [(\sum K_a \mathbf{E}_a + \sum K_a^0 \mathbf{E}_a^0)(\sum \overline{K}_a \mathbf{E}_a + \overline{K}_a^0 \mathbf{E}_a^0)] \, d\tau$$

$$= \frac{\epsilon}{2} \int \left[\sum_a \sum_b K_a \overline{K}_b \mathbf{E}_a . \mathbf{E}_b + \sum_a \sum_b K_a \overline{K}_b^0 \mathbf{E}_a . \mathbf{E}_b^0 \right] d\tau$$

$$= \frac{\epsilon}{2} \sum K_a . \overline{K}_a.$$

Thus the total maximum electric energy is the sum of the maximum electric energies in the modes as if each were present independently of the others. This is also equal to the total magnetic energy of the modes and to the total instantaneous stored energy which is constant. When ρ or $\mathbf{J}$ are not zero within the cavity, then it is necessary to employ further normalized functions in the expansions of $\mathbf{E}$ and $\mathbf{H}$.

These are generally chosen to be scalar functions ϕ_a, ϕ_b, etc., which are constant over the boundary and which satisfy the equations

$$\nabla^2\phi_a + k_a^2\phi_a = 0, \quad \nabla^2\phi_b + k_b^2\phi_b = 0. \tag{12}$$

It follows that

$$\int(\phi_a\nabla^2\phi_b - \phi_b\nabla^2\phi_a)\,d\tau$$

$$= (k_a^2 - k_b^2)\int\phi_a\phi_b\,d\tau$$

$$= \int_S(\phi_a\operatorname{grad}\phi_b - \phi_b\operatorname{grad}\phi_a)\,.\,d\mathbf{S} = 0,$$

if ϕ_a and ϕ_b are zero on S.

Thus
$$\int\phi_a\phi_b\,d\tau = 0 \quad (a \neq b).$$

The ϕ are normalized, so that

$$\int\phi_a\phi_b\,d\tau = 1 \quad (a = b).$$

It can be shown that the vectors $\mathbf{F}_a = \operatorname{grad}\phi_a$ and $\mathbf{F}_b = \operatorname{grad}\phi_b$ are mutually orthogonal and orthogonal to the $\mathbf{E}_a$, $\mathbf{E}_b$, $\mathbf{H}_a$, $\mathbf{H}_b$, etc.

Examples of ϕ_a would be $(\partial U_a/\partial z)$ in a cylindrical resonator, and (V_a/r) in a spherical resonator, multiplied by the appropriate normalizing constants.

7·18·4. Forced oscillations in cavity resonators

The orthogonal functions discussed in the previous section are of fundamental importance for the theory of forced oscillations in cavity resonators. This subject has been investigated by Hansen and Condon, and more completely by Slater, their treatments resembling similar analyses that occur in quantum mechanics.*

The following treatment of forced oscillations in cavity resonators is based on those of Condon and Slater but is intended to illustrate the method rather than to provide a rigorous discussion.

* E.g. E. U. Condon, 'Forced Oscillations in Cavity Resonators', *J. Appl. Phys.* vol. 12, pp. 129–32 (1941).

We suppose that total electric field **E**, total magnetic field **H** and current density **J** within the cavity are expanded in terms of the vectors $\mathbf{E}_a$, $\mathbf{H}_a$ and $\mathbf{F}_a = \operatorname{grad} \phi_a$ as follows:

$$\mathbf{E} = \Sigma K_a \mathbf{E}_a + \Sigma L_a \mathbf{F}_a, \quad \mathbf{H} = \Sigma M_a \mathbf{H}_a, \quad \mathbf{J} = \Sigma N_a \mathbf{E}_a + \Sigma T_a \mathbf{F}_a. \quad (1)$$

Here the coefficients K_a and M_a also include the previous K_a^0 and M_a^0. Further, **E** and **H** satisfy Maxwell's equations

$$\operatorname{curl} \mathbf{E} = -\mu \dot{\mathbf{H}}, \quad \operatorname{curl} \mathbf{H} = \epsilon \dot{\mathbf{E}} + \mathbf{J}, \quad \operatorname{div} \epsilon \mathbf{E} = \rho, \quad \operatorname{div} \mu \mathbf{H} = 0. \quad (2)$$

We wish to obtain formal solutions of (2) for a cavity, in terms of the $\mathbf{E}_a$, $\mathbf{H}_a$ and $\mathbf{F}_a$, but before we can proceed some preliminary results are required.

The electric field **E** and magnetic field **H** of a given characteristic mode are proportional to the normalized fields $\mathbf{E}_a$ and $\mathbf{H}_a$ of the mode, and we may write

$$\mathbf{E} = K_a \mathbf{E}_a, \quad \mathbf{H} = M_a \mathbf{H}_a. \quad (3)$$

Further **E** and **H** oscillate in quadrature. The total energy of the mode is

$$\frac{\epsilon |E|^2}{2} = \frac{\epsilon}{2} K_a \overline{K}_a \int \mathbf{E}_a . \mathbf{E}_a \, d\tau = \frac{\mu |H|^2}{2} = \frac{\mu}{2} M_a M_a \int \mathbf{H}_a . \mathbf{H}_a \, d\tau,$$

whence
$$\epsilon K_a \overline{K}_a = \mu M_a \overline{M}_a$$

or
$$|K_a| \sqrt{\epsilon} = |M_a| \sqrt{\mu}.$$

Since **E** and **H** oscillate in quadrature it follows that

$$K_a = \pm j \sqrt{\left(\frac{\mu}{\epsilon}\right)} M_a. \quad (4)$$

From (2) and (3), applied to this specific mode, we deduce (**J** = 0 and $d/dt \equiv j\omega$)

$$\left.\begin{aligned} K_a \operatorname{curl} \mathbf{E}_a &= -j\omega\mu M_a \mathbf{H}_a, \\ M_a \operatorname{curl} \mathbf{H}_a &= j\omega\epsilon K_a \mathbf{E}_a, \end{aligned}\right\} \quad (5)$$

or, from (4) and (5), with $k_a^2 = \omega^2 \mu \epsilon$,

$$\left.\begin{aligned} \operatorname{curl} \mathbf{E}_a &= -j\omega\mu \frac{M_a}{K_a} \mathbf{H}_a = \pm k_a \mathbf{H}_a, \\ \operatorname{curl} \mathbf{H}_a &= j\omega\epsilon \frac{K_a}{M_a} \mathbf{E}_a = \pm k_a \mathbf{E}_a. \end{aligned}\right\} \quad (6)$$

It is convenient to retain the $+$ sign before k_a, that is, the negative sign in (4).

Since, from (2), curl $\mathbf{E}$ is proportional to $\mathbf{H}$, we may, according to (1), expand curl $\mathbf{E}$ in terms of the $\mathbf{H}_a$,

$$\operatorname{curl} \mathbf{E} = \Sigma C_a \mathbf{H}_a, \tag{7}$$

where
$$C_a = \int (\mathbf{H}_a . \operatorname{curl} \mathbf{E}) \, d\tau.$$

We may express (7) as a surface integral, since

$$\operatorname{div} [\mathbf{E} \times \operatorname{curl} \mathbf{E}_a] = (\operatorname{curl} \mathbf{E}_a) . (\operatorname{curl} \mathbf{E}) - \mathbf{E} . (\operatorname{curl} \operatorname{curl} \mathbf{E}_a).$$

From (6) and 7·18·3 (5), this may be written

$$k_a \operatorname{div} [\mathbf{E} \times \mathbf{H}_a] = k_a \mathbf{H}_a . \operatorname{curl} \mathbf{E} - k_a^2 \mathbf{E} . \mathbf{E}_a. \tag{8}$$

Integrate both sides of (8) over the volume of the cavity and use Gauss's theorem,

$$k_a \int_{\text{surface}} (\mathbf{E} \times \mathbf{H}_a) . d\mathbf{S} = k_a C_a - k_a^2 K_a$$

or
$$C_a = k_a K_a + \int_{\text{surface}} (\mathbf{E} \times \mathbf{H}_a) . d\mathbf{S}. \tag{9}$$

The surface integral is taken over the boundary surface of the volume $\int d\tau$. When the cavity is empty this is its interior surface, but if metallic obstacles are present, then the surface integral covers their surfaces as well and $\int d\tau$ is the volume bounded by the obstacles and the closed boundary of the cavity. In this case k_a, C_a, K_a, etc., have different values from those of the corresponding mode in the empty cavity.

Thus, from (7) and (9),

$$\operatorname{curl} \mathbf{E} = \Sigma C_a \mathbf{H}_a = \Sigma k_a K_a \mathbf{H}_a + \Sigma \mathbf{H}_a \int (\mathbf{E} \times \mathbf{H}_a) . d\mathbf{S}. \tag{10}$$

Similarly, $$\operatorname{curl} \mathbf{H} = \Sigma D_a \mathbf{E}_a = \Sigma \mathbf{E}_a \int (\mathbf{E}_a . \operatorname{curl} \mathbf{H}) \, d\tau. \tag{11}$$

It may be shown, as above, that

$$D_a = \int (\mathbf{E}_a . \operatorname{curl} \mathbf{H}) \, d\tau = k_a M_a + \int (\mathbf{H} \times \mathbf{E}_a) . d\mathbf{S}. \tag{12}$$

Terms in $\mathbf{F}_a$ in the expansion of curl $\mathbf{H}$ are all found to have zero coefficients.

We now replace the vectors in Maxwell's equations (2) by their equivalent expansions in terms of the $\mathbf{E}_a$, $\mathbf{H}_a$ and $\mathbf{F}_a$. We have

$$\text{curl}\,\mathbf{E} + \mu\dot{\mathbf{H}} = 0,$$

whence, from (1) and (10),

$$\Sigma\mathbf{H}_a\left(k_a K_a + \int(\mathbf{E}\times\mathbf{H}_a).d\mathbf{S} + \mu\frac{dM_a}{dt}\right) = 0.$$

Since the field $(\mathbf{E}, \mathbf{H})$ is arbitrarily chosen, it follows that, for all a,

$$k_a K_a + \int(\mathbf{E}\times\mathbf{H}_a).d\mathbf{S} + \mu\frac{dM_a}{dt} = 0. \tag{13}$$

From $$\text{curl}\,\mathbf{H} - \epsilon\dot{\mathbf{E}} = \mathbf{J}$$

we obtain, using (1) and (12),

$$\Sigma\mathbf{E}_a\left(k_a M_a + \int(\mathbf{H}\times\mathbf{E}_a).d\mathbf{S} - \epsilon\frac{d}{dt}K_a - N_a\right)$$
$$+ \Sigma\mathbf{F}_a\left(-\epsilon\frac{d}{dt}L_a - T_a\right) = 0, \tag{14}$$

whence $$k_a M_a = \epsilon\frac{dK_a}{dt} - \int(\mathbf{H}\times\mathbf{E}_a).d\mathbf{S} + N_a, \tag{15}$$

$$\epsilon\frac{dL_a}{dt} + T_a = 0. \tag{16}$$

In an empty cavity ($\rho = 0$) div $\mathbf{E} = 0$ and $\mathbf{E}$ is represented by terms $\mathbf{E}_a$ with all the $\mathbf{F}_a$ equal to zero. This we shall assume to obtain in what follows. Equation (16) is then of no significance since $T_a = 0$. From (13) and (15) it follows that (with $\omega_a^2 = k_a^2/\mu\epsilon$)

$$\frac{dK_a^2}{dt} + \omega_a^2 K_a = \frac{1}{\epsilon}\frac{d}{dt}\left[\int(\mathbf{H}\times\mathbf{E}_a).d\mathbf{S} - N_a\right] - \frac{k_a}{\mu\epsilon}\int(\mathbf{E}\times\mathbf{H}_a).d\mathbf{S}. \tag{17}$$

Similarly,

$$\frac{d^2 M_a}{dt^2} + \omega_a^2 M_a = \frac{k_a}{\mu\epsilon}\left[N_a - \int(\mathbf{H}\times\mathbf{E}_a).d\mathbf{S}\right] - \frac{1}{\mu}\frac{d}{dt}\int(\mathbf{E}\times\mathbf{H}_a).d\mathbf{S}. \tag{18}$$

These equations (17) and (18), which are equivalent and consistent with (4) with the negative sign on the left, are similar to the equations of motion of an elastically bound particle moving under the influence of an impressed force. The ratio of the impressed force to the mass of the particle is here represented by the sum of the terms on the right-hand side. If, therefore, it is possible to evaluate the right-hand sides of (17) and (18), then, in principle, the differential equation for K_a can be solved and the complex amplitude of the corresponding natural mode obtained.

We consider the simple example of a resonator excited by a current in a small loop of thin wire fed from a coaxial transmission line. We suppose the outer conductor of the line to fit a circular window in the wall of the resonator and the central conductor to bend over within the cavity and to join the adjacent wall as a loop.

For simplicity, suppose the window and the loop to be so small compared with the dimensions of the resonator that the k_a, $\mathbf{E}_a$, $\mathbf{H}_a$ are effectively those of the empty resonator.

Let the current i in the loop oscillate at frequency $(\omega/2\pi)$. Consider the terms on the right-hand side of (18). By definition (from (1))

$$N_a = \int \mathbf{J} . \mathbf{E}_a \, d\tau.$$

Consequently, since $\mathbf{J} = 0$, $N_a = 0$.

Over the walls of the cavity $(\mathbf{H} \times \mathbf{E}_a)$ is tangential to the walls, and $(\mathbf{H} \times \mathbf{E}_a) . d\mathbf{S}$ is zero. Over the window $\mathbf{H}$ is effectively in the plane of the window and $(\mathbf{H} \times \mathbf{E}_a) . d\mathbf{S}$ is also zero.

The only contribution to $\int (\mathbf{H} \times \mathbf{E}_a) . d\mathbf{S}$ comes therefore from the surface of the loop. The total field $\mathbf{H}$ at the surface of the loop is tangential to the surface, and since the loop is a thin wire of radius a, carrying current i, we have $2\pi a |H| = i$.

Let $d\mathbf{l}$ be an element of length of the wire, then $d\mathbf{S}$ in the integral becomes $d\mathbf{S} = 2\pi a \, d\mathbf{l}$. The integrand $(\mathbf{H} \times \mathbf{E}_a) . d\mathbf{S} = (d\mathbf{S} \times \mathbf{H}) . \mathbf{E}_a$.

But $(d\mathbf{S} \times \mathbf{H})$ is a vector of magnitude $2\pi a \, dl |H|$ directed parallel to the axis of the wire in the direction of the current. Consequently,

$$(d\mathbf{S} \times \mathbf{H}) . \mathbf{E}_a = 2\pi a \, dl |H| |E_a| \cos\theta = i |E_a| \, dl \cos\theta,$$

where $\cos\theta$ is the angle between $\mathbf{E}_a$ and $d\mathbf{l}$ with $d\mathbf{l}$ pointing in the direction of the current. Consequently,

$$\int (\mathbf{H} \times \mathbf{E}_a) . d\mathbf{S} = i \int |E_a| \cos\theta \, dl$$
$$= i \times (\text{line integral of } \mathbf{E}_a \text{ around loop}).$$

If we form a closed path comprising the loop and a line on the wall of the cavity from the end of the loop to the end of the coaxial, the line integral of $\mathbf{E}_a$ along the line is zero since $\mathbf{E}_a$ is normal to the walls. Thus

$$\int | E_a | \cos \theta \, dl = \int_A \operatorname{curl} \mathbf{E}_a . d\mathbf{A},$$

where $\mathbf{A}$ is the area enclosed by the loop and the line mentioned above.

But from (6), $\operatorname{curl} \mathbf{E}_a = k_a \mathbf{H}_a$. Therefore

$$\int (\mathbf{H} \times \mathbf{E}_a) . d\mathbf{S} = k_a i \int \mathbf{H}_a . d\mathbf{A} = k_a i \psi_a, \qquad (19)$$

where ψ_a is the flux of $\mathbf{H}_a$ through the loop. ψ_a is the effective mutual coupling of the loop with the normalized mode vector $\mathbf{H}_a$.

It remains to discuss the third term on the right-hand side of (18), namely,

$$-\frac{1}{\mu} \frac{d}{dt} \int (\mathbf{E} \times \mathbf{H}_a) . d\mathbf{S}.$$

We obtain an approximation for this term which is sufficiently accurate in practice.

We note that the total magnetic field in the forced vibration in the cavity is, from (18),

$$\mathbf{H} = \Sigma M_a \mathbf{H}_a$$
$$= \Sigma \mathbf{H}_a \left(\frac{1}{(\omega_a^2 - \omega^2)} \left[\frac{k_a}{\mu \epsilon} (-k_a i \psi_a) - \frac{j\omega}{\mu} (\mathbf{E} \times \mathbf{H}_a) . d\mathbf{S} \right] \right). \qquad (20)$$

If, therefore, the resonant frequencies ω_a do not fall close together, the amplitude M_a is very large compared with all other coefficient M_b, M_c, etc., when $(\omega_a - \omega)$ is small.

In this case, the field $\mathbf{E}$ in the integrand of the integral under discussion is virtually the same as $\mathbf{E}_a$.

If the walls of the resonator are perfectly conducting then the total field $\mathbf{E}$ is normal to all metallic boundaries, and since $\mathbf{E} \times \mathbf{H}_a$ is tangential to the boundary the integral is zero over the walls and the loop. Over the window, however, $\mathbf{H}_a$ is virtually constant, and is tangential to the plane of the window. The tangential component of $\mathbf{E}$ in the plane of the window is approximately radial from the end of the centre conductor of the coaxial and the integral over the window is small or zero. Suppose, instead, the walls to be highly conducting but not infinitely conducting.

The integral over the metal boundary of the cavity is $\int (\mathbf{E} \times \mathbf{H}_a) \cdot d\mathbf{S}$.

Near resonance the total magnetic field is essentially $\mathbf{H}_a$, consequently the tangential component of $\mathbf{E}$ is that associated with a field $M_a \mathbf{H}_a$. The mean value of the integral $M_a \int (\mathbf{E} \times \mathbf{H}_a) \cdot d\mathbf{S}$ is the mean loss of power to the boundary and according to 7·14 (11) this is $\dfrac{M_a^2}{2\sigma\delta} \int \mathbf{H}_a^2 \, d\mathbf{S}$.

Since the other terms in (18) represent maximum values we require the maximum value of the integral, which is

$$\frac{M_a^2}{\sigma\delta} \int \mathbf{H}_a^2 \, d\mathbf{S}.$$

But according to 7·16 (2)

$$Q_a = \frac{2 \int M_a^2 \mathbf{H}_a^2 \, d\tau}{\delta \int M_a^2 \mathbf{H}_a^2 \, d\mathbf{S}} = \frac{2}{\delta \int \mathbf{H}_a^2 \, dS},$$

where Q_a is the Q-factor of this characteristic mode. Consequently,

$$\int \mathbf{H}_a^2 \, d\mathbf{S} = 2/\delta Q_a.$$

Thus the term $-\dfrac{1}{\mu} \dfrac{d}{dt} \int (\mathbf{E} \times \mathbf{H}_a) \cdot d\mathbf{S}$ in (18) becomes, using 7·14 (6),

$$-\frac{2}{\mu\sigma\delta^2 Q_a} \frac{dM_a}{dt} = \frac{\omega_a}{Q_a} \frac{dM_a}{dt} = \frac{j\omega\omega_a}{Q_a} M_a. \tag{21}$$

Equation (18) therefore reduces to

$$\left(\omega_a^2 + \frac{j\omega_a}{Q_a} \omega - \omega^2 \right) M_a = -\frac{k_a^2 \psi_a}{\mu\varepsilon} i$$

or

$$M_a = \frac{k_a^2 \psi_a i}{\mu\varepsilon \left(\omega_a^2 + \dfrac{j\omega_a}{Q_a} \omega - \omega^2 \right)}. \tag{22}$$

This is an expression for the coefficient of $\mathbf{H}_a$ in terms of the impressed current i and the angular frequency ω. The total magnetic field within the cavity is $\mathbf{H} = \sum_a M_a \mathbf{H}_a$. The quantity ψ_a is the coupling coefficient between the loop and the mode $\mathbf{H}_a$, that is, it is the flux $\int \mathbf{H}_a . \mathbf{dA}$ of $\mathbf{H}_a$ through the loop. The flux of magnetic induction due to this characteristic mode is $\mu \int M_a \mathbf{H}_a . \mathbf{dS} = \mu M_a \psi_a$, and the contribution of this flux to the input voltage is

$$v_a = -j\omega M_a \psi_a \mu,$$

which, according to (22), is

$$v_a = \frac{j\omega \psi_a^2 \mu i \omega_a^2}{\left(\omega_a^2 + \dfrac{j\omega_a}{Q_a}\omega - \omega^2\right)}.$$

The total input voltage is

$$v = \Sigma v_a = i\Sigma \frac{j\omega \psi_a^2 \omega_a^2 \mu}{\left[(\omega_a^2 - \omega^2) + \dfrac{j\omega\omega_a}{Q_a}\right]}. \tag{23}$$

The input impedance to the cavity is

$$Z = \frac{v}{i} = \Sigma \frac{j\omega \psi_a^2 \omega_a^2 \mu}{\left[(\omega_a^2 - \omega^2) + \dfrac{j\omega\omega_a}{Q_a}\right]}. \tag{24}$$

This expression is equivalent to that quoted in 6·11 (1) (with $M_k^2 = \mu\omega_a^2 \psi_a^2$).

When the field in the cavity is excited from a probe which is a short extension of the central conductor, equation (18) reduces to

$$\left[(\omega_a^2 - \omega^2) + \frac{j\omega_a \omega}{Q_a}\right] M_a = -\frac{k_a}{\mu\epsilon} \int (\mathbf{H} \times \mathbf{E}_a) . d\mathbf{S}. \tag{25}$$

As before the integral reduces to that over the surface of the probe, and it transforms to

$$-\frac{k_a}{\mu\epsilon} \int i \, | \, E_a \, | \cos\theta \, dl,$$

where i is the current at the element $\mathbf{dl}$ of the probe and θ is the angle between $\mathbf{E}_a$ and $\mathbf{dl}$.

In this case i varies along the probe, and when the latter is thin i is zero at the end. If the probe is short and straight as well as thin, then $| E_a |$ may be considered to be constant over the length of the probe. The integral then reduces to

$$-\frac{k_a}{\mu\epsilon} | E_a | \cos\theta \int i \, dl = -\frac{k_a}{\mu\epsilon} | E_a | \cos\theta \, l\bar{i},$$

where l is the length of the probe and $\bar{\imath}$ the average current along the probe.

The coupling coefficient χ_a of the probe with the ath mode is defined as

$$\chi_a = |E_a| \cos \theta l.$$

It follows from (25) that (writing $k_a = \omega_a \sqrt{(\mu \epsilon)}$)

$$M_a = \frac{-\omega_a \chi_a \bar{\imath}}{\sqrt{(\mu \epsilon)} \left[(\omega_a^2 - \omega^2) + \dfrac{j \omega \omega_a}{Q_a} \right]}. \qquad (26)$$

The e.m.f. induced in the probe by the ath mode is

$$
\begin{aligned}
v_a = K_a E_a l \cos \theta &= K_a \chi_a \\
&= -j \sqrt{\left(\frac{\mu}{\epsilon}\right)} M_a \chi_a \\
&= \frac{j \omega_a \chi_a^2 \bar{\imath}}{\epsilon \left[(\omega_a^2 - \omega^2) + \dfrac{j \omega \omega_a}{Q_a} \right]},
\end{aligned}
\qquad (27)
$$

and the input voltage is $v = \Sigma v_a.$

The relation between the average current $\bar{\imath}$ and the input current i depends on the length and thickness of the probe. When the probe is thin and short we shall make little error in assuming that the current falls linearly from i at the input to zero at the end of the probe. In this case $\bar{\imath} = \frac{1}{2}i$.

The input impedance then becomes

$$Z = \frac{v}{i} = \frac{j}{2\epsilon} \Sigma \frac{\omega_a \chi_a^2}{\left[(\omega_a^2 - \omega^2) + \dfrac{j \omega \omega_a}{Q_a} \right]}. \qquad (28)$$

It is evident that, to excite the ath mode strongly by means of a loop, ψ_a should be as large as possible, that is, the loop should be placed where $\mathbf{H}_a$ is a maximum and turned with its plane normal to $\mathbf{H}_a$, but with probe excitation the probe should be introduced where $\mathbf{E}_a$ is a maximum and set parallel to $\mathbf{E}_a$.

We consider the simple example of a rectangular resonator with sides a, b and c.

The normalized magnetic field $\mathbf{H}_a$ of the H_{011} mode is

$$H_{ay} = \frac{2}{c} \sqrt{\left(\frac{bc}{a(b^2+c^2)}\right)} \sin\frac{\pi y}{b} \cos\frac{\pi z}{c},$$

$$H_{az} = \frac{2}{b} \sqrt{\left(\frac{bc}{a(b^2+c^2)}\right)} \cos\frac{\pi y}{b} \sin\frac{\pi z}{c},$$

since these values give

$$\int_0^a \int_0^b \int_0^c [H_{ay}^2 + H_{az}^2]\,dx\,dy\,dz = 1 \quad \text{and} \quad \frac{H_{ay}}{H_{az}} = \left(\frac{H_y}{H_z}\right)_{H_{011}}.$$

Suppose a small loop of area A to be introduced so that it receives the maximum flux of H_{az}. Then (m.k.s. units)

$$\psi_{011} = (H_{az})_{\text{max}}.A = \frac{2}{b} \sqrt{\left(\frac{bc}{a(b^2+c^2)}\right)} A.$$

Thus, for a cube of side a,

$$\psi_{011} = \sqrt{\left(\frac{2}{a^3}\right)}A, \quad \omega_a^2 = \frac{k_a^2}{\mu\epsilon} = \frac{2}{\mu\epsilon}\left(\frac{\pi}{a}\right)^2.$$

The term $\omega_a^2 \chi_a^2 \mu$ in (24) here becomes

$$\frac{2}{\epsilon}\left(\frac{\pi}{a}\right)^2 \frac{2}{a^3} A^2 = \frac{4\pi^2 A^2}{\epsilon a^5} = \frac{36\pi \times 10^9 \times 4\pi^2 A^2}{K_e a^5}$$

$$= 4 \cdot 46 \times 10^{12} \frac{A^2}{K_e a^5}.$$

Normally $K_e = 1$. Since m.k.s. units have been used, A^2 is expressed in square metres and a in metres in this formula. The contribution of this H_{011} mode to the input impedance is then found from (24).

SHORT BIBLIOGRAPHY

BRILLOUIN, L. Theoretical Study of Dielectric Cables. *Electrical Communication*, April 1938.

COPSON, E. T. An integral equation method of solving plane diffraction problems (Babinet's Principle). *Proc. Roy. Soc.*, A, vol. 186, pp. 100–18, 1946.

FRANK, P. and VON MISES, R. *Die Differential und Integralgleichungen der Mechanik und Physik*, part II, §5.

LAMONT, R. L. *Wave Guides*. Methuen Monograph.

M.I.T. Radar School. *Principles of Radar*. McGraw-Hill Book Co.

Proceedings at the Radiolocation Convention, March–May 1946; *J. Instn Elect. Engrs*, vol. 93, part III A, nos. 1, 3 and 4, 1946.

SARBACHER and EDSON. *Hyper- and Ultra-High-Frequency Engineering.* John Wiley and Chapman and Hall.

SCHELKUNOFF, S. A. *Electromagnetie Waves*. Van Nostrand Co. Inc.

SLATER, J. C. *Microwave Transmission*. McGraw Hill Book Co.

STRATTON, J. A. *Electromagnetic Theory*. McGraw Hill Book Co.

WATSON, W. H. *The Physical Principles of Wave Guide Transmission and Antenna Systems*. Oxford: Clarendon Press.

LIST OF PRINCIPAL QUANTITIES AND THEIR SYMBOLS

Phase velocity of TEM-wave: $v = 1/\sqrt{(\mu\varepsilon)}$.
Phase velocity of wave-guide wave: v_g.
Wave-length of TEM-wave: λ.
Wave-length of wave-guide wave: λ_g.
Frequency f cyc./sec.
Angular frequency ω radians/sec.
Electric inductive capacity of medium: $\varepsilon = K_e \varepsilon_0$.
Magnetic inductive capacity of medium: $\mu = K_m \mu_0$.
Corresponding quantities for vacuum: ε_0 and μ_0;

$$\varepsilon_0 = 8 \cdot 854 \times 10^{-12} \doteqdot \frac{10^{-9}}{36\pi}, \quad \mu_0 = 4\pi \times 10^{-7}.$$

Dielectric constant or electric specific inductive capacity of medium: K_e.
Magnetic permeability or magnetic specific inductive capacity of medium: K_m.
Electric field strength: **E**.
Magnetic field strength: **H**.
Electric induction $\mathbf{D} = \varepsilon\mathbf{E}$.
Magnetic induction $\mathbf{B} = \mu\mathbf{H}$.
Specific conductivity: σ.
Skin depth: δ.
Surface resistance: R.

Voltage: V; current: i.
Characteristic impedance: Z_0.
Intrinsic or wave impedance:

(a) TEM-wave: $\sqrt{\dfrac{\mu}{\varepsilon}}$.

(b) E- or TM-wave: $\sqrt{\left(\dfrac{\mu}{\varepsilon}\right)}\dfrac{\lambda}{\lambda_y} = Z_E$.

(c) H- or TE-wave: $\sqrt{\left(\dfrac{\mu}{\varepsilon}\right)}\dfrac{\lambda_g}{\lambda} = Z_H$.

Impedance: $Z = R + jX$.
Admittance: $Y = G + jB$.
Normalized impedance: $z = Z/Z_0 = r + jx$.
Normalized admittance: $y = YZ_0 = Y/Y_0 = g + jb$.
Reflexion coefficient $\rho = |\rho|\,e^{j\phi}$.
Conjugate complex of ρ is: $\bar{\rho} = |\rho|\,e^{-j\phi}$.
Phase advance on reflexion: ϕ.
Scattering coefficient $h = |h|\,e^{j\theta}$.
Wave-guide dimensions:

(a) Rectangular wave guides: a and b.
(b) Circular-radius: a.

Coordinates:

(a) Cartesian: x, y, z.
(b) Cylindrical: r, θ, z.
(c) Spherical polar coordinates: r, θ, ϕ
(d) Curvilinear coordinates: u_1, u_2, u_3.

Differential multipliers: h_1, h_2, h_3.
Unit vectors: $\mathbf{i}_1$, $\mathbf{i}_2$, $\mathbf{i}_3$.
Standing wave ratio: $S = 1/s$.
Propagation constant of TEM-wave: $k = \omega\sqrt{(\mu\varepsilon)} = (2\pi/\lambda)$.
Propagation constant of TE- or TM-wave: $\gamma = (2\pi/\lambda_g)$.
Cut-off wave-length: λ_c.
Cut-off frequency: f_c.
Attenuation coefficient: α.
Mode subscripts: m, n, p.
Electromagnetic potentials:

(a) Electric Hertz vector: $\mathbf{\Pi}$.
(b) Magnetic Hertz vector: $\mathbf{\Pi}^0$.
(c) Single component Hertz vectors: $\mathbf{\Pi} = \mathbf{i}_1\,U$; $\mathbf{\Pi}_0 = \mathbf{i}_1\,V$.

General symbol for scalar magnitudes of Hertz vectors U and V: T.
Quality factor of a resonator: Q.
Complementary screens: S and C.
Charge density: ρ.
Current density: $\mathbf{J}$.
$\mathbf{E}_a$, $\mathbf{H}_a$: normalized characteristic modes.
S-band: $\lambda = 8\cdot9 - 10\cdot7$ cm.
X-band: $\lambda = 3 - 3\cdot3$ cm.

Subject Index

Section references in clarendon type; page references in normal type

Printed in the United States
By Bookmasters